T0203295

IET SECURITY SERIES 04

User-Centric Privacy and Security in Biometrics

IET Book Series in Advances in Biometrics – Call for authors
Book Series Editor: Michael Fairhurst, University of Kent, UK

This Book Series provides the foundation on which to build a valuable library of reference volumes on the topic of Biometrics. *Iris and Periocular Biometric Recognition, Mobile Biometrics, User-centric Privacy and Security in Biometrics*, and *Hand-based Biometrics* are the first volumes in preparation, with further titles currently being commissioned. Proposals for coherently integrated, multi-author edited contributions are welcome for consideration. Please email your proposal to the Book Series Editor, Professor Michael Fairhurst, at: m.c.fairhurst@kent.ac.uk, or to the IET at: author_support@theiet.org.

Other Titles in the Series include:

Iris and Periocular Biometric Recognition (Christian Rathgeb and Christoph Busch, Eds.): Iris recognition is already widely deployed in large-scale applications, achieving impressive performance. More recently, periocular recognition has been used to augment biometric performance of iris in unconstrained environments where only the ocular region is present in the image. This book addresses the state of the art in this important emerging area.

Mobile Biometrics (Guodong Guo and Harry Wechsler, Eds.): Mobile biometrics aim to achieve conventional functionality and robustness while also supporting portability and mobility, bringing greater convenience and opportunity for deployment in a wide range of operational environments. However, achieving these aims brings new challenges, stimulating a new body of research in recent years, and this is the focus of this timely book.

Hand-based Biometric Methods and Technologies (Martin Drahanský, Ed.): This book provides a unique integrated analysis of current issues related to a wide range of hand phenomena relevant to biometrics. Generally treated separately, this book brings together the latest insights into 2D/3D hand shape, fingerprints, palmprints, and vein patterns, offering a new perspective on these important biometric modalities.

User-Centric Privacy and Security in Biometrics

Edited by
Claus Vielhauer

The Institution of Engineering and Technology

Published by The Institution of Engineering and Technology, London, United Kingdom

The Institution of Engineering and Technology is registered as a Charity in England & Wales (no. 211014) and Scotland (no. SC038698).

The Institution of Engineering and Technology
Michael Faraday House
Six Hills Way, Stevenage
Herts, SG1 2AY, United Kingdom

www.theiet.org

British Library Cataloguing in Publication Data
A catalogue record for this product is available from the British Library

ISBN 978-1-78561-207-7 (hardback)
ISBN 978-1-78561-208-4 (PDF)

Typeset in India by MPS Ltd
Printed in the UK by CPI Group (UK) Ltd, Croydon

Contents

14 Secure cognitive recognition: brain-based biometric cryptosystems using EEG   **325**
Emanuele Maiorana and Patrizio Campisi

15 A multidisciplinary analysis of the implementation of biometric systems and their implications in society   **353**
Vassiliki Andronikou, Stefanos Xefteris, Theodora Varvarigou, and Panagiotis Bamidis

Preface

This book is the latest in the recently established "IET Advances in Biometrics" Book Series. In fact, the history of this series dates back to the publication by the IET in late 2013 of "Age factors in biometric processing", which provided the impetus and set the pattern for an on-going series of books, each of which focuses on a key topic in biometrics. Each individual volume will bring together different perspectives and state of the art thinking in its topic area, shedding light on academic research, industrial practice, societal concerns and so on, and providing new insights to illuminate and integrate both specific and broader issues of relevance and importance.

Strategies for dealing with system security and personal data protection, topics of rapidly increasing importance in modern society, are – perhaps unsurprisingly – varied and very wide-ranging, and many different approaches have emerged over the years to improve the way in which we protect vital systems and sensitive data. Issues of importance in this context range from basic authentication to questions about confidentiality and privacy, and compiling a book which covers such a diverse landscape is both timely and challenging.

This book will explore specifically the way in which developments in biometrics have influenced how we regard and address privacy and security issues. Although, originally, the primary aim of biometrics was seen as offering an alternative to less robust ways of establishing or verifying the identity of individuals, the role of biometrics more recently has developed significantly, leading to a whole array of security approaches, encompassing cryptographic processes, multibiometrics, soft biometrics, privacy and many other areas. This sort of work has been greatly stimulated by the emergence of complex problems arising from, for example, the management of very large databases, distributed system configurations, mobile computing platforms, reliability requirements and an increasing desire to put the user at the centre of system design and control. This book therefore provides a collection of state-of-the-art contributions which survey, evaluate and provide new insights about the range of ways in which biometric techniques can now enhance and increase the reliability of security strategies in the diverse range of applications encountered in the modern world.

The contributors come from a variety of backgrounds, and the volume overall represents an integration of views from across the spectrum of stakeholders, including, of course, academia and industry. We hope that the reader will find this a stimulating and informative approach, and that this book will take its place in the emerging series as a valuable and important resource which will support the development of influential work in this area for some time to come.

Other books in the series are in production, and we look forward to adding regularly new titles to inform and guide the biometrics community as we continue to grapple with fundamental technical issues and continue to support the transfer of the best ideas from the research laboratory to practical application. It is hoped that this book series will prove to be an on-going primary reference source for researchers, for system users, for students and for anyone who has an interest in the fascinating world of biometrics where innovation is able to shine a light on topics where new work can promote better understanding and stimulate practical improvements. To achieve real progress in any field requires that we understand where we have come from, where we are now, and where we are heading. This is exactly what this book and, indeed, all the volumes in this series aim to provide.

Michael Fairhurst
Series Editor, The IET Book Series on Advances in Biometrics

Part I

Introduction and interdisciplinary approaches

Chapter 1
The interplay of privacy, security and user-determination in biometrics
Claus Vielhauer[1,2]

On a general scale, doubtlessly system security has become a major challenge in modern society. Along with the enormous spreading of applications of technologies such as biometrics, cloud computing and Internet-of-Things (IoT), at the same time, the range and intensity of attacks to such systems has dramatically increased. Consequently, information technology (IT) security as the discipline of actively defending against such threats is a very active and flourishing domain in computer sciences, in which many different strategies have emerged over the years as how to improve the way in which we protect vital systems and sensitive data throughout the broad variety of technologies. The common goals in IT security research are the preservation of confidentiality, authenticity, integrity and privacy of information during the processing of data in the presence of existence of threats such as targeted attacks.

1.1 The technological view

With regards to biometrics, the rapid emergence of reliable biometric technologies has brought up a new dimension to this area of research. The utilisation of human's biometric traits, based on physiological and behavioural properties of natural persons towards automatic recognition of their identities promises accurate, convenient and efficient means for authentication of users in various day-to-day situations such as border crossings, access to buildings or personal devices such as smartphones. Being one piece of a larger puzzle of security infrastructures, biometrics can therefore be considered as developments towards new approaches to embedding security into systems and processes, and providing opportunities for integrating new elements into an overall typical security chain.

[1]Department of Informatics & Media, Brandenburg University of Applied Sciences, Germany
[2]Faculty of Computer Science, Otto-von-Guericke University of Magdeburg, Germany

However, as in most emerging technologies, there are downsides, which can be embraced for example by three observations:

- Biometrics is inherently highly individual and unique to persons, as its purpose is to robustly identify these across larger systems, geographical areas and time spans. This implicitly enables linkability and profiling of biometric data across system borders. For example, Minder and Bernstein study the potential to use face recognition methods in combination with text-based attributes towards social network aggregation, i.e. recognising identities across different social networks robustly [1].
- For the very same reasons, biometric data is highly sensitive data, as it may unveil information about health conditions, medications, intoxication or emotional states. Faundez-Zanuy and Mekyska for example describe in their chapter of this book, how symptoms of Parkinson's and/or Alzheimer's disease can be derived from online handwriting signals, as used in signature biometrics (see Chapter 2). Of course, this kind of observations raises strong privacy concerns.
- Further biometrics, being a concept to increase overall security, are vulnerable in itself against attacks in their design, implementation and operation. The study of potential and practical attack scenarios has been increasingly subject to research over the past view years and one way to comprehensive the broadness of potential attacks is to structure these as threat vectors[1] to a generic biometric system layout. Roberts, for example, identifies a total of 18 main threat vectors, as summarised and explained in Table 1.1 [2]. These are reaching from physical attacks to sensor functionality, with the goal to achieve denial of service of the authentication system, over spoofing scenarios such as fake physical/digital data presentations for impersonation, all the way to attacks on system level, for example by overriding system internal information such as decision signals or template data.

In addition, Voloshynovskiy *et al.* identify 12 imposter attack scenarios, which are common to biometric and physical object identification systems in Chapter 4 of this book. They categorise into three cases, where attacks are (i) derived directly from available biometrics, (ii) from traces left behind and (iii) based on system level alterations.

In the presence of these three problematic and challenging areas, and inherently to the very concept of biometrics, at the same time there has been a continuous trend over the past years towards distributed, ubiquitous computing concepts such as cloud computing and technologies exploring huge amounts of data for processing and classification tasks, i.e. big data analysis and deep learning. In view of this light, it needs to be considered also that there is a tendency for biometric systems to move from local identification systems, to such distributed, ubiquitous and packed structures. This implies that the entire biometric infrastructure, including sensor, network, server and data security, which used to be under control and responsibility of the operators, is now also exposed to the cloud.

---

[1]Threat vectors describe the path and/or method utilised used by a threat to reach a target. It is a term widely common in IT security analysis.

Table 1.1   Biometric system threat vector categories [2] and their short explanations

Threat vector according to Roberts [2]	Short explanation
Denial of service	Attack targets to make biometric systems/services unavailable to legitimate users in part or as whole
False enrolment	Fake production of identities during the registration process. Requires attacker to creep in as legitimate user during enrolment phase at least once
Fake physical biometric	Physical generation of fake biometric samples to circumvent the system by impersonation
Fake digital biometric	Ditto, but in digital domain. Can be achieved for example by signal injection
Latent print reactivation	Copying, enhancement and re-activation of residues left behind on biometric sensors (e.g. latent fingerprints) for impersonation
Reuse of residuals	Misuse of left behind biometric data in the computer memory of local systems
Replay attacks/false data inject	Based in intercepting legitimate data transmission in the biometric pipeline and re-injecting it at a later point in time
Synthesised feature vector	Generation and injection of feature vectors into the biometric processing pipeline
Override feature extraction	Deactivation of the original feature extraction module within a system and overriding the feature extraction result by a version modified by the attacker
System parameter override/ modification	Modification or overriding of system parameters allowing illegitimate use. For example, a decision threshold can be modified in order to increase false-acceptances by relaxation of the similarity requirement
Match override/false match	Functional overriding of the matching score-level result within the biometric system to grant illegitimate access by falsely increased modified matching scores for example
Storage channel intercept and data inject	Interaction and modification of the transmission between the template storage and the comparison process. This can be achieved for example by injecting faked templates prior to the matching
Unauthorised template modification	Modification of references in the data storage itself, for example a database, towards the attacking goals
Template reconstruction	Attempts to reconstruct original biometric data or even physical samples by information recovered from biometric templates. For example, an artificial fingerprint can be constructed from minutiae-based template data
Decision override/false accept	Completely override the overall system decision at the last stage of the biometric authentication process
Modify access rights	Modification of access rights to user data such way, that data required for the biometric authentication can no longer be accessed, thus resulting in a denial-of-service

(Continues)

Table 1.1   (Continued)

Threat vector according to Roberts [2]	Short explanation
System interconnections	In case the biometric system is integrated in a larger infrastructure of multiple components, the intercommunication between components may be attacked toward various attack goals, such as denial-of-service
System vulnerabilities	Exploitation of vulnerabilities of the platform(s) on which the biometric system is executed. This includes, but is not limited to, operating system and device driver implementation vulnerabilities

With regards to sensor technology, there have been a dramatic developments regarding the acquisition potentials. On the one side, for example biometric sensory nowadays can be considered almost ubiquitously available due to the fact that practically all new personal devices such as smartphones, tablets or notebooks are equipped with high-resolution cameras, microphones and an increasing fraction of these even with fingerprint sensors. In addition, the significantly improved resolution and image quality of camera sensors (typically charge-coupled-devices) allow innovative acquisition concepts. To give just one example, Venugopalan *et al.* propose a long range iris acquisition sensor, built from off-the-shelf components, which allows the capturing of iris images with resolutions of 200 pixels in diameter from a distance of 8 m between sensor and subject. Even at a distance of 13 m, the system can still allow taking iris images with a resolution of 150 pixels [3].

1.2   Some societal, ethical and legal views

From a societal perspective, in order to protect citizens from misuse of their private data, many countries are increasingly adopting and specifying more precisely their legal privacy laws. For example, the European Union has recently specified a common data protection directive, which legally regulates the collection, storage and processing of personal data such as biometrics across all member states [4].

In Chapter 15 of this book, Andronikou *et al.* discuss the actual landscape of biometrics in various scales and in the presence of IoT, cloud and ubiquitous computing. They further illuminate impacts of very practical aspects of large-scale biometric implementations such as the Indian Unique ID project and reflect on social and secondary impacts of all these observations. Discussions relating to social impacts can further be structured by their effects to individuals, societies as a general or especially marginalised groups. Gelb and Clark for example report various cases of exclusion of person groups due to failure to enrol to fingerprint systems, determined due to worn out fingers of tobacco planters for example. Other concerns addressed are the potential exclusion of children or people having religious concerns in taking their biometrics, for example facial images [5].

In summary, Pato and Millett conclude the challenges arising from the interplay of broad biometric application and the technical, legal and social developments as follows [6]:

> *Although biometric systems can be beneficial, the potentially lifelong association of biometric traits with an individual, their potential use for remote detection, and their connection with identity records may raise social, cultural, and legal concerns. Such issues can affect a system's acceptance by users, its performance, or the decision on whether to use it in the first place. Biometric recognition also raises important legal issues of remediation, authority, and reliability, and, of course, privacy. Ultimately, social, cultural, and legal factors are critical and should be taken into account in the design, development, and deployment of biometric recognition systems.*

These perceptions impose additional security and privacy challenges, including:

- better and more accurate biometrics (e.g. iris from a distance, allowing fusion of face and iris biometrics)
- fusion of soft-, hard biometrics and other personal profiling information (e.g. smartphone localisation)
- ubiquitous and seamless data collections, both of biometric and non-biometric data (e.g. video and telecommunication surveillance)
- vast and widely uncontrolled/unregulated collection, aggregation and deep analysis of personal, private data, including biometric data, beyond the control of the owner of it
- limited awareness of the risks of unsuspective to naive authorisation of user's for web services to utilise their own private data
- limited control of individuals of their personal credentials and biometric data in circulation

This chapter is an effort to suggest a structure for all these terms and aspects by means of a rather simplistic taxonomical proposal. The purpose of it is to allow a mapping of this theoretical frame to the variety of topics addressed in the remaining chapter contributions of this book in order to provide thematic guidance to the readership.

1.3   A taxonomical approach for discussions

There are various ways to systematically reflect the privacy and security considerations of biometrics. For example, of course there are technical categories, which may group the variety of protection schemes for biometric data by means of their underlying technologies, such as cryptography or mathematical projections for securing biometric templates or blurring for de-identification of biometric information.

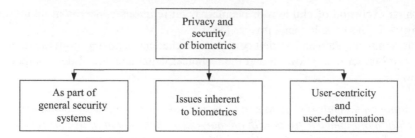

Figure 1.1   Three main aspects of privacy and security from a holistic view

When trying to organise the considerations from a rather holistic view on how biometrics affect privacy and security of large-scale, complex, dynamic and mobile IT infrastructures, including both the technical and societal/ethical considerations introduced in the previous part of this chapter, a possible 3-fold structure can be derived, as illustrated in Figure 1.1.

The **first aspect** focuses on biometrics as one **complementary authentication concept being part of broader security systems**: Initially, biometrics was seen primarily as offering a more or less isolated alternative to less robust ways of ensuring the identity of individuals engaging in important and sensitive transactions in digital domain. More recently, the role of biometrics has developed significantly and has spawned a whole new array of security approaches, including convergence with cryptographic processes (i.e. biometric cryptosystems), multi-biometrics, multi-factorial authentication and in presence of ubiquitous smart devices. Particularly, the widespread use of smartphones allow for the aggregation of all kind of personal-and-non-personal data on the device itself and also in connected cloud systems. This makes data generated or stored on such devices potentially much more prone to theft, cyberattacks and malware infection. The trend towards complete cloud storage of data allows for an increasing tendency of nomadic users, and it comes together with various new technologies of high complexity, like big data analysis and deep learning. In view of this, an essential question is how to protect and ensure the trust in and security of sensitive and private biometric data in potentially untrusted, semi-honest settings and environments?

Privacy-enhancing technologies (PET) have been introduced recently to provide technical solutions to these kind of challenges. For example, the concept of oblivious information retrieval has been suggested by Huang *et al.* to realise new backtracking protocols allowing for comparing and matching biometric references between two parties [7], whereas Nagar *et al.* discuss the security of PET on the example of key-based transformation methodology a vulnerability analysis on two concepts (Biohashing and cancellable fingerprints) [8].

In attempting to categorise the scopes of all these works, one could think of the title 'Privacy and security of biometrics as part of general security systems', i.e. discussions on impacts arising from the integration of biometric concepts into larger, holistic and rather complex infrastructures to provide security.

The **second aspect** involves **security and privacy** challenges **inherent to biometric systems**. From this **'within-biometrics' perspective**, due to the continuous and ubiquitous propagation of practical biometric recognition techniques on a large scale, increasingly an additional manifold of new requirements and challenges arise. Of course, there are many issues on performance, accuracy, interoperability, standardisation, etc. But in addition to these rather metric-oriented optimisations, there obviously are challenges with regards to the privacy of biometric data in context of (very large) databases, soft biometric profiling, biometric recognition of persons across distributed systems and in nomadic scenarios, as well as the convergence between user convenience, usability and authentication reliability. To give a few examples, the impact of biometric recognition accuracy in case of sensor-interoperability is an area of actual research, as studied by Ross and Jain [9]. Part of this 'within-biometrics' aspect is also the importantly relevant research on template protection, i.e. protection of the reference data within biometric systems against unauthorised access and misuse by design. Rathgeb and Busch for example present a very comprehensive survey on this field in their chapter 'Biometric template protection: state-of-the-art, issues and challenges' (see Chapter 8). Another important challenge is the active detection of biometric spoofing attacks, for example by replay in the analogue domain in front of biometrics sensors. In this book, for the modality of speech, Korshunov and Marcel reflect recent advances in detection of such presentation attacks by means of signal processing and pattern recognition, as well as reflections on integrating such detection mechanisms within automatic speaker recognition systems (see Chapter 10). To sum up, this second perspective covers challenges such as de-identification, template and data protection, etc. and can be titled as 'Security and privacy issues inherent to biometrics'.

A **third perspective** in the proposed taxonomy addresses **user-centric aspects** in biometrics **to ensure privacy**. All privacy legislation is established on the idea of individuals having the civil right to keep track of and control the acquisition, usage purpose, storage, collection and deletion of personal data. For this purpose, legislation of most of modern societies and countries derive the essential concepts of obliging institutions to take measures to protect any kind of personal data against misuses and at the same time granting citizens the right to self-determination with regards to their sensitive data. Increasingly, in order to reliably achieve the aforementioned obligations of institutions and citizen rights, the concept of **privacy by design and by default** is being deployed to IT systems, which includes the following main requirements:

- technical and organisational measures need to be taken throughout the entire processing life cycle,
- the application design must be in such way that personal data is only processed when required for a specific purpose,
- mechanisms for users are integrated and activated in such way that they are able to actively access and control their personal data in order to empower them to grant, modify or revoke their informed consent to processing of their data, and

• data protection mechanisms should be enabled at maximum level always by default, enabling users to grant exclusively by their explicit will ('opt-in') based on their informed consent.

The societal and ethical needs are strongly supported by a publication of Fischer-Hübner *et al.* They sketch guidelines for self-determination of personal data and identities in the Internet, under considerations of the aforementioned European legal doctrines and they strongly claim the need for transparency enhancing tools (TETs), allowing end users to achieve transparency, visibility and user-centricity in the processing of their private data [10].

   Of course, all this should have an impact not only to IT systems involving biometric authentication, but also to all other applications, which are processing personal data. For example, social networks and also operating systems provide reluctantly, yet increasingly means to end users to configure access to, and use of their private data by default. Although there are good arguments saying that the consideration of privacy still is insufficient to some degree, the fact that commercial services provide such control mechanisms – which might appear somewhat contradictory to their underlying business model – indicates a stronger push towards the acceptance of claiming personal rights from legal, social and increasingly industrial principals.

   For the specific areas of subject of this book, biometrics, it can be foreseen that privacy by design, PET and TET could form the technological basis for future user-centric, self-determined management of sensitive biometric data. It seems that three concepts, as suggested by Barocas and Nissenbaum, could play an important role for user-centric privacy in biometrics:

• **Informed consent attempts**, i.e. collection, handling and processing of data into matters of individual choice,
• **Anonymisation**, i.e. rendering of privacy concerns irrelevant by decoupling data from identifiable subjects and
• **Fair information practice principles**, i.e. general, common and broad concepts that form a set of principles that constitute data protection with regards to their substantive (e.g. data quality, use limitation) and procedural (e.g. consent, access) collection, storage and processing [11].

Although in their publication [12], Barocas and Nissenbaum focus on privacy management challenges in Big Data scenarios close to the domain of biomedicine, it can be assumed that they are of high relevance for biometrics well.

   Regarding potential PET realisations, prior to this book, Vielhauer *et al.* suggest a stringent system design along the biometric pattern recognition pipeline, which considers each of the necessary (sub-)processes in biometric reference generation and authentication, respectively. Potential relevant PET as suggested in literature is discussed and an exemplary study on how PET for biometrics can be achieved by the Homomorphic Encryption methods indicates a possible roadmap [13]. Other ways to implement PET could be methods of de-identification, which allow to adjust or limit the amount of information in media date, which can be exploited for biometric

recognition. Within their chapter on 'De-identification for privacy protection in biometrics' as part of this book, Ribaric and Pavesic reflect on recent advances within
this kind technology (see Chapter 13). Other potential technological ways towards
user-centric management of biometric data could be paved by transformation-based
biometric cryptosystems. As shown by Jassim in another chapter contribution of this
works, random projections could be another key concept towards de-identification
and cancellability of biometric templates and protection of the private biometric data.
In the Jassim approach, pools of random orthonormal sparse matrices are to be used
for de-identifying biometric data whereby only the user stores the transformation
parameters while service providers only store the transformed biometric template
captured at enrolment (see Chapter 12).

The relevance of user-centric privacy can also be seen by the establishment
of research networks such as AMBER ('enhAnced Mobile BiomEtRics', Marie
Sklodowska-Curie Innovative Training Network) as part of the EU Horizon 2020
Research & Innovation programme, which aims to address a range of research issues.
This includes protection of data and the management privacy of personal information
within mobile biometric systems, to enhance confidence in their usage [14].

1.4   Contributions of this book

As outlined in the taxonomical discussions in the previous sections, it can be expected
that privacy and security aspects to biometrics will be amongst the more critical
concerns in upcoming technologies, and this book will therefore try to explore the
way in which developments in biometrics will address various security aspects in
relation to a consideration of privacy. The book is organised in to four parts:

In Part I, 'Introduction and interdisciplinary approaches', the editor attempts to
set the scene for readership by proposing the taxonomic view. The three perspective
branches developed here shall be the blueprint for the further three parts, whereas
Part I objects to further complement these three areas by a space for discussions on
an interdisciplinary scope. To this end, in this part, two interdisciplinary chapters
present works, which appear to bridge gaps to two related research domain with interesting links to biometrics: as already referred to in the introduction to this chapter,
Faundez-Zanuy and Mekyska present interesting insights in the link between biometrics, privacy and medical analysis on the example of handwriting in their chapter
'Online handwritten analysis for biomedical applications' (see Chapter 2). In the
chapter 'Privacy concepts in biometrics: lessons learned from forensics', Dittmann
and Kraetzer present relevant insights to potential impacts of forensic analysis to the
conceptional design of biometrics in future. This includes relevant findings regarding privacy-preserving analysis of fingerprints from the domain of digitised crime
scene forensics. One example here are considerations regarding de-identification of
fingerprint images by resolution limitation, in a sense that data is acquired in such
way, that while specific properties (e.g. the age of a fingerprint) can be estimated
based on de-identified biometric or forensic samples, there is no possibility to use
the sample for identification. Another bias of this chapter is to study the potential

of face morphing with regards to attacks to biometric systems and their privacy (see Chapter 3).

Following the taxonomy suggested by Figure 1.1, Part II, is titled 'Privacy and security of biometrics within general security systems'. In the contributions to this part, the reader may find interesting views on how to relate biometrics to physical layer security concepts in the chapter 'Physical layer security: biometrics vs object fingerprinting'. Voloshynovskiy, Holotyak and Diephuis here present biometrics from the viewpoint, that it can be considered as one specialised variation of a more generalised problem of automatically recognising individual physical objects or phenomena. Besides detailed reflections on common signal processing/pattern recognition models for this, authors address security issues of such concepts. They do so by performing an attack analysis, as already referred to Chapter 4. Sanchez-Reillo focuses in his chapter 'Biometric systems in unsupervised environments and smart cards: conceptual advances on privacy & security' on impacts of practical uses of biometrics in unsupervised scenarios by categorising possible attacks to use cases: Hereby, he categorises depending on the question if these use cases can be considered as supervised or non-supervised situations. Further, the chapter presents a discussion on possible technical approaches to secure against such threats by utilising smart-card technology. For this purpose, a technical introduction with special regards to application biometric context is included (Chapter 5). As the title of the third chapter in the second part suggests – 'Inverse biometrics and privacy', Gomez-Barrero and Galbally present a thorough state-of-the art review to various methods to reconstruct or generate biometric data under various assumptions and in different application scenarios. They identify inverse biometrics being one specific branch of synthesising biometric data, and they expand on ways as how to achieve this under consideration of different levels of knowledge about an actual biometric (sub-)system. Further, this chapter discusses concepts as how to evaluate such inversion schemes (see Chapter 6). In presenting an own original method to achieve multi-layer security by combining secret sharing, biometrics and steganography, Tran, Wang, Ou and Hu contribute one example as how to achieve privacy protection in use cases to combine photo imagery with biometric properties of users or owners. In their chapter 'Double-layer secret sharing system involving privacy preserving biometric authentication', they exemplify one specific showcase to what can be considered as 'General security systems' (see Chapter 7).

To set the scene for the contributions in the third part 'Security and privacy issues inherent to biometrics', Rathgeb and Busch present a comprehensive review of today's technological potential for the protection of biometric reference data, as already mentioned earlier in this chapter. In their contribution 'Biometric template protection: state-of-the-art, issues and challenges', they review methods from three categories: biometric cryptosystems, cancellable biometrics and multi-biometric template protection. In their survey part, authors present an enormous number of references to the most relevant works, which may of high interest for the readership. At least of the same level of interest seem their reflections on issues and challenges in this area. They identify three sub-domains: performance decrease under presence of protection schemes, data representation and feature alignment, as well as standardisation and deployment issues (see Chapter 8). Feature analysis and feature selection methods are believed to be important practices for optimising recognition accuracy in pattern recognition in general, but also in biometrics. Specifically, for one biometric modality, namely the on-line handwriting, Scheidat addresses this research domain by suggesting empirical optimisation of biometric algorithms in his chapter 'Handwriting biometrics – feature-based optimisation'. He suggests to apply eight different feature analysis/selection methods from the categories of wrappers and filters for a given training set and presents experimental studies on the impact of such optimisations by comparative analysis of the recognition accuracy. In this chapter, the outcome of the feature selection process to two reference algorithms is validated. It turns out that significant reduction in error rates go along with feature space reduction by a factor of around five (see Chapter 9). Korshunov and Marcel's contribution to this book 'Presentation attack detection in voice biometrics', which also has been mentioned before in this chapter, also focuses on one single modality, in this case the voice. In analysing various attack scenarios for this on a general scale, they focus on attacks at the first stage of a speech-based biometric system: the capturing/speech acquisition, which are commonly referred to as presentation attacks. They elaborate on the experimental analysis of such attacks to voice biometrics and motivate the concept of actively implementing presentation attack detection and integrating these in the biometric process pipelines (see Chapter 10). The validation of originality of biometric images and data is also the goal of the chapter 'Benford's law for classification of biometric images' contributed by Iorliam, Ho, Poh, Zhao and Xia. They explore the utilisation of Bendford's law, also known as the first digit law, for classification of given biometric images by their sensor source. While this idea had already been pursued earlier for forensic imagery, authors extent the concept towards biometric imagery taken from renown biometric databases for face and fingerprint modality. Experimental results included in this chapter indicate that their method, based on a Neural-Network classifier and features derived from Bendford's law, is capable to correctly classify the sensor source with accuracies of 90% and higher (see Chapter 11).

The concluding part of this book shall combine the user-centric aspects, as introduced in the taxonomy suggested in this chapter, with some considerations regarding the future of biometrics. Introducing to this part 'User-centricity and the future' of this book, Jassim recapitulates on the evolution of biometrics over the past 25 years, not only for biometrics, but also keeping the recent developments in communication technologies, Cloud computing and IoT in view. This general reflection may give valuable impulses to future considerations and as summarised earlier, the application of using Random Projection methods to overcome some of the future challenges is proposed in addition in the chapter 'Random projections for increased privacy' (see Chapter 12). As mentioned earlier in this chapter, de-identification may be one of the most promising concepts for user-centric, transparent protection of privacy in the application of biometrics, as summarised by the chapter on 'De-identification for privacy protection in biometrics', authored by Ribaric and Pavesic. Here, a clarification of terminology is presented with a thorough survey of technical methods which have been suggested so far to achieve de-identification for physiological, behavioural and soft-biometric modalities (see Chapter 13). A possible future approach towards user-centricity based on electroencephalography (EEG) measurements is discussed by Campisi and Maiorana in their chapter 'Secure cognitive recognition: brain-based biometric cryptosystems using EEG'. They follow the idea of Cognitive Biometric Cryptosystems, i.e. analysing the impact of cognitive tasks to measurable brain activity and deriving biometric information from it in such way, that it is appropriate for further inclusion in cryptographic concepts. Besides detailed technical explanations of their own method, authors present results from experimental studies based on EEG activity measurements from subjects in two setups (eyes-closed resting and eyes-open following a light point on a screen), which indicate the concept works at a current level of approximately 5% Half-Total-Error-Rate, as reported in Chapter 14.

In the concluding two chapters of this book, authors reflect on social, ethical, historical and philosophical perspectives and developments of identity, identification of humans in general. Further, these two contributions round up the book by linking such general perceptions to biometric technologies with regards to their privacy and security considerations.

The chapter 'A multidisciplinary analysis of the implementation of biometric systems and their implications in society' by Andronikou, Xefteris, Varvarigou and Bamidis allows the reader to enable again another holistic mode of thinking, with reflections on the social impacts of biometrics. Authors here present an interdisciplinary discourse about the practical and social effects of large-to-huge scale application of biometrics in presence of other emerging trends in IT technology. By suggesting another very interesting line of thinking in this chapter, the potential to use biometrics for cross-linking and information aggregation in social media, as well as other secondary exploitation or even misuse of biometrics for purposes other than the intended are also discussed (see Chapter 15). Finally, Mordini steps back in history in emphasising the importance of identity for the development of personality form the ancient times to what we understand as a modern society. In his chapter 'Biometrics, identity, recognition and the private sphere: where we are, where we go', a brief introduction to the terms 'Identity' and 'Recognition' from a philosophical point of view is provided, and the author then recapitulates major milestones in the history of personal recognition from the ages of Neolithic Revolution to date. From this, perspectives are derived, how biometrics may influence privacy, person and human dignity in future from a philosophical viewpoint rather from a legal, such as for example data protection regulations. To illustrate these prospects, Mordini stresses two currently highly relevant challenges in global societies: refugees and digital economies (see Chapter 16).

1.5 Proposed reading and acknowledgements

As illustrated in this chapter, the structure of this book is intending to allow readers to explore the various aspects of privacy and security of and within biometrics in a top-down-top manner. With the proposed taxonomy and their mapping to the individual chapters of the book, readers should be able to step down into the individual parts of the book, to identify the scopes of each of it and also to understand the interlinks of biometrics to related disciplines such as, for example, medical computing or cryptography. On one side, this should allow – depending on the professional or academic background and the particular interest of readers, specific and targeted reading of chapters. On the other side, in following the structure by the proposed parts, this book can also be accessed as a text book. This way, hopefully, the book will allow readers such as students, teachers, researchers or professionals to explore the various poles of biometrics as part of broader security concepts and objectives, including security and privacy within and for biometrics, protection of biometric data, concepts towards informed consent of data usage, transparency on biometric data and biometric data fraud prevention.

Of course, this book project has been a team effort and could be successful only because of the strong support by many protagonists, to which the editor would like to express his great thankfulness. In first place, the authors of the remaining 15 chapters certainly have the greatest share in the overall merit. The quality of their individual presentations, their thematic interlinks, as well as the compilation of all of them as a whole, constitute the core scientific value of this publication. Second, the editorial team of IET has been of enormous help during the entire process of planning, compiling, typesetting (and, of course, sometimes error correcting), and completing this book. Jennifer Grace, Olivia Wilkins and Paul Deards have been extraordinarily supportive, motivating and always constructive in their advises. The editor would thank Mike Fairhurst for igniting the idea of a book series on biometrics published by IET, for his motivating words when starting this particular book project and for his valuable advises here and there during the many months of the editorial process. There have been anonymous reviewers who helped in producing constructive advises to authors, which the editor and author of this first chapter is very thankful for and last not least, the editor would express his gratitude to his family, which has contributed the most valuable asset to the book: a great share of time.

References

[1] Minder, P., Bernstein, A.: 'Social Network Aggregation Using Face-Recognition', Proc. Fourth Int. Workshop on Social Data on the Web Workshop (SDoW2011), Bonn, Germany, October 2011.
[2] Roberts, C.: 'Biometric Attack Vectors and Defences', Computers & Security, 2007, 26, pp. 14–25.
[3] Venugopalan, S., Prasad, U., Harun K., et al., 'Long Range Iris Acquisition System for Stationary and Mobile Subjects', Proc. 2011 International Joint Conference on Biometrics, October 11–13, 2011, pp. 1–8.
[4] The European Parliament and the Council: 'Directive (EU) 2016/680 of the European Parliament and of the Council of 27 April 2016 on the protection of natural persons with regard to the processing of personal data by competent authorities for the purposes of the prevention, investigation, detection or prosecution of criminal offences or the execution of criminal penalties, and on the free movement of such data, and repealing

Council Framework Decision 2008/977/JHA', http://eur-lex.europa.eu/legal-content/EN/TXT/?uri=uriserv:OJ.L_.2016.119.01.0089.01.ENG&toc=OJ:L:2016:119:TOC, accessed May 2017.

[5]   Gelb, A., Clark, J.: 'Identification for Development: The Biometrics Revolution', Center for Global Development, Working Paper 315, Jan. 2013, pp. 4–49, https://www.cgdev.org/sites/default/files/1426862_file_Biometric_ID_for_Development.pdf, accessed May 2017.

[6]   Pato, J.N., Millett, L.I. (Ed.): 'Biometric Recognition: Challenges and Opportunities', (Washington, D.C.: The National Academic Press, 2010), pp. 85–115.

[7]   Huang, Y., Malka, L., Evans, D., Katz, J.: 'Efficient Privacy-Preserving Biometric Identification', Proc. Network and Distributed System Security, 2011.

[8]   Nagar, A., Nandakumar, K., Jain, A.: 'Biometric Template Transformation: A Security Analysis', Proc. SPIE 7541, Media Forensics and Security II, January 2010.

[9]   Ross, A., Jain, A.: 'Biometric Sensor Interoperability: A Case Study in Fingerprints', Proc. ECCV Workshop BioAW, 2004, pp. 134–145.

[10] Fischer-Hübner, S., Hoofnagle, C., Krontiris, J., Rannenberg, K., Waidner, M.: 'Online Privacy: Towards Informational Self-Determination on the Internet', Dagstuhl Manifestos 11061, Vol. 1, Issue 1, Dagstuhl Publishing, 2011, pp. 1–20.

[11] Cate, F.H.: 'The Failure of Fair Information Practice Principles'. Consumer Protection in the Age of the Information Economy, (2006), pp. 1–2, available at SSRN: https://ssrn.com/abstract=1156972.

[12] Barocas, S., Nissenbaum, H.: 'Big Data's End Run Around Procedural Privacy Protections', Communications of the ACM, 2014, 57.11, pp. 31–33.

[13] Vielhauer, C., Dittmann, J., Katzenbeisser, S.: 'Design Aspects of Secure Biometric Systems and Biometrics in the Encrypted Domain', In: Campisi P. (eds) 'Security and Privacy in Biometrics.' (London: Springer, 2013), pp. 25–43.

[14] AMBER: enhAnced Mobile BiomEtRics, https://www.amber-biometrics.eu/, accessed 16 October 2017.

Chapter 2

Privacy of online handwriting biometrics related to biomedical analysis

Marcos Faundez-Zanuy[1] and Jiri Mekyska[2]

Online handwritten signals analysis for biomedical applications has received lesser attention from the international scientific community than other biometric signals such as electroencephalogram (EEG), electrocardiogram (ECG), magnetic resonance imaging signals (MRI), speech, etc. However, handwritten signals are useful for biometric security applications, especially in the case of signature, but to support pathology diagnose/monitoring as well. Obviously, while utilising handwriting in one field, there are implications in the other one and privacy concerns can arise. A good example is a biometric security system that stores the whole biometric template. It is desirable to reduce the template to the relevant information required for security, removing those characteristics that can permit the identification of pathologies.

In this paper, we summarize the main aspects of handwritten signals with special emphasis on medical applications (Alzheimer's disease, Parkinson's disease, mild cognitive impairment, essential tremor, depression, dysgraphia, etc.) and security. In addition, it is important to remark that health and security issues cannot be easily isolated, and an application in one field should take care of the other.

2.1 Introduction

Online handwritten biometrics belongs to behavioural biometrics because it is based on an action performed by a user. This is opposed to morphological biometrics, which is based on direct measurements of physical traits of the human body. From human behaviour and health condition point of view, it appears more appealing than other hard biometrics such as fingerprint or iris. Although health applications based on online handwriting today have not been deeply explored, there is a nice set of possibilities that will probably grow in the future, such as diagnosis/monitoring of depression, neurological diseases, drug abuse, etc. It can be noted that nowadays,

[1]Pompeu Fabra University, Spain
[2]Department of Telecommunications, Brno University of Technology, Brno, Czech Republic

most of the published research in biometric signal processing is based on image and speech, reasons for which can be that these signals are easier to acquire and cheaper than online handwriting tasks. The price of a webcam or a microphone has been low since the past century, while digitizing devices for online handwritten tasks was by far more expensive. Fortunately, in the recent years, tactile screens have become more popular and online handwritten signals are more present in the society than a few years ago. This has permitted a reduction in the cost of acquiring devices. Thus, nowadays, the price of the acquisition device is not a drawback anymore. We can forecast a growing in applications in this field, and we should take care of privacy issues. In this chapter, we will present an introduction to online handwritten signals and discuss several applications of them in the medical field, which we consider relevant for the biometric community.

This chapter is written for signal-processing engineers devoted to security biometric applications. Even if readers have a background in speech and/or image but are not familiar with online handwritten signals, they will find an explanation including fundamentals of the acquisition process as a starting point. However, and even more challenging, this part of the book is also written for people outside the biometric community, including the audience of medical doctors, willing to enter into this topic and collaborate with engineers. Today, it seems hard to establish collaborations between engineers and medical doctors. Quite often, we do not understand each other due to our different background. Thus, we tried to write the chapter in an easy-to-read way. Breaking innovations are hardly produced in the core of a knowledge area, and the main contribution is seen rather in terms of focussing on the borders between different areas.

The structure of this chapter is as follows: Section 2.2 introduces to the properties and characteristics of the acquisition devices as well as the online handwritten signal. Section 2.3 is devoted to examples of implications between both fields, security and health, with special emphasis on those situations where the privacy of the user can be compromised, and the authentication task is performed under pressure or without consciousness of the users (e.g. suffering a severe disease). Section 2.4 summarizes the chapter.

2.2   Online handwritten signals – an introduction

Online handwritten signals acquisition consists of dynamic acquisition of various properties of the moving pen during the writing process in real time, whereas the digital representation of the signals is typically given by time-stamped sequences of measurement points/tupels. For instance, using a digitizing tablet, smartphone, etc., which typically acquires information listed in Table 2.1.

Using this set of dynamic data, further information can be inferred by analytical computation, which is usually more suitable for certain applications (e.g. handwriting velocity, duration, width, height). This results in what is usually called feature sets, being similar to the use of body mass index for overweight classification. Body mass

Table 2.1   Information acquired from a digitizing tablet

Abbreviation	Description
x	Position of pen tip in x axis
y	Position of pen tip in y axis
s/a	On-surface/in-air pen position information
p	Pressure applied by the pen tip
az	Azimuth angle of the pen with respect to the tablet's surface (see Figure 2.1)
al	Altitude angle (sometimes called tilt) of the pen with respect to the tablet's surface (see Figure 2.1)
t	Timestamp

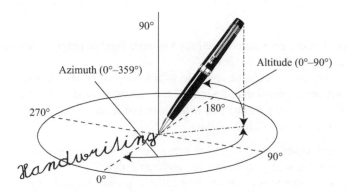

Figure 2.1   Handwriting online information acquired in typical cases (x and y position, pressure, azimuth, altitude)

index is not a direct measure. In fact, it is based on weight and height but it is more useful than body/weight alone.

2.2.1   In-air and on-surface movements

Some digitizing devices, such as Intuos Wacom Tablet™, Samsung Galaxy Note™, etc., are able to track the pen-tip movement even when it is not touching the surface. Thus, it is possible to record the x and y coordinates of in-air movements when pressure is equal to zero. Unfortunately, this is only possible when the distance between the tip of the pen and the surface is less or equal to approximately 1 cm, otherwise the tracking is lost. Nevertheless, the time spent in air is still known because the acquisition device provides a timestamp of each sample. By looking at the difference between consecutive samples, it is possible to know the exact amount of time spent in-air,

Figure 2.2   Illustration of the distance from pen tip to surface

although the x and y coordinates are only known when the height is smaller or equal to 1 cm (see Figure 2.2).

While some devices can be operated with a sheet of paper and a special ink pen, others do not permit this kind of pen, and the handwriting must be directly done on the tablet's surface using plastic pen without an immediate visual feedback.

Thus, we know three kinds of data:

1. Movement on-surface: typically provides the five features described in the previous section (x, y, pressure, azimuth, altitude).
2. Movement in-air at short distance to surface: provides x and y position, azimuth and altitude.
3. Movement in-air at long distances to surface: when distance is higher than approximately 1 cm, we only know the time spent in-air, as no samples are acquired.

Figure 2.3 shows the aspect of raw samples acquired by a digitizer. For each sampling instance, a set of features is acquired: x coordinate; y coordinate; timestamp t provided by the machine; surface/air bit s/a, which is equal to zero when there is no contact between tip of pen and surface, and one where there is contact; pressure value p; azimuth az and altitude al. In this example, we may observe some samples in-air at short distance plus some time in-air (between $t = 11{,}253{,}657$ and $11{,}253{,}827$), with a subsequent measurement at long distance. This can be observed because the jump in timestamp between $t = 11{,}253{,}827$ and $11{,}253{,}843$ is higher than the usual sampling rate for on-surface samples. For the later, the time-stamp progress in t is 10 units, while for the last sample in-air at short distance, it is 16 time units. Time in-air at long distance can appear after in-air at short distance before touching again the surface. For most of the users and tasks, this time is negligible, because movements between strokes tend to be short.

Looking at Figure 2.3, we observe that raw data provided by digitizing tablet is really simple in structure and thus can be processed in a straightforward way, even

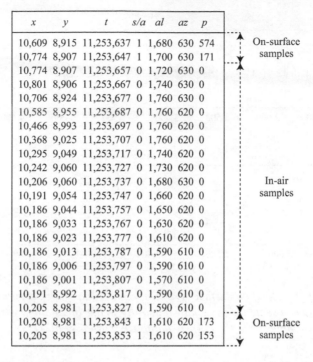

x	y	t	s/a	al	az	p	
10,609	8,915	11,253,637	1	1,680	630	574	On-surface samples
10,774	8,907	11,253,647	1	1,700	630	171	
10,774	8,907	11,253,657	0	1,720	630	0	
10,801	8,906	11,253,667	0	1,740	630	0	
10,706	8,924	11,253,677	0	1,760	630	0	
10,585	8,955	11,253,687	0	1,760	620	0	
10,466	8,993	11,253,697	0	1,760	620	0	
10,368	9,025	11,253,707	0	1,760	620	0	
10,295	9,049	11,253,717	0	1,740	620	0	
10,242	9,060	11,253,727	0	1,730	620	0	
10,206	9,060	11,253,737	0	1,680	630	0	In-air samples
10,191	9,054	11,253,747	0	1,660	620	0	
10,186	9,044	11,253,757	0	1,650	620	0	
10,186	9,033	11,253,767	0	1,630	620	0	
10,186	9,023	11,253,777	0	1,610	620	0	
10,186	9,013	11,253,787	0	1,590	610	0	
10,186	9,006	11,253,797	0	1,590	610	0	
10,186	9,001	11,253,807	0	1,570	610	0	
10,191	8,992	11,253,817	0	1,590	610	0	
10,205	8,981	11,253,827	0	1,590	610	0	
10,205	8,981	11,253,843	1	1,610	620	173	On-surface samples
10,205	8,981	11,253,853	1	1,610	620	153	

Figure 2.3   *Example of digital representation of samples acquired with digitizer in two scenarios: on-surface, in-air. x – x position, y – y position, t – timestamp, s/a – on-surface/in-air pen position information, p – pressure, az – azimuth, al – altitude*

by people without programming skills. For instance, it can be easily imported in any standard spreadsheet software and processed there to extract simple and useful statistics such as mean time on-surface/in-air, variation in pressure, etc.

Although most of the many existing works related to handwritten signals in biometrics and handwriting recognition have been based on surface movements (see e.g. [1]), there are evidences of the importance of in-air movements as well. Sesa-Nogueras *et al.* [2] presented an analysis of in-air and on-surface signals from an information theory point of view. They performed the entropy analysis of handwriting samples acquired in a group of 100 people (see the BiosercurID database for more information [3]) and observed that both types of movements contain approximately the same amount of information. Moreover, based on the values of mutual information, these movements appear to be notably non-redundant. This property has been advantageously used in several fields of science. For instance, Drotar *et al.* [4,5] proved that in-air movement increases the accuracy of Parkinsonic dysgraphia identification. Specifically, when classifying the Parkinsonic dysgraphia by support vector machine (SVM) in combination with the in-air features, they reached 84% accuracy

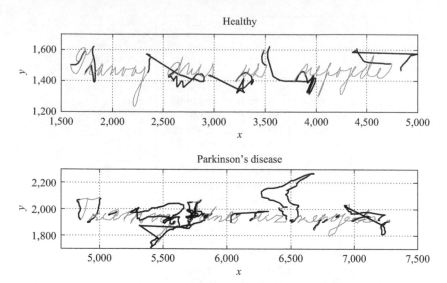

*Figure 2.4   Example of on-surface (grey line) and in-air (black line) movement.
Czech sentence written by a healthy writer and patient with PD
(samples from the PaHaW database, Drotar et al. [11])*

which is by 6% higher in comparison to classification based on the on-surface features
only. When combining both feature sets, they observed 86% classification accuracy.
Faundez-Zanuy *et al.* [6] reported that the in-air movement supports diagnosis of
Alzheimer's disease (AD). They observed that patients with AD spend seven times
longer in-air when comparing to a control group. In the case of on-surface movement, it is only three times longer. Similarly, Rosenblum *et al.* [7] found out that
the in-air duration can be a good measure for performance analysis of children with
high-functioning autism spectrum disorder. The in-air movement has also been used
for identification and quantitative analysis of developmental dysgraphia in children
population [8–10]. Mekyska *et al.* [8] proved that kinematic features derived from
this kind of movement (especially jerk, which is rate at which the acceleration of a
pen changes with time) provide good discrimination power between children with
dysgraphia and control group.

Figure 2.4 contains an example of Czech sentence written by a healthy writer and
writer with Parkinson's disease (PD). As can be seen, the in-air movement (transition
between strokes plotted in black bold) is in the case of PD writer very unsmooth and
irregular. We can see that the writer spent a lot of time in-air before he initiated the
writing of next word. This is tightly related to cognitive functions, the writer has to
think about the next movement, and sometimes, he forgets what to write. We wouldn't
be able to objectively describe these cognitive processes without the in-air movement.

2.3 Handwriting signals from biometrics to medical applications

The analysis of handwriting in security applications, i.e. for the automated identification or verification of subjects by means of biometric methods, today appears to be a well-studied domain. We thus in this section discuss this part very briefly, with reviews of some relevant works. Further, we expand the views to metadata analysis (also referred to as Soft Biometrics) with a brief review of selected works published. Finally, we bridge the gap towards the analysis of handwriting signals for medical purposes, for example to support diagnostics of some diseases. These aspects will be the main focus of discussions in the following subsections.

2.3.1 Biometric security applications

Biometric security applications based on handwritten tasks are mainly based on signatures. Several international competitions summarize the state of the art achieved by dozens of teams, such as Houmani *et al.* [12], signature verification competition (SVC) [13] and SigWiComp (competitions on signature verification and writer identification for on- and offline skilled forgeries) [14]. Although less known, there are also some works where biometric recognition is based on handwritten text, either text-dependent or independent.

The individuality of handwriting has been demonstrated by several authors. Srihari *et al.* [15] assessed the individuality of handwriting in the off-line case. They collected a database of 1,500 writers selected to be representative of the US population and conducted experiments on identification and verification. Regarding identification, they reached accuracy of about 83% at the word level (88% at the paragraph-level and 98% at the document-level). These results allowed the authors to conclude that the individuality hypothesis, with respect to the target population, was true with a 95% confidence level. Zhang and Srihari [16] complemented the previous work of [15]. They analysed the individuality of four handwritten words (*been*, *Cohen*, *Medical* and *referred*) taken from 1,027 US individuals, who wrote each word three times. The combination of the four words yielded an identification accuracy of about 83% and a verification accuracy of about 91%.

With regard to the online case, some authors have addressed the issue of individuality of single words and short sentences. Hook *et al.* [17] showed that single words (the German words *auch*, *oder*, *bitte* and *weit*) and the short sentence *Guten Morgen* exhibit both considerable reproducibility and uniqueness (i.e. equal items written by the same person match well while equal items written by different people match far less well). They used a small database consisting of 15 writers that produced, in a single session, ten repetitions of each item captured by a prototype of a digitizing pen. Chapran [18] used the English words *February*, *January*, *November*, *October* and *September* (25 repetitions of each word donated by 45 writers). The identification rate reached 95%. In Sesa and Faundez-Zanuy [19], a writer identification rate of 92.38% and a minimum of detection cost function [20] of 0.046 (4.6%) was achieved with 370 users using just one word written in capital letters. Results were improved up to 96.46% and 0.033 (3.3%) when combining two words.

2.3.2   Metadata applications

Behavioural biometrics, in addition to security and health applications, can provide a set of additional information, known as metadata. Sometimes also referred to as Soft Biometrics, it can be based on system hardware specifics (technical metadata) and on the other side on personal attributes (non-technical metadata) [21,22]. System-related metadata represent physical characteristics of biometric sensors and are essential for ensuring comparable quality of the biometric raw signals. Previous work in personal related metadata has shown that it is possible to estimate some metadata like script language, dialect, origin, gender and age by statistically analysing human handwriting. In this section, we will summarize some non-technical metadata applications.

Gender recognition attempts to classify the writer as a male or a female. In [23] using only four repetitions of a single uppercase word, the average rate of well-classified writers is 68%; with 16 words, the rate rises to an average of 72.6%. Statistical analysis reveals that the aforementioned rates are highly significant. In order to explore the classification potential of the in-air strokes, these are also considered. Although in this case, results are not conclusive, and an outstanding average of 74% of well-classified writers is obtained when information from in-air strokes is combined with information from on-surface ones. This rate is slightly better than the one achieved by calligraphic experts. However, we should keep in mind that this is a two-class problem and even by pure chance (for instance, flipping a coin) we would get 50% accuracy.

Bandi *et al.* [24] proposed a system that classifies handwritings into demographic categories using measurements such as pen pressure, writing movement, stroke formation and word proportion. The authors reported classification accuracies of 77.5%, 86.6% and 74.4% for gender, age and handedness classification, respectively. In this study, all the writers produced the same letter. Liwicki *et al.* [25] also addressed the classification of gender and handedness in the on-line mode. The authors used a set of 29 features extracted from both on-line information and its off-line representation and applied support vector machines and Gaussian mixture models to perform the classification. The authors reported an accuracy of 67.06% for gender classification and 84.66% for handedness classification. In [26], the authors separately reported the performance of the offline mode, the on-line mode and their combination. The accuracy reported for the off-line mode was 55.39%.

Emotional states, such as anxiety, depression and stress, can be assessed by the depression anxiety stress scales (DASS) questionnaire. Likforman-Sulem *et al.* [27] presents a new database that relates emotional states to handwriting and drawing tasks acquired with a digitizing tablet. Experimental results show that anxiety and stress recognition perform better than depression recognition. This database includes samples of 129 participants whose emotional states are assessed by the DASS questionnaire and is freely distributed for those interested in researching in this line.

2.3.3   Biometric health applications

As to be seen from the example on emotional states and the reasons for emotional changes, the transition from metadata to medical analysis is somewhat fluent. In this

Basal	6 Months	12 Months	18 Months

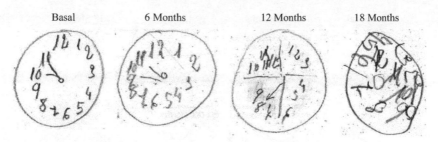

Figure 2.5   *Clock drawing test of ACE-R for a person with AD, showing initial
baseline on the left, and then from left to right, samples from the same
person after 6, 12 and 18 months*

section, we focus on selected analysis for the latter case, with regards to handwriting modality. While signature and handwritten script samples are also useful for health issues, we focus on a set of probably more interesting tasks such as drawings or sketches. These kinds of signals can also be used for biometric recognition, although they are not as usual in real life as handwriting or signature (some examples can be found in [28]).

One important unsolved problem is how the dementia syndrome is associated with diseases such as Parkinson's and Alzheimer's, etc. In the case of Alzheimer's, it is estimated that the cost per year for a single patient is 35,000 USD in the USA. One in ten patients is below 60 years old. The incidence of Alzheimer's is doubled for every 5 years after 65, and beyond 85 years old the incidence is between one-third and half of the amount of population. If a solution is not found, this problem will be unbearable for society. Consequently, a relevant issue related to dementia is its diagnostic procedure. For example, AD is the most common type of dementia, and it has been pointed out that early detection and diagnosis may confer several benefits. However, intensive research efforts to develop a valid and reliable biomarker with enough accuracy to detect AD in the very mild stages or even in pre-symptomatic stages of the disease have not been conclusive. Nowadays, the diagnostic procedure includes the assessment of cognitive functions by using psychometric instruments such as general or specific tests that assess several cognitive functions. A typical test for AD is the clock drawing test (CDT) [29] that consists of drawing a circle and distributing the 12 hours inside. An example of this is shown in Figure 2.5. The initial result produced by a person (baseline) is shown on the left, and on the right, several samples of the same person after 6, 12 and 18 months of being damaged are also shown. This same test has also been used for detecting drug abuse, depression, etc. Figure 2.6 shows a similar situation when copying two interlinking pentagons, which is one of the tasks of the mini-mental state examination (MMSE) [30]. The MMSE or Folstein test is a brief 30-point questionnaire test that is used to screen for cognitive impairment. It is also used to estimate the severity of cognitive impairment at a specific time and to follow the course of cognitive changes in an individual over time, thus making it an effective way to document an individual's response to treatment.

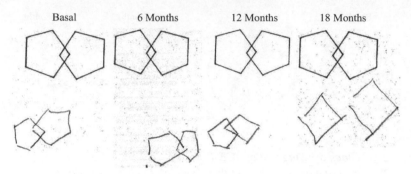

*Figure 2.6   Pentagons of MMSE for a person with AD, showing initial baseline on
the left, and then from left to right, samples from the same person after
6, 12 and 18 months. A pentagon template copied by the patients
appears in the top*

Figure 2.7 presents a house drawing that includes perspective notions (3D aspect
[31]). The first two rows are performed by individuals with AD of different clinical
severity. The visual inspection of the on-surface image suggests a progressive degree
of impairment, where drawing becomes more disorganized and the three dimensional
effect is only achieved in the second row (mild case). The visual information provided
by the in-air drawing between AD individuals also indicates a progressive impairment
and disorganization when the individuals try to plan the drawing. It is also important
to note that the comparison of the on-surface drawing between the mild case of AD and
the control (third and fourth rows) also shows important differences. Even in the case
when the drawing is performed with the non-dominant hand. Besides the increased
time in-air, there is an increased number of hand movements before writers decide to
put the pen on the surface to drawn. We consider that these graphomotor measures
applied to the analysis of drawing and writing functions may be a useful alternative
to study the precise nature and progression of the drawing and writing disorders
associated with several neurodegenerative diseases [6,31]. Figure 2.7 illustrates the
potential of in-air information, which is neglected when medical doctors use the
classical ink pen system in off-line mode.

Generally, in the area of diagnostics in medical context, drawings are widely
used. In summary, some common drawings and their potential usage in medical field
(included the cases described already above) are

1.   *Pentagon test* – used in the MMSE to assess cognitive impairment [30] (see
     Figure 2.6). A template provided to the patient appears in the top row. The second
     row is the result produced by the patient when trying to copy the template, which
     is always the same.
2.   *CDT* – can be utilized as a precursory measure to indicate the likelihood
     of further/future cognitive deficits. It is used in the Addenbrooke's cognitive
     examination-revised (ACE-R) test [32] (see Figure 2.5). As in the previous case,

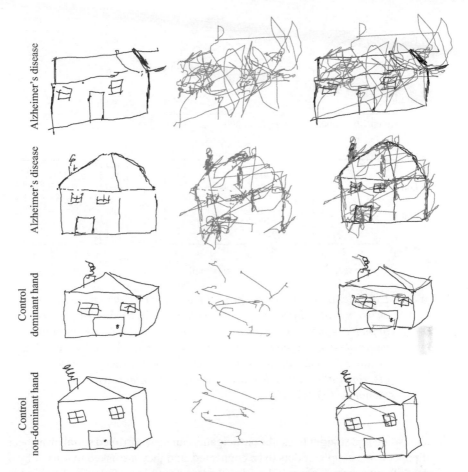

Figure 2.7   House drawing performed by four individuals. First and second rows
correspond to individuals with Alzheimer's disease (one per row). Third
and fourth rows correspond to a healthy user when drawing with
dominant and non-dominant hand. First column – on-surface
movement, second column – in-air movement, third column –
combination of both. Extracted and adapted with permission from
Reference [33]

the different clocks (from left to right) are produced by the same patient passing
6 months.

3. *House drawing copy* – used for identification of AD [6,33] (see Figure 2.7).
    Patients have to copy a shown image of a house sketch.
4. *Archimedes spiral and straight line (drawing between points)* – useful to discrim-
    inate between PD and essential tremor; diagnose mild cognitive impairment, AD,
    dysgraphia, etc. [11,34–38] (see Figure 2.8). In the case of the Archimedes spiral

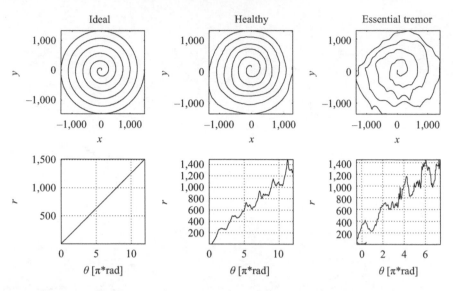

Figure 2.8   Examples of Archimedean spiral: ideal, spiral written by a healthy writer, spiral written by a patient with essential tremor. The second row of picture represents an unwrapped version, where x axis contains angle of each sample on spiral with respect to its centre, and y axis a distance from this centre. This representation is usually used for discrimination among healthy controls and patients with essential tremor or PD [43–45]

acquisition and straight lines, the participants can have a printed spiral on a sheet of paper and a couple of dots to be connected, and they are asked to trace it by a pen without touching the spiral neither the bars (see Figure 2.9). Or, the spiral is shown to them on a template, and they are asked to replicate it on a blank sheet of paper. Similarly, the straight lines can be acquired. In addition, the participants can be asked to connect printed points.

5. *Overlapped circles (ellipses)* – can be used for quantitative analysis of schizophrenia or PD [39–41]. See Figures 2.10 and 2.11, which represents some simple kinematic features that can be used for an effective diagnosis.

6. *Rey–Osterrieth complex figure test* – developed in 1941 and further consists of copying a complex drawing [42]. It is frequently used to further explain any secondary effect of brain injury in neurological patients, to test for the presence of dementia or to study the degree of cognitive development in children. In this task, patients have to memorize an image, and later, they have to replicate it without looking at the example.

Changes in handwriting are usually among the first manifestations of the second most common neurodegenerative disorder – PD [46]. PD patients are usually associated with several motor features, such as tremor in rest, rigidity (resistance to passive

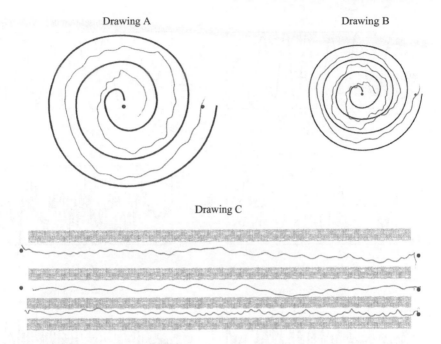

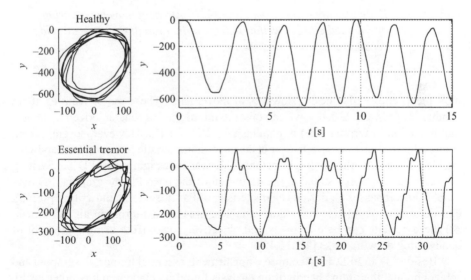

Figure 2.9   Spiral and straight lines test. In the spiral test, the user has to trace a spiral without touching the walls of the traced spiral. In the line test, he has to connect the dots with a straight line without touching the upper and lower bars

Figure 2.10   Examples of overlapped circles (ellipses): healthy subject and patient with essential tremor. The right part of figure represents vertical movement in time

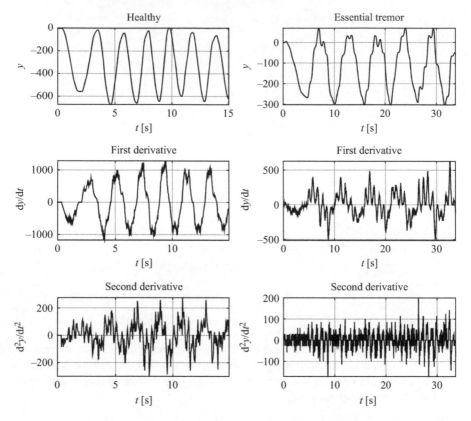

Figure 2.11   Examples of signals from Figure 2.10 (vertical movement during overlapped ellipses writing) after application of first (speed) and second (acceleration) derivative

movement) and bradykinesia (slowness of movement). These motor features affect handwriting as well. Usually, we can observe micrographia (characterized by abnormal reduction in writing size) in patients with PD [46,47]. However, several recent studies showed that micrographia is not the most characteristic feature of PD handwriting [48]. The availability of graphical tablets allowed investigating the PD handwriting in great detail, which resulted in the definition of the term PD dysgraphia. PD dysgraphia encompasses all deficits characteristic of Parkinsonian handwriting [46,48], e.g. deficits in geometry, kinematic variables, pressure patterns, in-air movement. All these deficits can be effectively and objectively quantified using a protocol of handwriting/drawing tasks [4,5,11,49].

Beside AD and PD, mild cognitive impairment, essential tremor, dysgraphia and schizophrenia, the online handwriting analysis found its place in many other health applications, e.g. analysis of depression [50], autism [51] or obsessive–compulsive disorder [52].

2.4 Security–health implications and concerns

Security and health applications have been described as isolated fields to each other in previous sections. They indeed are separated for most of the scientific community. However, they are not separated in real life.

Engineers can develop a technological solution for security or health issues. In general, they tend to be experts in just one of the two fields, and they do not take care of the other field. This solution can be based on biometrics but hardly can we consider that security is not related to health and the opposite. In some cases, both should be considered jointly. In other ones, we cannot isolate one from the other one. In the next subsections, we will describe several examples.

2.4.1 Security where health aspects influence biometric security

Most biometric security applications only try to determine the identity of a user or to verify if he is who claims to be. However, in the common knowledge that automated biometric systems are subject to erroneous classifications, it is important to extract some additional information in context of the acquisition of biometric signals. In the following, we summarise three of such possible scenarios, motivated by the questions as follows:

1. *Is the user under stress?* It is not the same to put a finger in a sensor, confirm the identity, and open a door if, for example, his heart is beating at 70 beats per minute (bpm) than if it is beating at 120 bpm. If his heart is much more accelerated than expected, some suspicious activity can be happening. To solve this, some fingerprint systems have a mechanism called duress finger, which is a way to notify security guards or the police about the threatening situation without letting the threatening person notice it. To do that the user enrols at least two fingers. Both will open the door but one of them will activate a silent alarm. This concept is known as duress detection. See for instance [53]. Some examples are the commercialized fingerprint products by Fingertec[1] and Suprema[2]. This is a simple example to illustrate the idea, but obviously, it is quite simple for the threatening person to force the user to use the specific finger that does not activate the alarm. This knowledge can be obtained just looking how the user interacts with the sensor in previous days. Similarly, the user can enrol a couple of different signatures, one for duress recognition and the other one for normal operation system. Again, it would be possible for a third party to be familiar with the genuine signature that does not activate any silent alarm and force the user to use that signature.

   A robust biometric security system should be able to detect the stress situation based on characteristics that cannot be easily controlled by the user. Detection of user stress from signature or handwriting is a challenging research topic that can indeed improve security systems.

---

[1] http://www.fingertec.com/ver2/english/vd-duressfingerprint.html
[2] https://www.supremainc.com/es/node/613

Figure 2.12   On the left, documents signed in 1985 (hesitated) and 1986. On the
right, some signatures on blank sheets, when the elder woman was
suffering dementia. Extracted with permission from Reference [33]

2. *Is he suffering any disease that makes him unable to understand the real implica-*
    tion on his acts? In [33], we presented the case of a woman affected by AD. This
    case was presented to us by Viñals and described in his book [54]. In this case,
    several women made an elder woman sign her name on blank sheets of paper (see
    Figure 2.12). Theoretically, due to some issues related to medicines. When the
    elder person died, the women took advantage of the signed sheets in order to write
    a rental agreement. The declared date of this agreement was 1985 (Figure 2.12
    on the bottom left), but several documents signed in 1986 (Figure 2.12 on the
    top left) showed better control of calligraphic movements. In fact, the hesitantly
    written signature document signed in 1985 was closer in appearance to the blank
    sheets signed when the elder woman had dementia than to the 1986 document.
    Thus, it was demonstrated that in fact the rental document was not signed in
    1985. It was signed later.
        Another possibility is to be affected by depression. Heinik *et al.* [55] used
    drawings for analysing depressive disorders in older people.
        These two examples indicate that while even if in the context of biometric
    signature verification one can conclude that the signature is genuine, this may be
    not enough. One should in addition take into account aspects such as the health
    state of the user. Considering both aspects (identity, i.e. degree of signature
    matching and health), one can conclude in doubt of a good health condition of

the subject the signature, from a legal point of view, may not be valid. In these cases, the biometric authentication of the individual does not solve the problem, and some additional considerations should be taken. This is not just related to health. Another similar situation where a genuine biometric sample is used in a fraudulent way is a replay attack. In a replay attack, the biometric signal is usually genuine, but it was acquired/recorded in the past and presented again and should be considered as a fake attempt.

3. *Is he temporarily affected by drug substances abuse?* References [56,57] found changes in handwriting due to alcohol. Tucha *et al.* [58] detected the effects of caffeine on handwriting. Foley and Miller [59] performed similar experiments about the effects of marijuana and alcohol. While this consumption could be hardly detected by fingerprint analysis, for instance, this is not the case with biometric signals such as handwriting/signature and speech.

2.4.2 Situations where health information can be extracted from security applications

One of the main concerns of biometrics applied to security is about privacy issues [33]. Technological advances let to store, gather and compare a wide range of information on people. Using identifiers such as name, address, passport or social security number, institutions can search databases for individuals' information. This information can be related to salary, employment, sexual preferences, religion, consumption habits, medical history, etc. This information can be collected with the consent of the user, but in some cases it could also be extracted from biometric samples without the knowledge of the user. Thus, the user could ignore that some additional and private information can be extracted from his biometric samples.

Though in most of the scenarios there should be no problem, there is a potential risk. Let us think, for instance, in sharing medical information. Obviously, in case of emergency, this sharing between hospitals would be beneficial. On the contrary, if this information is transferred to a personal insurance company or a prospective employer, the insurance or the job application can be denied. The situation is especially dramatic when biometric data collection is intended for security biometric recognition to grant access to a facility or information but a third party tries to infer the health condition of the subject. For instance, in the case of retina and iris recognition, an expert can determine that a patient suffers from diabetes, arteriosclerosis, hypertension, etc.

For any biometric identifier, there is a portion of population for which it is possible to extract relevant information about their health, with similar implications to the ones described in the previous paragraph. This is not a specific problem of handwritten signals. Some other biometric signals exhibit the same potential problems. For example, speech disorders, hair or skin colour problems, etc. An important question is what exactly is disclosed when biometric scanning is used. In some cases, additional information not related to identification might be obtained. One possible scenario could be a company where an attendance sheet must be signed each day. The main purpose of this task could be to check if the worker is at his workplace all the labouring days. However, once the handwriting is provided, the company could decide to analyse the signature to detect some pathologies or drugs abuse and to fire out those workers who do not show a good health. And last but not the least, once we provide our biometric samples, they can last in a database for dozens of years, and due to technological advances, they can be used in a simple way in the future to extract additional information that was not intended during acquisition. For this reason, we should think about technical solutions to preserve privacy and legal regulations to avoid that.

2.4.3   Situations where the identity information must be removed

Sometimes, the situation is just opposite to that one mentioned in the previous section. With the growth of eHealth and telemedicine fields, scientists started to develop automatic handwriting analysis systems that can be used for disease diagnosis, rating or monitoring. However, to introduce a robust analysis system, it is necessary to develop it using a large database consisting of hundreds or thousands of subjects. This could be problematic, especially when acquiring patients with rare diseases or patients with cognitive deficits (e.g. patients with dementia). In these cases, it is difficult to find enough samples and explain the handwriting tasks, respectively.

One possibility to overcome the lack of data is to fuse databases acquired by different research or clinical teams around the world, i.e. make the data publicly available (or at least for research purposes). But this is usually not allowed by the local ethics committee or by the participants themselves. Just a few people would make their health data available when containing identity information. Therefore, during last few years, scientists started to develop de-identification methods, that would remove this information, but that would still keep the information about pathology (see the next paragraph). Usually, this is done using a sophisticated parameterization process. For example, in future datasets used for analysis of handwriting in patients with PD, it would be enough to keep and disseminate kinematic, in-air, tremor and pressure characteristics.

In this field, there was a European Cooperation in Science and Technology action devoted to de-identification for privacy protection in multimedia content. De-identification in multimedia content can be defined as the process of concealing the identities of individuals captured in a given set of data (images, video, audio, text), for the purpose of protecting their privacy. This will provide an effective means for supporting the EU's Data Protection Directive (95/46/EC), which is concerned with the introduction of appropriate measures for the protection of personal data. The fact that a person can be identified by such features as face, voice, silhouette and gait, indicates the de-identification process as an interdisciplinary challenge, involving such scientific areas as image processing, speech analysis, video tracking and biometrics. This action aims to facilitate coordinated interdisciplinary efforts (related to scientific, legal, ethical and societal aspects) in the introduction of person de-identification and reversible de-identification in multimedia content by networking relevant European experts and organisations.

2.5 Summary and conclusions

In this chapter, we have described the main characteristics of online handwritten signals as well as their applications on biometric recognition and health. We have emphasized the importance of taking into account that security and health should not be isolated to each other. Care must be taken to protect privacy in health applications (in some studies, the identity should not be revealed), and vice versa, it is important to preserve health state privacy in security ones.

To sum up, some background of both fields is desirable although we are working only in the field of security or health. In addition, privacy aspects must be carefully considered in our technological solutions.

References

[1] Plamondon R., Shirari S. 'On-line and off-line handwriting recognition: a comprehensive survey'. IEEE Transactions on Pattern Analysis and Machine Intelligence. 2000;22(1):63–84.

[2] Sesa-Nogueras E., Faundez-Zanuy M., Mekyska J. 'An information analysis of in-air and on-surface trajectories in online handwriting'. Cognitive Computation. 2012;4(2):195–205.

[3] Fierrez-Aguilar J., Galbally J., Ortega-Garcia J., *et al.* 'BiosecurID: a multimodal biometric database'. Pattern Analysis and Applications. 2010;13(2):235–246.

[4] Drotar P., Mekyska J., Rektorova I., Masarova L., Smekal Z., Faundez-Zanuy M. 'A new modality for quantitative evaluation of Parkinson's disease: In-air movement'. Proceedings of the IEEE 13th International Conference on Bioinformatics and Bioengineering; Chania, GR, Nov. 2013. Chania: IEEE; 2013, pp. 1–4.

[5] Drotar P., Mekyska J., Rektorova I., Masarova L., Smekal Z., Faundez-Zanuy M. 'Analysis of in-air movement in handwriting: a novel marker for Parkinson's disease'. Computer Methods and Programs in Biomedicine. 2014;117(3): 405–411.

[6] Faundez-Zanuy M., Sesa-Nogueras E., Roure-Alcobé J., Garré-Olmo J., Lopez-de-Ipiña K., Solé-Casals J. 'Online drawings for dementia diagnose: in-air and pressure information analysis'. Proceedings of the 8th Mediterranean Conference on Medical and Biological Engineering and Computing; Seville, ES, Sep. 2013. New York: Springer; 2013, pp. 567–570.

[7] Rosenblum S., Simhon H. A. B., Gal E. 'Unique handwriting performance characteristics of children with high-functioning autism spectrum disorder'. Research in Autism Spectrum Disorders. 2016;23:235–244.

[8] Mekyska J., Faundez-Zanuy M., Mzourek Z., Galaz Z., Smekal Z., Rosenblum S. 'Identification and rating of developmental dysgrahia by handwriting analysis'. IEEE Transactions on Human-Machine Systems. 2017;47(2), pp. 235–248. doi: 10.1109/THMS.2016.2586605.

[9]   Rosenblum S., Parush S., Weiss P. L. 'Computerized temporal handwriting characteristics of proficient and non-proficient handwriters'. American Journal of Occupational Therapy. 2003;57(2):129–138.

[10]   Rosenblum S., Dvorkin A. Y., Weiss P. L. 'Automatic segmentation as a tool for examining the handwriting process of children with dysgraphic and proficient handwriting'. Human Movement Science. 2006;25(45):608–621.

[11]   Drotar P., Mekyska J., Rektorova I., Masarova L., Smekal Z., Faundez-Zanuy M. 'Evaluation of handwriting kinematics and pressure for differential diagnosis of Parkinson's disease'. Artificial Intelligence in Medicine. 2016;67:39–46.

[12]   Houmani N., Mayoue A., Garcia-Salicetti S., *et al.* 'BioSecure Signature Evaluation Campaign (BSEC'2009): evaluating online signature algorithms depending on the quality of signatures'. Pattern Recognition. 2012;45:993–1003.

[13]   Yeung D. Y., Chang H., Xiong Y., *et al.* 'SVC2004: first international signature verification competition'. Proc. of Intl. Conf. on Biometric Authentication, ICBA, LNCS-3072. New York: Springer, 2004, pp. 16–22.

[14]   Malik M. I., Ahmed S., Marcelli A., *et al.* 'ICDAR2015 competition on signature verification and writer identification for on- and off-line skilled forgeries (SigWIcomp2015)'. IEEE International Conference on Document Analysis and Recognition; Nancy, France, 23–26 Aug. 2015.

[15]   Srihari S., Sung-Hyuk C., Sangjik L. 'Establishing handwriting individuality using pattern recognition techniques'. Proceedings of the Sixth International Conference on Document Analysis and Recognition; 2001, pp. 1195–1204.

[16]   Zhang B., Srihari S. 'Analysis of handwriting individuality using word features'. Proceedings of the Seventh International Conference on Document Analysis and Recognition; 2003, pp. 1142–1146.

[17]   Hook C., Kempf J., Scharfenberg G. 'A novel digitizing pen for the analysis of pen pressure and inclination in handwriting biometrics'. Biometric Authentication Workshop, Prague 2004, Lecture Notes in Computer Science. 2004;3087:283–294.

[18]   Chapran J. 'Biometric writer identification: feature analysis and classification'. International Journal of Pattern Recognition & Artificial Intelligence. 2006;20:483–503.

[19]   Sesa-Nogueras E., Faundez-Zanuy M. 'Biometric recognition using online uppercase handwritten text'. Pattern Recognition. 2012;45:128–144.

[20]   Martin A. A. F., Doddington G., Kamm T., Ordowski M., Przybocki M., 'The DET curve in assessment of detection task performance'. Proc. Eurospeech '97; Rhodes, Greece, Sept. 1997;4:1899–1903.

[21]   Vielhauer C., Basu T., Dittmann J., Dutta P. K. 'Finding meta data in speech and handwriting biometrics'. Proceedings of the SPIE. 2005;5681: 504–515.

[22]   Scheidat T., Wolf F., Vielhauer C. 'Analyzing handwriting biometrics in metadata context'. Proceedings of the SPIE. 2006;6072:60720H1–60720H12.

[23]   Sesa-Nogueras E., Faundez-Zanuy M., Roure-Alcobe J. 'Gender classification by means of online uppercase handwriting: a text-dependent allographic approach'. Cognitive Computation. 2016;8(1):15–29.

[24]   Bandi K. R., Srihari S. N. 'Writer demographic identification using bagging and boosting'. Proceedings of the 12th Conference of the International Graphonomics Society (IGS); Salerno, IT, Jun. 2005. Salerno: IGS; 2005, pp. 133–137.

[25]   Liwicki M., Schlapbach A., Loretan P., Bunke H. 'Automatic detection of gender and handedness from on-line handwriting'. Proceedings of the 13th Conference of the International Graphonomics Society (IGS); Melbourne, AU, Jun 2007. Melbourne: IGS; 2007, pp. 179–183.

[26]   Liwicki M., Schlapbach A., Loretan P., Bunke H. 'Automatic gender detection using on-line and off-line information'. Pattern Analysis and Applications. 2011;14:87–92.

[27]   Likforman-Sulem L., Esposito A., Faundez-Zanuy M., Clemençon S., Cordasco G. 'EMOTHAW: a novel database for emotional state recognition from handwriting and drawing'. IEEE Transactions on Human-Machine Systems. 2017;47(2):273–284. doi: 10.1109/THMS.2016.2635441.

[28]   Vielhauer C. 'Processing of handwriting and sketching dynamics' in Thiran J. P., Marqués F., Bourlard H. (ed.). Multimodal Signal Processing: Theory and Applications for Human–Computer Interaction. Oxford: Academic Press; 2009. pp. 119–142.

[29]   Sunderland T., Hill J. L., Mellow A. M., *et al.*, 'Clock drawing in Alzheimer's disease: a novel measure of dementia severity'. Journal of the American Geriatrics Society. 1989;37:725–729.

[30]   Folstein M. F., Folstein S. E., McHugh P. R. 'Mini-mental state: a practical method for grading the cognitive state of patients for the clinician'. Journal of Psychiatric Research. 1975;12(3):189–198.

[31]   Faundez-Zanuy M., Sesa-Nogueras E., Roure-Alcobe J., *et al.* 'A preliminary study of online drawings and dementia diagnose'. Neural Nets and Surroundings. Volume 19 of the series Smart Innovation, Systems and Technologies; 2013, pp. 367–374, Springer.

[32]   Larner A. J. 'Addenbrooke's cognitive examination-revised (ACE-R) in day-to-day clinical practice'. Age Ageing. 2007;36(6):685–686.

[33]   Faundez-Zanuy M., Hussain A., Mekyska J., *et al.* 'Biometric applications related to human beings: there is life beyond security'. Cognitive Computation. 2013;5(1):136–151.

[34]   Albert D., Opwis K., Regard M. 'Effect of drawing hand and age on figural fluency: a graphomotor study with the five-point test in children'. Child Neuropsychology. 2009;16(1):32–41.

[35]   Louis E. D., Yu Q., Floyd A. G., Moskowitz C., Pullman S. L. 'Axis is a feature of handwritten spirals in essential tremor'. Movement Disorders. 2006;21(8):1294–1295.

[36]   Popovic M. B., Dzoljic E., Kostic V. 'A method to assess hand motor blocks in Parkinson's disease with digitizing tablet'. Tohoku Journal of Experimental Medicine. 2008;216(4):317–324.

[37]   Pullman S. L. 'Spiral analysis: a new technique for measuring tremor with a digitizing tablet'. Movement Disorders. 1998;13(Suppl. 3):85–89.

[38]   Thenganatt M. A., Louis E. D. 'Distinguishing essential tremor from Parkinson's disease: bedside tests and laboratory evaluations'. Expert Review of Neurotherapeutics. 2012;12(6):687–696.

[39]   Yan J. H., Rountree S., Massman P., Doody R. S., Li H. 'Alzheimer's disease and mild cognitive impairment deteriorate fine movement control'. Journal of Psychiatric Research. 2008;42(14):1203–1212.

[40]   Caligiuri M. P., Teulings H. L., Dean C. E., Lohr J. B. 'A quantitative measure of handwriting dysfluency for assessing tardive dyskinesia'. Journal of Clinical Psychopharmacology. 2015;35(2):168–174.

[41]   Eichhorn T. E., Gasser T., Mai N., *et al.* 'Computational analysis of open loop handwriting movements in Parkinson's disease: a rapid method to detect dopamimetic effects'. Movement Disorders. 1996;11(3):289–297.

[42]   Lepelley M. C., Thullier F., Bolmont B., Lestienne F. G. 'Age-related differences in sensorimotor representation of space in drawing by hand'. Clinical Neurophysiology. 2010;121(11):1890–1897.

[43]   Rey A. 'The psychological examination in cases of traumatic encepholopathy. Problems'. Archives de Psychologie. 1941;28:215–285.

[44]   Saunders-Pullman R., Derby C., Stanley K., *et al.* 'Validity of spiral analysis in early Parkinson's disease'. Movement Disorders. 2008;23(4):531–537.

[45]   Wang H., Yu Q., Kurtis M. M., Floyd A. G., Smith W. A., Pullman S. L. 'Spiral analysis-improved clinical utility with center detection'. Journal of Neuroscience Methods. 2008;171(2):264–270.

[46]   Pinto S., Velay J. L. 'Handwriting as a marker for PD progression: a shift in paradigm'. Neurodegenerative Disease Management. 2015;5(5):367–369.

[47]   Rosenblum S., Samuel M., Zlotnik S., Erikh I., Schlesinger I. 'Handwriting as an objective tool for Parkinson's disease diagnosis'. Journal of Neurology. 2013;260(9):2357–2361.

[48]   Letanneux A., Danna J., Velay J. L., Viallet F., Pinto S. 'From micrographia to Parkinson's disease dysgraphia'. Movement Disorders. 2014;29(12):1467–1475.

[49]   Drotar P., Mekyska J., Rektorova I., Masarova L., Smekal Z., Faundez-Zanuy M. 'Decision support framework for Parkinson's disease based on novel handwriting markers'. IEEE Transactions on Neural Systems and Rehabilitation Engineering. 2015;23(3):508–516.

[50]   Schröter A., Mergl R., Bürger K., Hampel H., Möller H. J., Hegerl U. 'Kinematic analysis of handwriting movements in patients with Alzheimer's disease, mild cognitive impairment, depression and healthy subjects'. Dementia and Geriatric Cognitive Disorders. 2003;15(3):132–142.

[51]   Johnson B. P., Phillips J. G., Papadopoulos N., Fielding J., Tonge B., Rinehart N. J. 'Understanding macrographia in children with autism spectrum disorders'. Research in Developmental Disabilities. 2013;34(9):2917–2926.

[52] Mavrogiorgou P., Mergl R., Tigges P., *et al.* 'Kinematic analysis of handwriting movements in patients with obsessive–compulsive disorder'. Journal of Neurology, Neurosurgery & Psychiatry. 2001;70(5):605–612.

[53] Martin C., Oh E., Addy K., Eskildsen K. 'Biometric verification and duress detection system and method'. Patent US 20070198850 A1, 2007.

[54] Viñals F., Puente M. L. 'Alteraciones neurológicas y biológicas' in Herder (ed.). Grafología criminal. 2009. Chapter 3.

[55] Heinik J., Werner P., Dekel T., Gurevitz I., Rosenblum S. 'Computerized kinematic analysis of the clock drawing task in elderly people with mild major depressive disorder: an exploratory study'. International Psychogeriatrics. 2010;22(3):479–488.

[56] Faruk A., Turan N. 'Handwritten changes under the effect of alcohol'. Forensic Science International. 2003;132(3):201–210.

[57] Phillips J. G., Ogeil R. P., Muller F. 'Alcohol consumption and handwriting: a kinematic analysis'. Human Movement Science. 2009;28:619–632.

[58] Tucha O., Mecklinger L., Walitza S., Lange K. W. 'The effect of caffeine on handwriting movements in skilled writers'. Human Movement Science. 2006;25:523–535.

[59] Foley R. G., Miller L. 'The effects of marijuana and alcohol usage on handwriting'. Forensic Science International. 1979;14(3):159–164.

Chapter 3

Privacy concepts in biometrics: lessons learned from forensics

Jana Dittmann[1] and Christian Kraetzer[1]

This chapter discusses lessons that can be learned in biometrics from the field of the forensic sciences. It acknowledges the fact that biometrics and forensics are both very old research disciplines which have a very different perspective in life: While work in biometrics is mostly focused on application issues, like achieving certain error levels, forensics need a very thorough backing to achieve the ultimate goal in this field, admissability in court. This automatically results in high standards for methods that exceed simple performance issues by far. One aspect that is used in this chapter as the focus of the discussions is the matter of privacy. In the first half of the chapter it is illustrated by example how current research work in one digitized forensics field, here digitised dactyloscopy (i.e. the science of forensic analysis of fingerprint traces), influences the current view on fingerprint biometrics and which lessons in regards to privacy can be derived. In the second half, the ever popular field of face biometrics is addressed as an example of an widely used biometric modality in desperate need of not only digital image forensics but also guidelines for privacy preserving methods.

3.1 Introduction: forensic science and selected privacy concepts

As widely known, forensic science can be defined as the study and application of sciences to support solving legal problems or crimes by finding the truth or an understandable version of the truth following scientific methods and processes (see e.g. in [1]). Forensic science contains, for example, crime scene investigation aspects such as crime scene management and trace (evidence) collection, processing and documentation. It supports investigators to understand and interpret traces and derive information, for example, to determine the identity of an unknown suspect. The process of evidence collection, its analysis, argumentation and presentation in court cases needs to follow scientific and legal rules (see, e.g. [2]).

[1]Department of Computer Science (FIN), Otto-von-Guericke University Magdeburg, Germany

The basic principles of processing personal and person-related data in forensics, as well as in general handling of such data, can be derived from the EU Directive 2016/680, see in [3], dated 27 April 2016. This directive focusses on the protection of natural persons with regard to the processing of personal data by competent authorities for the purposes of the prevention, investigation, detection or prosecution of criminal offences or the execution of criminal penalties, and on the free movement of such data, and repealing EU Council Framework Decision 2008/977/JHA, for example, in [3] by stating, 'Any processing of personal data must be lawful, fair and transparent in relation to the natural persons concerned, and only processed for specific purposes laid down by law'.

These general principles are closely related to the general privacy principles established in the field of biometrics, such as limitation to the defined purpose, minimisation of the data which is collected, relevance principle, as well as proportionality, see, for example, in the EU Data Protection Reform [4].

In the following two sections of this chapter, first, selected findings from digitised dactyloscopy (i.e. the science of forensic analysis of fingerprint traces), as classical crime scene traces, are summarised, which have been identified during practical research work (see German national projects Digi-Dak and DigiDak+, e.g. in [5]), and privacy rules are derived from these research efforts for fingerprint biometrics. In particular, privacy interests of subjects and the overall interest of the society in solving crime cases need to be fairly balanced. To address both goals, in both mentioned projects, sensor acquisition guidelines have been derived and proposed for crime scene trace acquisition procedures along with privacy-conform testing procedures to allow privacy-conform benchmarking of forensic methods. In Section 3.2, these two project outcomes are summarised and guidelines for the field of biometrics are derived.

Second, the ever popular field of face biometrics is addressed as an example of a biometric modality in need of digital image forensics and guidelines for privacy-preserving methods. In particular, we focus on known face-morphing manipulations designed to cheat passport-based authentication systems. This specific class of face image attacks has been introduced in [6] as a threat to border-control systems. In this chapter, an analysis based on established forensic methodology is presented, and selected results from the German national project ANANAS are summarised to also derive lessons learned for the application of morphing for privacy preserving face biometrics, as suggested by Sweeney [7], Newton *et al.* [8], Driessen and Dürmuth [9], and others. It shows that the lessons learned in forensics can help to increase the maturity in applied privacy preserving face biometrics.

3.2   Privacy concepts – findings from digitised forensics of latent fingerprints

In digitised dactyloscopy, covering the digitised processing and analysis of forensic fingerprints, finger marks or traces from the contact with fingers are acquired, digitised and pre-processed to find patterns and features for a forensic analysis. The goal is to find and identify individual characteristics in the trace as potential basis to derive evidence for subject identification or verification. As the crime scene is searched for traces broadly and intensively, in most cases also traces are acquired which are not related to the actual crime case (such as prints from uninvolved and innocent subjects). Therefore, it is essential to follow the principles that the traces are only processed for the specific forensic purpose and to ensure the data protection by design and by default requirement (Article 20) from [3]. Here, Article 20 requires appropriate technical and organisational measures supporting data protection principles, such as data minimisation, to protect the rights of subjects. To this end, Section 3.2.1 summarises a selection of privacy-supporting measures for fingerprint recognition and pre-selection, such as sensor acquisition findings from coarse scans about potential privacy-preserving dot distances and resolutions, spatial dimensions and sample part selection as guideline for biometric de-identification, selected biometric sub-tasks such as soft biometrics (print age determination) or sensor-quality assessment.

In order to support a privacy-preserving testing, a benchmarking concept using data and artefact simulations is summarised in Section 3.2.2. It is essential to ensure quality aspects such as accuracy of the acquired and processed trace data, as requested from EU Directive 2016/680 [3], Article 7 (2). Further, it is necessary to consider quality aspects for the applied processing methods from other legal perspectives and to become accepted as evidence, as for example required by the well-known Daubert standard [2]). Therefore, testing of forensic methods is very essential and strictly required. Here, scientific testimony and certification procedures (see, e.g. ISO/IEC 17025:2005 – General requirements for the competence of testing and calibration laboratories) have been suggested and are regularly performed to ensure the quality in forensic processes.

The acquisition devices in [5] use contactless, optical, high-resolution sensing principles (hyperspectral imaging in the ultraviolet (UV) to near-infrared (NIR) range, chromatic white light sensing and confocal laser scanning microscopy) with point or area measurements. In comparison to the most used acquisition techniques in biometrics domain, these devices are slow in its processing.

3.2.1 Sensor-acquisition-related privacy-preserving guidelines

Here, three different sensor-related privacy preserving approaches are introduced and discussed to derive recommendations for the biometrics field

3.2.1.1 Localisation of fingerprints using a coarse scan: findings for privacy-preserving dot distances and resolutions

At crime scene forensics, one important step is the search and localisation of potential traces. In [5], the so-called coarse scan was introduced for such a search followed by a so-called fine scan for a detailed forensic analysis. In particular, in [10], the contactless, optical acquisition sensors are investigated, and a privacy-preserving coarse scan for fingerprint localisation is introduced. The investigated overall acquisition dot distances and resulting resolutions of acquired data can ensure that the fingerprint details for detailed expert analysis are not collected in the required quality, whereas the quality/resolution is still sufficient for the detection of traces. This allows for a case-dependent coarse analysis to determine which and where fingerprint traces are present and might be case relevant – before the relevant parts are further scanned in detail with smaller dot distances and resulting higher resolutions.

The first assessment in [10] of fingerprint traces acquired with a Chromatic White Light CWL600 sensor determined a suitable dot distance for such coarse scans of 400 µm resulting in a 63.5-dpi resolution. In further work, [11] showed that an acquisition dot distance of approximately 200 µm resulting in 10×10 pixels for blocks of 2×2 mm can also be used for trace detection in a coarse scan scenario, by investigating smaller areas and timebased features (trace changes over time, e.g. due to degradation; called persistency), but which would require a multiple sample acquisition (multiple samples over time series).

Recommendation for biometrics: From the proposed coarse scan, we can derive privacy-preserving strategies for biometrics, for example in applications of fingerprint biometrics on touchscreens. It should keep resolution as low as possible do prevent verification and identification on the acquired data, while at the same time, allowing for a precise localisation as basis for further processing steps. Furthermore, the findings can be used for de-identification also: the identified dot distance boundaries and resolution settings can help in ensuring good de-identification degrees, for example, by appropriate downscaling of original biometric images.

3.2.1.2   Small area scans on the example of age detection: findings for privacy-preserving spatial dimensions

In [12], it was investigated how the contactless, optical sensors behave for trace ageing analysis to determine, for example, the freshness of a trace. The age of the trace is often important for forensic investigations and court hearings (to link a print to the time of a crime) but also allows for a pre-selection of case-relevant data to ensure its limitation to the defined purpose and compliance to the relevance principle. The work showed that, for a Chromatic White Light CWL600 sensor, a measured area of 4×4 mm of the latent fingerprint region with a lateral dot distance of 20 µm (200 $\times$ 200 pixels) is a reliable lower boundary for age detection, whereas not sufficiently detailed for verification.

Recommendation for biometrics: For biometrics application fields, it might be of interest to investigate and use the findings that, for specific sub-tasks (e.g. age-related analysis), the required analysis might also be performed successfully on small sample regions only. This would also allow for a privacy-preserving analysis by only using a minimum amount of data. With the *StirTrace* tool [13], different sample sizes can be assessed and tested easily to determine the impact on biometric samples, e.g. on the biometric task of subject–age-detection–error-rate-determination. Furthermore, for sensor-quality assessment also, small sample sizes might be sufficient too and would allow a privacy-preserving testing.

3.2.1.3   Structural and semantical selection: findings for privacy-preserving sample part selection

In the mentioned German national projects, Digi-Dak and DigiDak+, fingerprint characteristics and the possibility to select parts from a fingerprint to generate a privacypreserving data set were discussed with the project partners. The joint discussion considered the three levels of fingerprint features known in the literature, see for example in [14] general pattern, the minutiae and the third-level details. Here, in [14], the first-level features are summarised as features describing the morphology and general shape of the friction ridge flow. It is an overall global structural characteristics and relevant as a kind of first-level matching of plausibility of the global structural appearance. The second-level features as summarised in [14] are derived from the individual characteristics along the papillary ridges and defined as minutiae with type (three basic types: ridge ending, bifurcation, and dot), location and orientation and spatial arrangement. The known third-level features are summarised in [14] as friction ridge details by characterising ridge contours or edges, shape of the minutiae, as well as shape and position of the pores. Reference [14] also identifies further known individual characteristics such as scars, incipient ridges and warts which might influence one or more parts of the three-level features.

These three known features levels were studied jointly with the project partners in the DigiDak project, and guidelines (rules) were derived on how parts from a whole fingerprint can be de-identified and used as privacy-preserving data set. As result, the following rules were jointly identified (based on partial prints with a maximum size of 5×5 mm and not more than two crops per complete fingerprint):

- **Level 1**: The fingerprint part considered for an analysis shall not show any global, general pattern. Core parts (such as inner sections) should be avoided in the selection, and from the selected part no global, general pattern shall be identifiable – with the only exemption, when level 2 requirements are all fulfilled.
- **Level 2**: The selected part shall not contain more than 3–5 minutiae depending on level 3 details: if level 3 features are present, only 3 minutiae shall be visible in the selected part. Further individual characteristics such as scars, incipient ridges and warts shall be avoided and not part of the selection [15].
- **Level 3**: If level 3 features are present, parts with non-common characteristics (such as scares) need to be avoided.

Recommendation for biometrics: For data anonymisation, the rules summarised for latent fingerprint traces can be directly adopted to generate a privacy-conform test set, if only parts are required from the whole fingerprint. Such test sets might be valuable, for example, for sensor acquisition quality testing and reproducibility testing or also for biometric subtasks, such as the age analysis outlined before. If the whole fingerprint is required, protection mechanisms such as encryption or de-identification techniques can use the rules too, to achieve such a protection. The proposed privacy-preserving strategy can be adopted and used also for other biometric modalities. Table 3.1 summarises the recommendations.

3.2.2 Privacy-preserving-benchmarking concepts and guidelines

Forensic sciences developed over the ages a methodology (e.g. in the form of compliance to the Daubert standard [2] or similar regulations) which, in its elaborateness and accurateness, by far exceeds the scope of current best practices in most fields of applied computing, like biometrics. The main reason for this fact is the realisation,

Table 3.1   Summary of sensor-acquisition-related privacy-preserving guidelines

Tasks	Recommendation	Example settings
Forensic task: localisation of fingerprints using a coarse scan Biometric task: localisation such as on touchscreens, de-identification	Keep resolution as low as possible	e.g. CWL600 coarse scans of 400 μm resulting in a 63.5-dpi resolution or approximately 200 μm resulting in 10 × 10 pixels for blocks of 2 × 2 mm
Forensic task: small area scans on the example of age detection Biometric task: specific biometric sub-tasks such as age determination of sensor-quality assessment	Selection of small sample regions only	CWL600 sensor, a measured area of 4 × 4 mm with a lateral dot distance of 20 μm (200 × 200 pixels)
Forensic task: structural and semantical selection for privacy-conform test sets Biometric task: privacy-conform test set generation, de-identification, requirements and selection criteria for protection	Use small areas and follow rules for selection of fingerprint parts by analysing the levels 1–3 for structural and semantical specifics, partial prints with a maximum size of 5 × 5 mm and not more than two crops per complete fingerprint	Areas settings similar for the age detection by additionally selecting parts with no structural or semantical outstanding characteristics

that the ultimate benchmark for any forensic technique is usually its admissibility in court, with all the severe consequences that can be attached including imprisonment (or in some countries even death sentence) of innocent. This subsection summarises lessons learned in quality control and testing of forensic methods (forensics benchmarking) on the example of fingerprints, which can help to increase the maturity in applied biometrics by also ensuring privacy issues in testing. First, data set issues are discussed, second, known simulation approaches are summarised.

3.2.2.1   Privacy-preserving data sets for benchmarking (testing): artificial sweat-printed computer-generated fingerprints

To compare the performances and reproducibility of forensic methods of fingerprint acquisition, detection and analysis, [16] proposed to create latent fingerprints with artificial sweat (an amino acid solution) using a bubble jet printer allowing a fine granular determination of amino acid as simulated artificial sweat. This idea was further used by [17] and combined with artificial fingerprint patterns generated using SFinGe [18]. This concept proposes to produce computer-generated fingerprint images which printed by an inkjet printer by using artificial sweat. These latent, artificial sweat prints can be further developed and acquired with traditional treatments known in dactyloscopy or acquired with contactless sensors and further digitally processed. The acquired latent fingerprints can also be directly compared and evaluated with the digitally generated counterpart. This comparison allows several quality evaluations such as for sensor acquisition quality and sensor reproducibility, the evaluation of the impact of treatment methods or for different algorithm-benchmarking purposes. This approach enables a creation of privacy-conform test sets which can be exchanged publicly. In the field of contactless, optical sensors, test sets with samples from [17] exist and contain 50 fingerprints printed with a Canon Pixma iP4950 inkjet printer on a compact disk platter surface, acquired with a Keyence VK-x110 series confocal laser scanning microscope. Here, the topography and intensity data is captured (laser and colour data by using a charge-coupled device (CCD) camera, colour-intensity image by combining the colour data and the laser data). Based on this approach, further printed fingerprints have been generated and captured using a FRT CWL 600 sensor with 500 and 1,000 ppi resulting in 24 fingerprints for each resolution. This sample data set consists of 48 contactless acquired computer generated and artificial sweat printed latent fingerprints generated using SFinGe [18] and printed with Canon Pixma iP4600 inkjet printer to an overhead foil, available from the AMSL websites [19].

Recommendation for biometrics: The approach to use SFinGe [18] is already known and commonly applied in privacy-preserving biometrics, nevertheless the combination with artificial sweat printing could be a valuable extension to further include sweat-related artefacts simulated in during the printing and further acquisition processes. The existing data sets might be an extension to the test sets used in biometrics, or biometric test sets can be easily enhanced by additionally printing the data sets and acquiring samples in different conditions.

3.2.2.2 Simulation framework: from *StirMark* to *StirTrace* for artefact simulations in forensics and biometrics

StirTrace introduced in [20] and enhanced in [21] simulates a selection of expected typical artefacts occurring during trace acquisition at the crime scene. It is derived from the known *StirMark* tool, originating from the research field of digital watermarking [22] and its application in [23] to assess exemplary fingerprint sensor acquisition quality in biometrics domain. From the forensic application of forgery detection, specific simulations for artificial sweat-printed fingerprints (printing artefacts) and further trace acquisition artefacts with various parameters (assumed potential disturbances during sensor acquisition) are included. In summary, eight original *StirMark* techniques have been identified as appropriate. In addition, banding simulation (to test printing defects caused by blocked nozzles), trace acquisition bit depth and random Gaussian noise (as sensor characteristics simulations), surface characteristic simulation (for substrate characteristics), salt and pepper noise (as simulation of sensor scan artefacts) and tilting (as acquisition condition) were defined and included. Furthermore, combinations of artefacts are foreseen in the test concepts and were investigated.

Recommendation for biometrics: The simulation of artefact in the biometrics domain is already applied as suggested in [23]. In addition, from the forensics work, acquisition bit depth, further noise types and artefact combinations might also be valuable to be considered in biometrics testing. As shown in forensics, the simulations can be easily applied to privacy-preserving data sets, such as on computer-generated fingerprints or privacy-preserved partial data or de-identified data sets.

3.3   Privacy concepts – findings from digital forensics of face-morphing detection in face authentication systems

The concept of identity theft is as old as the concept of user authentication. Even though the intention of biometrics is to generate a strong (i.e. hard to steal) link between a person and an authentication token, also biometric authentication scenarios are in the focus of various of identity theft schemes. In their paper, 'The Magic Passport' [6], Ferrara *et al.* present a relevant, novel identity theft scenario for face biometrics that, in its potential consequences, by far outperforms simplistic spoofing or presentation attacks (see e.g. [24]) in biometric contexts. Their face-morphing attacks aim at the creation of valid (authentic and un-tampered) documents (e.g. biometric passports) that could be used by multiple persons. Summarising the work of Ferrara *et al.*, on one hand, the potential outcome of their attack could be a 'clean', authentic and perfectly regular passport, issued by official authorities, for a wanted criminal. This document will therefore pass all optical and electronic authenticity and integrity checks. On the other hand, virtually no security mechanisms for detecting the attack in the document-issuing process seem to be currently foreseen in any of the stages of the document life cycle (application, generation, release, usage, etc.), despite the fact that it is currently a hot research topic [to discuss the individual contributions in the targeted detectors against face-morphing attacks (e.g. [25]) is outside the scope of this chapter; for such an overview, the authors refer to [26]].

Besides such novel challenges in face biometrics, the already existing attack scenarios like presentation attacks (as described in [24] for biometrics in ATMs) or more complex information leakage scenarios – i.e. the degradation of the security level of face-based authentication systems under face image disclosure scenarios (e.g. [27]) – still provide interesting application scenarios for forensic applications.

Inherently linked to attacks against face recognition systems is the natural wish to preserve the personal privacy. In contrast to specific attacks on biometric authentication, this introduces general concerns against any form of potentially un-wanted commercial exploitation or discrimination of any kind (e.g. racial, age, etc.) of disclosed face data. A good work summarising such concerns in the context of the usage of face images in online social media would be [28] looking at scenarios of misuse of face images taken from social media websites. Such work on person tracking and techniques invasive of privacy motivates security researchers to focus on the design and implementation of de-identification techniques. Works like Sweeney [7], Newton *et al.* [8], Driessen and Dürmuth [9], etc. show that, amongst others, the same face-morphing techniques that have above been discussed as a method for attacking a biometric authentication system can at the same time help to achieve a certain wanted degree of anonymity in various application scenarios such as profiling of users, preparation of a training base for identification or the preparation of identity theft attempts. Comparing different face de-identification techniques, research work like [29] focusses on requirements as well as quality criteria. In [29], these are summarised as the degree of personally identifiable information removed from an image and the data utility or intelligibility retained (i.e. whether the de-identification output is still fit for the purpose intended by its generator). A chapter by Ribarić & Pavešsić in this book is devoted to the broad aspects of de-identification in biometrics.

In comparison to the aforementioned forensic challenges arising from attacks against biometric authentication systems, such face de-identification provides further relevant challenges to media forensics. Here, usually even a quick check would easily reveal enough hints on the usage of media manipulations and therefore raise suspicion in many scenarios, so the user would have to check carefully (or forensically) that the level of anonymity achieved by the chosen de-identification or protection method actually meets the expectations. For a thorough discussion on differentiable levels of anonymity (identifiability, pseudonymity, anonymity, unlinkability, unobservability, etc.) that are achievable see, Pfitzmann *et al.* in [30].

The claim that the need for establishing suitable privacy-protection mechanisms for images is not only an academic trend, as to be seen by a call for proposals that has been published by ISO/IEC 19566 (JPEG Systems) WG1 on 'JPEG Privacy and Security' in January 2017. The goal of the call is summarised in [31] as 'This activity aims at developing a standard for realising secure image information sharing, which is capable of ensuring privacy, maintaining data integrity, and protecting intellectual property rights. This activity is not only intended to protect private information carried by images [...] but also to provide degrees of trust while sharing image content and metadata based on individual preference'. Amongst other things, the call is looking into 'system-level solutions for privacy and security protection of images encoded by JPEG coding technologies to ensure backward and forward compatibility', which would, to some extent, be synonymous to the de-identification considerations addressed here.

In the following subsections, the image manipulation strategy of face morphings is discussed in two different contexts to pay tribute to its dual-use nature: (a) as an identity theft scheme against face biometrics and (b) for privacy protection/ de-identification purposes. As a foundation of the practical considerations made here, the face-morphing methods summarised by Ferrara *et al.* in 'The Magic Passport' [6] are referred to.

Based on the methodology for forgery detection introduced in [32] for context analysis-based digitised forensics, in Section 3.1, a generalised attack procedure and privacy implications for this image manipulation strategy are discussed. Further, in Section 3.2, the attached forensic implications and the potential interplay between biometrics, privacy and forensics are discussed for this exemplary attack.

3.3.1 Face-morphing attacks – generalised attack procedure and privacy implications

The general concept for morphing attacks is best described in [6] as a face-morphing operation combining input images from different persons into one image to be submitted into the document generation process for an ID card or similar document. At

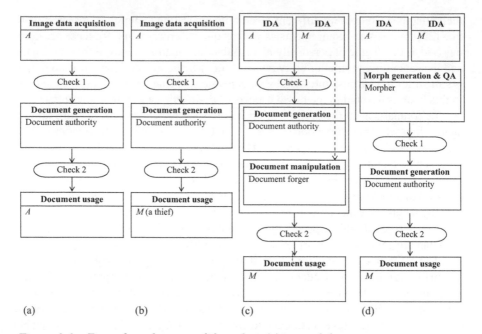

Figure 3.1 Exemplary document life cycles: (a) normal document usage,
(b) document is stolen and presented by the thief (presentation attack),
(c) existing passport is tampered by a forger, (d) face-morphing attack

the time of verification, the face can be successfully matched with the face of two
or more different subjects. As a result, two or more subjects can share the same document, which would allow the criminal exploitation and the overcoming of security
controls.

In order to understand the severity of face-morphing attacks, it has to be put
in contrast to other attacks on photo-ID documents. Figure 3.1(a) introduces a
generalised life-cycle model that is valid for a wide range of (paper as well as radio-
frequency identification (RFID)-enabled) documents including passports (eMRTDs),
national ID cards, drivers' licences, company IDs, etc.

The generalised process for generation and usage of a face-image-based authentication token (e.g. a passport, driver's licence or ID card) can be reduced to three
core operations: image data acquisition (IDA), document generation and document
usage. First, the face image is newly generated or acquired for re-usage. Second, a
person applies for a document and all administrative and authentication steps that
are necessary for the commissioning of the document creation are performed by a
corresponding authority. Furthermore, the document is created and picked up by or
delivered to the applicant. Third, the document is used in authentication scenarios like
border control, authentication at the desk of a car rental agency, etc. Besides the operations, the model includes another important component – the checks that connect two
operations in the pipeline. In these checks, various document-specific characteristics

are evaluated, including, amongst others, authenticity (document as well as entity authenticity) and integrity checks as well as tests on compliance to standards. In these checks, media forensic considerations are mostly neglected so far, even though most of the example processes discussed in this paper already contain automated check components that could strongly benefit from media forensic detectors.

In Figure 3.1(a), the generalised legitimate document usage is shown representing a first exemplary document life-cycle use case. In operation 1, the applicant Alice (*A*) asks a photographer to produce a face image for the purpose of applying for a document (e.g. a passport).

Prior to the commissioning, the document that *A* applies for in operation 2, the passport authority verifies in Check 1 that all process requirements are met. At present, such a check contains the following components: Alice is authenticated using an already existing *ID* document (e.g. old passport, national identity card or birth certificate), the face on the image is verified against the presented face and the image properties are checked against the technical requirements of the document-issuing authority [e.g. compliance to International Civil Aviation Organization (ICAO) standards for eMRTD images [33]]. If the check is successful, in operation 2, the photograph is accepted by the authorities for document generation. Notice that this submission is strongly regulated in each country. Therefore, both digital and analogue submissions (the latter in form of a developed, hardcopy photograph) have to be foreseen. On the basis of the face image and the additional ID information provided, the document generation is commissioned.

Prior to document usage (e.g. for authentication of *A*), the document is checked in Check 2. Such checks verify the document integrity ('Does it look tampered?') as well as additional characteristics ('Is the document still valid?', etc.). If positive, the document is used, by a border-control agent, for authenticating the presenting person.

Figure 3.1(b)–(d) shows three different attack scenarios: simple presentation attack, document tampering/forgery and face-morphing attack.

In Figure 3.1(b), a simplistic presentation attack is shown. Here, a thief Mallory (*M*) steals the passport from its owner Alice (*A*) and tries to use it at a border control. This would be the most simplistic form of presentation attack [34] but might succeed if *A* and *M* look similar (and the other characteristics manifested into the document also seem plausible). The modelling of this attack looks similar to the example of legitimate document usage presented in Figure 3.1(a). Differences are only found in operation 3. The outcome of Check 2 is that no anomaly for the document could be found (i.e. the document was created by an authorised entity and was not tampered), but the ID and the presenting subject show a mismatch. If this mismatch is not detected in Check 2, *M* could successfully use the document for illicit authentication.

Figure 3.1(c) shows an attack scenario where an existing passport is modified by a forger. After the photos of *A* and *M* are taken in two parallel IDA operations, the forger takes a valid passport from *A* and replaces the image in the document with a photo of *M*. Admittedly, this is a very unlikely scenario for modern day ICAO compliant travel documents, but for other photo-ID documents, company ID cards, this might still be a valid attack. As shown in Figure 3.1(c), the 'document generation' step in the pipeline is extended by the 'document manipulation' step. As a result, the corresponding modelling of the attack is identical to the previous two examples for operation 1 and Check 1. For operation 2, the original operation by the document authority is seeing an extension by forging. The forger tries to transfer another ID into the document by replacing the photograph. After this manipulation, the ID and the photo show a mismatch, i.e. the document was tampered. In Check 2, the documents integrity check should detect the tampering, otherwise M could successfully use the document for illicit authentication in operation 3, since the acting entity in this step (e.g. a customs agent) only compares the face of person presenting the document with the image available in the document.

In Figure 3.1(d), a face-morphing attack is shown. In contrast to all previous examples, operation 1 is extended by an illegitimate addition – the morph generation and the morph quality assurance (QA). The photographer takes pictures of two persons [Alice (A; an innocent accomplice) and Mallory (M; assumedly a wanted criminal who wants a new document that can be used, e.g. to cross a border] in two parallel IDAs. The morpher creates one face image from two inputs. Literature indicates that the morph creation can happen as manual-, semi-automated- or automated processes, yielding potentially different quality levels of morphs (see e.g. [35]). The output image is visually as well as biometrically similar to A and M and is used by (previously innocent) accomplice A to apply for a new passport. If Check 1 does not detect the face-morphing operation (i.e. a suitable media forensic detector is missing), a valid document is generated in operation 2. Since M is assumedly a wanted criminal, she asks A to do the application and document pickup. Check 2 performs the same analyses as in the previous examples. If no traces of the morphing operation are found (i.e. a suitable media forensic detector is missing), M could successfully use the document for illicit authentication in operation 3. Notice that it might be relevant to distinguish between being a cooperating accomplice or an innocent victim of an ID theft attempt by face-morphing attack. This would have consequences for the IDA step and might influence the quality if the morphs, i.e. in case of a cooperating accomplice, the photos for M and A could be taken with the same camera, at the same location and identical environmental conditions, while for A the camera, etc. might strongly differ.

The difference of the face-morphing attacks to the other attacks can be summarised in the fact that the attack happens before the document generation step and that it enables the (presumably criminal) attacker to obtain a 'clean', valid and un-tampered document issued by the official authority.

Besides these attack use cases, a wide range of other black hat scenarios could be discussed on the basis of the exemplary generalised document life cycle presented in Figure 3.1. But the main point can already be made with the few presented examples: they show that the current realisations of Check 1 and Check 2 are sufficient to handle the established threats like presentation attacks or document forgery attempts but are ill-designed to handle face-morphing attacks. The clear message of our paper is that media forensics methods must be integrated into the established processes to enable a higher level of protection also including this novel threat.

Recommendations for biometrics: Biometric use cases with non-controlled enrolments (e.g. a person submitting an own image for passport creation) or with templates that are in the possession of the user (e.g. a passport) should incorporate forensic detectors to validate the input or template integrity.

Similar to the black hat use cases presented above, white hat use cases could be discussed here, using the same modelling – life-cycle modelling. Figure 3.2 shows three exemplary selected cases of face-image usage for social networks and the potential checks, Check 1 and Check 2 for de-identification purposes. Figure 3.2(a) shows a normal usage in contrast to an ID-theft scenario in Figure 3.2(b). Figure 3.2(c) illustrates the white hat application use case of face morphing for privacy concerns (i.e. for legitimate privacy protection/de-identification purposes). Here, the Check 1 and Check 2 aim at establishing whether the users intended degree of anonymity have been achieved on the image itself (Check 1) and on the social network the image is intended to be used for (Check 2). Especially, the latter is of importance because being the only user using de-identified images in a network would be an individualising characteristic for a user, making him easily traceable, which is exactly the opposite of the original intention of de-identification.

Using the methodology for forgery analysis and detection introduced in [32] for context analysis-based digitised forensics, a modelling of this attack scenario has to identify the attack procedure, the context properties (CP), the detection properties (DP) and the context anomaly properties (CAP). The application of this methodology to morphing attacks is challenging because here we would need to distinguish different classes of influences to the image: allowed image manipulations (certain croppings, scalings, compressions, etc.), non-allowed but non-malicious image manipulations (e.g. the common 'beautification') as well as malicious manipulations.

The description of the attackers operations presented in [6] can be extended into the generalised attack procedure shown in Figure 3.3 (please note: the CP, DP and CAP are discussed in detail in Section 3.2).

- In **step 1**, the attacker acquires the (two or more) face images needed as input for the morphing operations. Ideally, the attacker would be able to take the photos of all faces in a fixed, condition-optimised environment. But if this cannot be achieved, also other means of acquisition can be used, e.g. by taking face image inputs from social media websites or other databases. Here, the attacker/morpher takes the first step to achieve an visually imperceptible, format compliant, biometrically sound and forensically hard to detect morph, but at the same time, this is the step where the first CP are formed (like a PRNU-based camera fingerprint [36], the colour filter array pattern or chromatic aberration artefacts) that might be later used to detect morphs.
- In **step 2**, the attacker performs pre-morphing image-processing operations, like cropping, scaling and alignment. Obviously, these operations are required to produce visually imperceptible, format compliant, biometrically sound morphs, but they introduce further CP/CAP (like influences to the camera fingerprint or the colour filter array pattern) that might help a media forensic analysis later on.
- In **step 3**, an iterative sequence of morphing and image-quality-control operations is performed by the attacker until all quality criteria [perceptual quality, face-matching scores against a reference database (containing, amongst others,

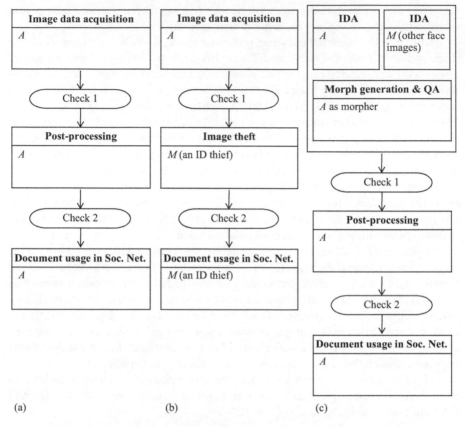

Figure 3.2   Exemplary image usage for social network scenarios, based on the concept of the document life cycle shown in Figure 3.1: (a) normal face-image usage, (b) image is stolen and presented by an ID thief (very common case for dating websites); (c) white hat scenario of face morphing for privacy concerns

images of the persons to be morphed together), standards compliance on the output images (e.g. to ICAO standards for eMRTD images, etc.)] are met. It has to be assumed that, regardless of which face-morphing method is used (e.g. complete morphs, splicing morphs, etc. – for details on possible realisations of face-morphing attacks we refer to [35]), this step introduces characteristic CP/CAP to the output that could be transferred into DP for this specific image manipulation since a morphing usually performs (at least) a series of geometric transformations of image regions to correct the face geometry together with a colour interpolation to correct the face colour and texture. Therefore, it has to be assumed that the morpher might also use counter-forensics methods to disguise the attack efforts, leading to an arms-race cycle between attacker and detector.

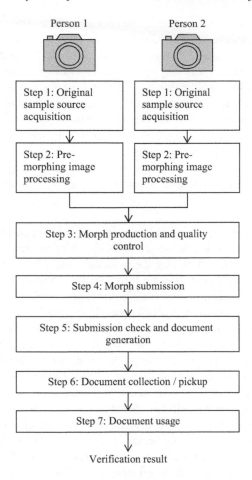

Figure 3.3   Generalised morphing attack pipeline

- In **step 4**, the generated morph image is submitted to the authorities for document generation. Here, it has to be mentioned that this submission is strongly regulated in each country but that digital as well as analogue submissions (the latter in form of a developed, hardcopy photograph) have to be foreseen, as they are practically used widely. The latter might strongly influence media forensic investigations, because in since the required scan, many CP, DP and CAP, that have been introduced in the previous steps, are assumedly eliminated by the digitisation process (i.e. the analogue–digital conversion).
- During submission, in **step 5**, the image is analysed to see whether it meets the quality requirements of the document-issuing authority (e.g. compliance with ICAO standards for eMRTD images) and if this check is successful, the document generation will be commissioned. Since the exact technical details of the document creation process are kept secret in many countries, it can only be assumed

that additional image manipulations are taking place here that could either foster or hinder a later media forensic analysis of the face image used.

- In **step 6**, the generated document is obtained from the authorities by one of the participants in the morphing operation (most likely an innocent one). Here, usually a short form of biometric verification is performed before the document is handed over. In step 7, the document is used, most likely in criminal exploitation, e.g. at a border-control station. Here, usually biometric verification and a document verification take place in different scenarios (by a border-control agent, an automated border-control gate, etc.). In some of these scenarios, the detection of face morphings could be supported, e.g. by using the additional face images created in an automated border-control gate for biometric thresholding analyses.

Obviously, if the perspective is shifted from face morphings as a black-hat-identity-theft scheme against face biometrics towards its white hat application for privacy protection/de-identification purposes, the procedure has to be slightly amended. The main deviations can be found in steps 5 to 7 because the authority for document generation is here replaced by some usually less constrained and constraining entity, like the image-upload service of a social network website. There might be checks in place, but these are usually less strict than official standardisation processes like ICAO conformity request. In such scenarios, steps 6 and 7 are then usually website requests or similar access mechanisms for the images provided for de-identification purposes. Especially, if the example of de-identification of face images for the intended usage in social network websites is used, the fact that those images are thereafter globally, openly available proposes a significant threat for the detection of the application of de-identification techniques. The images could be viewed, checked or analysed by anyone even with the help with dedicated image manipulation detection methods, which would very likely reveal the morphing. Therefore, it is safe to assume that the morpher also aims at creating visually imperceptible, format compliant, biometrically sound and forensically hard to detect morphs leaving as few as possible traces (or CAP/DP) in the image as possible.

When looking at privacy implications, for the black-hat-attack scenario described above – as in most identity-theft scenarios for face biometrics – the attacker is in need to obtain several face images. Obviously, such images are required for the persons to be morphed, these images are obtained in step 1 of the attack outlined above and for the quality control operations he performs in step 3 (where the morpher would have to assure the person intending to apply for a document that the morph generated is up to the intended task and is very likely not to backfire in steps 5 to 7 of the attack). Ferrara *et al.* assume in [6] that all images to be used in the actual morphing are taken in the same environment, limiting the impact of the photo studio, lighting conditions, background, etc., but this is not strictly necessary. With the current trend of specialised services like document forgeries becoming darknet services, instead of in-house jobs, it is more likely that different acquisition environments have to be considered. Therefore, for a targeted morph, any face image source, even a social media profile page, can potentially provide the input to be used. The lessons to be learned here might be unpopular and opposing current trends, but the prevention of the disclosure of unnecessary information (in this case face images) is very likely to make such attacks more difficult. This fact is, from the point of more simplistic presentation attacks, more thoroughly discussed in [27].

For the investigation processes that are supposed to happen in steps 5 and 7 of the procedure described above, it would be beneficial for the operator at the application desk or at an authentication station to have ample of image material of the person shown in a face image to test a morphing-attack hypothesis. However, in the worst case scenario, the operator has only the face image on the document and the face presented. In situations more beneficial for the operator, more image sources, like additional images from still and video cameras, from additional databases, etc., are available. In such a case, biometric face matchers as well as media forensic methods could help in supporting or rejecting a morphing-attack hypothesis. Here, the amount of material that might be available and could be used is in most cases clearly regulated by privacy and data protection legislation which strongly varies between the different countries worldwide. To assess the boundaries imposed in this field for relevant application scenarios like document application or border control, authentication is outside the scope of this work, but such an assessment would highlight the restrictions imposed by privacy regulations.

Privacy considerations for the white hat de-identification scenario are here, on the one hand, the question whether the intended level of anonymity (i.e. the layer of added privacy) has been achieved in practice (e.g. if only one user using de-identified face images in a social network this will stick out), but this question leads to complex analyses since the questions of cross-system information leakage and traceablity/linkability would have to be addressed. On the other, the interplay with the privacy terms and terms of usage of the target service have to be considered, including the question whether a face image that underwent de-identification still has to be considered to be a person-related information protected by the data-protection acts in various countries.

Recommendations for biometrics: As a consequence for biometrics, the level of public exposure of a person – here, the amount of biometric data such as face images available – can be considered to be a good indicator for this person regarding his/her potential vulnerability against targeted attacks. This hypothesis can easily be justified for simplistic spoofing/presentation attacks, but it also holds true in more complex scenarios like the aforementioned face-morphing attack. Therefore, it seems to be a good recommendation to minimise the amount of publicly available face images to the absolute minimum and consider the usage of de-identification methods if face images must be provided for any reason. Biometrics and privacy concerns are two strongly connected topics. As a result, the field of face biometrics has to acknowledge the dual-use nature of face morphings and similar techniques and identify resulting threats as well as security (i.e. privacy)-enhancing benefits that derive from this processing method. In case, de-identification methods for biometrics are actually used, it is recommended to evaluate their performance in the chosen application scenario to establish whether the intended degree of anonymity is actually achieved.

3.3.2   Media forensic investigations and biometrics on the example of face-morphing attacks

The full detailed analysis, based on media forensic techniques, of the black hat as well as white hat use cases of face morphings using the model of CP, DP and CAP from [32] would go beyond the scope of this chapter. Nevertheless, this concept shall be used here, to link CP and derived DP with the CAP that might raise suspicion or even allow (in steps 5, 6 or 7 of the attack procedure outlined in Figure 1.1) the detection of the manipulation effects or anomalies introduced in steps 1, 2, 3 and 4. In particular, a systematisation and a high-level overview on a potential media forensic investigation in a morphing attack investigation are presented.

The statements of recent ICAO documents list the following allowed post-acquisition-processing operations for images created for the purpose of obtaining MRTD/eMRTD state [33]: in-plane rotation, cropping, down sampling, white balance adjustment, colour management transformation and/or compression. With focus on compression, it is said in [33]: '*It shall not be compressed beyond a ratio creating visible artefacts* [...]'. The statements by other document-regulating authorities, the US Department of State (as the authority issuing visa documents for the U.S.A.), are at the same time stricter and less technical on the issue of image modifications [37]: '*Photos must not be digitally enhanced or altered to change your appearance in any way*'. None of these prohibitions are backed by objective detection mechanisms, which could easily be turned into forensic methods. On the contrary, it seems to be common practice by photographers to apply 'beautification' to face images intended for the creation of documents.

In regard to the identified steps from the generalised attack procedure identified in Section 3.1 and for the black hat application scenario, three different points are evident, where a forensic investigation on the morphed image is very likely going to happen: the first would be step 3 'morph production and quality control' where the creator of the morphings will have to convince the person intending to use the morph of the quality of the work. The second investigation is the step 5 'submission check and document generation' where the receiving authority will run checks on the submitted objects. The third investigation is the document verification in step 7 'document usage'. Obviously, in all these three points, an inspection will take place to find, amongst others, traces of document forgery or manipulation. For presentation attacks, it has to be assumed that current, high-quality documents like eMRTDs have achieved a security level which prevents such attacks. But for face-morphing attacks hardly anything is currently in place to aid such investigations. In practice, a media forensic investigation that would be required here on an face image would have to distinguish between allowed image manipulations (e.g. cropping and compression) as well as non-malicious image manipulations ('beautification') on one hand and malicious manipulations (e.g. in a morphing attack) on the other hand. The following systematisation gives a limited-scale overview over relevant CP, DP and CAP for the face-morphing attack scenario. Here, four different classes of CP with corresponding derived DP should be distinguished:

1. **Standards compliance**: The accepted files have to comply the defined standards for the corresponding scenario (e.g. ICAO standards for eMRTDs). A failure to such compliance criteria could be interpreted as DP/CAP. In step 3 'morph production and quality control', it has to be assumed that the attacker/creator of the morphed images is well aware of the requirements imposed by the target authorities and also has some insights into verification methods that are used in steps 5 and 7. Thus, such an attacker can perform own QA work on high-quality versions of the image material and could introduce counter-forensics to actively manipulate the outcome of the later forensic analyses. Therefore, it is very unlikely that such non-compliance artefacts occur and can therefore be used to detect face morphs.

   In step 5 'submission check and document generation', the issuing agency will run its checks to enforce standards compliance on the input, but here the amount of data as well as the knowledge of the processing history of a specific image are of course smaller than in step 3. Since the procedures for these checks are usually well standardised (e.g. by ICAO), it should be assumed that they are known to the morpher and, as a result, it is very unlikely that artefacts are introduced here that can be used for morph detection.

   In step 7, any failure in compliance with existing document standards would have to be handled accordingly, but since the document that is the intended output of the face-morphing attack has neither been forged nor tampered with, it should pass any such checks successfully.

2. **Media forensic traces:** At the steps 3, 5 and 7, in the attack procedure, the image(s) will undergo media forensic analyses to check traces or artefacts from steps 1 to 2 and from 3 itself as well as the method of submission chosen by the attacker and the consequences to the image to be used. As already mentioned earlier for the check on standards compliance, the attacker usually also has the best knowledge and the best material to work on, while the data (and knowledge on the processing history of the image) become more scarce in the later steps, i.e. a successful detection becomes harder. While it can be assumed that the attacker will take precautions and counter-forensics to disguise the tampering with the image to be submitted, an image is a complex object which is hard to manipulate without leaving any trace. So here an arms race between the attacker and the detector has to be assumed as long as the attacker can control the steps 1 to 4 of the procedure. The only way to break out of this arms race (and to make the overall attack way more difficult for the attacker) would be to forbid the submission of images to the document creation process – instead, the photos would have to be taken directly in the application office, as it is already custom in some countries.

   In this black hat application scenario for face morphings introduced in [6], the image content and context of the submitted image has to undergo a thorough media forensic analysis distinguishing between CP, DP and CAP from different classes of manipulations in the steps prior to the image submission. These classes are the artefacts introduced in steps 1, 2, 3 and 4 by allowed manipulations (certain croppings, scalings, compressions, etc.), non-allowed but non-malicious image manipulations (e.g. 'beautification') as well as malicious manipulations. DP/CAP for all three classes could be generated by a multitude of source authentication as well as integrity verification methods currently discussed in the media forensics communities. In general, the range of CP, DP and CAP that could be utilised here would be very long. Summary documents on image-manipulation detections, [38], would provide extensive, but never complete lists including methods like:

    i.   violation of **sensor-intrinsic footprints**,
    ii. **re-sampling & splicing detection** (based on general intrinsic footprints, based on lens distortion or inconsistencies in lighting/shadows or the geometry/perspective of a scene, based on coding artefacts or based on camera artefacts),
   iii. **copy-move detection**,
   iv. **enhancement/beautification detection**,
    v. **seam-carving detection**, etc.

Specific DP/CAP for the detection of the face-morphing attacks would very likely focus on the manipulations that are intrinsic to the morphing operation like the geometrical interpolation of certain areas in the image (in case of splicing morphs) or the whole image (in case of complete morphs) in combination with colour value interpolation and double-compression artefacts. Recent articles [39] or [25] are already discussing such targeted detection scenarios for face-morphing attacks.

One imminent problem for the application of media forensic methods in this black hat application scenario is the fact that in many countries also the analogue submission of face images (i.e. of old-fashioned hardcopy photographs) is still allowed in document generation. In this case, the scanning which has to take place makes the analysis task much harder due to the additional information loss and noise addition in the digitalisation process.

3. **Perceptual quality:** A human eye is very well trained to observe implausibilities in human faces and face images. Therefore, a human inspection (of submitted face images in step 5 or the document in step 7) might help to identify morphs, in averse situations (e.g. bearded men – because morphing algorithms are notoriously bad for highly textured areas) or when image splicing techniques have been used. But here, the 'morph production and quality control' in step 3 is very likely to filter out morphs that look suspicious. Therefore, it is unlikely that visual artefacts from steps 1, 2 and 3 will remain in the image submitted by the morpher and therefore might be used as DP/CAP for morphing attacks in later media forensic analyses. This claim is supported by an experiment performed at the author's institute, the Advanced Multimedia and Security Lab (AMSL) at Otto-von-Guericke University Magdeburg. In this experiment, with 40 test candidates, the test persons were asked to classify real face images and automatically generated face morphs printed in passport-image size (3.5×4.5 cm), from sets of

morphed and original images. Summarising the results, it can be said that none of the test candidates achieved reliable detection rates for the face morphs – the average result in the experiment was equivalent to a coin toss (i.e. guessing in this two-class problem). Details on this experiment are presented in [35].

4. **Biometric thresholding:** In steps 3, 5 and 7, the discriminatory power of a face image against a database of human faces can be measured. Such a measurement would in general give an indication how strong the specimen is clustering after the projection into the feature space and would therefore give a rough estimate on the number of persons that could be suitably represented using this feature space (i.e. biometric system). Besides this generalised aspect, such a discriminatory power for an individual image could be used for QA in document generation. Here, also a DP/CAP might be generated from this value, because it has to be assumed that the morphing operations in step 3 (usually a series of geometric transformations of image regions to correct the face geometry together with a colour interpolation to correct the face colour and texture) will result in a face image which will have a lower discriminatory value in such a biometric thresholding. Especially, in step 7, a detection of potential face morphings might benefit from additional cameras recording face images at an authentication point (e.g. a border-control station).

Obviously, no single DP can be used to build a detection mechanism that could hope to counter such an sophisticated threat as posed by face-morphing attacks. Instead a fusion of multiple DPs would be necessary to reduce the attacks threat.

For the white hat application scenario of de-identification, similar considerations can be made for the interplay between creating non-suspicious privacy-enhanced-de-identified face images and the detection of such images. It has to be assumed that in this scenario the detection is harder to implement, since the face images to be used for social media websites are hardly supposed to follow strict guidelines (like e.g. ICAO compliance). On the other hand, the detector could benefit from cross-system information-leakage effects which might strongly decrease the level of privacy achieved by the de-identification, particularly, if both de-identified and original versions of such images are in circulation for a particular subject.

Recommendation for biometrics: In upcoming biometric application schemes, the integration of media forensics as well as biometric thresholding operations have to be foreseen to achieve a higher resilience against attacks like the face-morphing scenario considered here. Both would enable the usage of DP for security verification in presentation attacks as well as in scenarios targeting the template space. This is especially relevant if the authentication template is handed over to the person to be authenticated (e.g. in the case of eMRTD) which might pass them on to 'similar looking' persons for usage, or when the templates are created from sensor data, which are not captured in a protected, controllable environment (e.g. in the case where it is allowed to submit a photograph to the application office as basis for the commission of an eMRTD). The integration of media forensics as well as biometric thresholding into existing biometric schemes could reduce or even close security vulnerabilities in such systems.

Regarding the white hat application scenarios for face-morphing-based de-identification, it has to be acknowledged that the morpher has in media forensics a strong adversary. Successfully ensuring the highest level of privacy would also require the morpher to eliminate as much of the artefacts introduced in steps 1, 2 and 3 of the procedure as possible. But this would require the person applying such techniques to understand what is actually done (on a signal-level) and be aware of the current state of the art in image manipulation detection – which would be a huge prize to pay for anonymity.

Acknowledgements

The authors would like to thank the editor Claus Vielhauer for his motivation and providing us with possibility to write this book chapter. Further, we thank Mario Hildebrandt and Ronny Merkel for their very valuable joint work and their support in rephrasing the text and the possibility to work with them. In addition, we would like to thank the co-authors of all papers that led to the results summarised in this chapter, such as Stefan Kiltz, Ronny Merkel and Claus Vielhauer, for the inspiring joint work on our ideas and findings and their valuable help in the implementation and experimental validation.

Special thanks are given to Michael Ulrich motivating the application field and for his input to the findings from the crime scene forensics field.

The work presented in this chapter has been funded in part by the German Federal Ministry of Education and Research (BMBF) through the Research Programme ANANAS under Contract No. FKZ: 16KIS0509K and through the project Digi-Dak and the Research Program 'DigiDak+ Sicherheits-Forschungskolleg Digitale Formspuren' under Contract No. FKZ: 13N10816.

Thanks also to all Digi-Dak and DigiDak+ as well as ANANAS project staff and the project partners and students working in the projects for their help, input and discussions in the field.

References

[1]   National Institute of Justice (NIJ). Forensic Sciences [online]. 2017. Available from http://www.nij.gov/topics/forensics/pages/welcome.aspx. [Accessed 2 November 2017].

[2]   National Forensic Science Technology Center (NFSTC). *The Daubert Standard: Court Acceptance of Expert Testimony* [online]. 2016. Available from http://www.forensicsciencesimplified.org/legal/daubert.html [Accessed 25 November 2016].

[3]   *EU DIRECTIVE (EU) 2016/680 OF THE EUROPEAN PARLIAMENT AND OF THE COUNCIL of 27 April 2016 on the protection of natural persons with regard to the processing of personal data by competent authorities*

for the purposes of the prevention, investigation, detection or prosecution of criminal offences or the execution of criminal penalties, and on the free movement of such data, and repealing Council Framework Decision 2008/977/JHA [online]. 2016. Available from http://ec.europa.eu/justice/data-protection/reform/files/directive_oj_en.pdf [Accessed 6 Sept 2016].

[4]   *EU Reform of EU data protection rules* [online]. 2016. Available from http://ec.europa.eu/justice/data-protection/reform/index_en.htm [Accessed 6 Sept 2016].

[5]   Kiltz S. *Digi-Dak Homepage* [online]. 2017. Available from https://omen.cs.uni-magdeburg.de/digi-dak-plus/cms/front_content.php?idcat=2 [Accessed 20 Jan 2017].

[6]   Ferrara M., Franco A., Maltoni D. 'The Magic Passport'. *Proceedings International Joint Conference on Biometrics (IJCB)*; Clearwater, Florida, USA, Oct. 2014. IEEE; 2014. pp. 1–7.

[7]   Sweeney L. '*k*-Anonymity: A Model for Protecting Privacy'. *International Journal of Uncertainty, Fuzziness and Knowledge-Based Systems*. 2002;10(5); pp. 557–570.

[8]   Newton E.M., Sweeney L., Malin B. 'Preserving Privacy by De-Identifying Face Images'. *IEEE Transactions on Knowledge and Data Engineering*. 2005;17(2); pp. 232–243.

[9]   Driessen B., Dürmuth M. 'Achieving Anonymity Against Major Face Recognition Algorithms'. *Proceedings Communications and Multimedia Security (CMS2013)*; Magdeburg, Germany, September 2013. Berlin Heidelberg: Springer; 2013. pp. 18–33.

[10]   Hildebrandt M., Dittmann J., Pocs M., Ulrich M., Merkel R., Fries T. 'Privacy Preserving Challenges: New Design Aspects for Latent Fingerprint Detection Systems with Contact-Less Sensors for Future Preventive Applications in Airport Luggage Handling'. *Proceedings of Biometrics and ID Management: COST 2101 European Workshop* (BioID 2011); Brandenburg (Havel), Germany, March 2011. Berlin Heidelberg: Springer; 2011. pp. 286–298.

[11]   Merkel R., Hildebrandt M., Dittmann J. 'Latent Fingerprint Persistence: A New Temporal Feature Space For Forensic Trace Evidence Analysis'. *Proceedings of the IEEE International Conference on Image Processing* (ICIP); Paris, France, Oct. 2014. IEEE; pp. 4952–4956.

[12]   Merkel R., Hildebrandt M., Dittmann J. 'Application of StirTrace Benchmarking for The Evaluation of Latent Fingerprint Age Estimation Robustness'. *Proceedings of the 3rd IEEE International Workshop on Biometrics and Forensics* (IWBF2015); Gjovik, Norway, March 2015. IEEE; 2015. pp. 1–6.

[13]   Hildebrandt M. *StirTrace Website* [online]. 2017 Available from https://sourceforge.net/projects/stirtrace/ [Accessed 6 Sept 2016].

[14]   Meuwly D. 'Forensic Use of Fingerprints and Fingermarks' in Li S.Z., Jain A.K. (eds.). *Encyclopedia of Biometrics*. New York: Springer Science+Business Media; 2014. pp. 1–15.

[15] Roßnagel A., Jandt S., Desoi M., Stach B. *Kurzgutachten: Fingerabdrücke in wissenschaftlichen Datenbanken.* Report Projektgruppe verfassungsverträgliche Technikgestaltung (provet), Kassel, Germany, 2013.

[16] Schwarz L. 'An Amino Acid Model for Latent Fingerprints on Porous Surfaces'. *Journal of Forensic Sciences.* 2009;54(6); pp. 1323–1326.

[17] Hildebrandt M., Sturm J., Dittmann J., Vielhauer C. 'Creation of a Public Corpus of Contact-Less Acquired Latent Fingerprints without Privacy Implications'. *Proceedings Communications and Multimedia Security (CMS2013)*; Magdeburg, Germany, September 2013. Berlin, Heidelberg: Springer; 2013. pp. 204–206.

[18] Maltoni D., Maio D., Jain A.K., Prabhakar S. *Handbook of Fingerprint Recognition.* 2nd edn. London: Springer; 2009.

[19] Advanced Multimedia and Security Lab (AMSL). *Public Printed Fingerprint Data Set – Chromatic White Light Sensor – Basic Set V1.0* [online]. 2016. Available from https://omen.cs.uni-magdeburg.de/digi-dakplus/cms/front_content.php?idcat=47 [Accessed 23 May 2017].

[20] Hildebrandt M., Dittmann J. 'From StirMark to StirTrace: Benchmarking Pattern Recognition Based Printed Fingerprint Detection'. *Proceedings of the 2nd ACM Workshop on Information Hiding and Multimedia Security (IH&MMSec'14)*; Salzburg, Austria, June 2014. New York: ACM; 2014. pp. 71–76.

[21] Hildebrandt M., Dittmann J. 'StirTraceV2.0: Enhanced Benchmarking and Tuning of Printed Fingerprint Detection'. *IEEE Journal Transactions on Information Forensics and Security.* 2015;10(4); pp. 833–848.

[22] Petitcolas F., Anderson R., Kuhn M. 'Attacks on copyright marking systems' in *International Workshop on Information Hiding* (IH98). Volume 1525 of Lecture Notes in Computer Science. Berlin, Heidelberg: Springer; 1998. pp. 218–238.

[23] Hämmerle-Uhl J., Pober M., Uhl A. 'Towards A Standardised Testsuite to Assess Fingerprint Matching Robustness: The Stirmark Toolkit – Cross-Feature Type Comparisons'. *Proceedings of Communications and Multimedia Security (CMS2013)*; Magdeburg, Germany, September 2013. Volume 8099 of Lecture Notes in Computer Science. Berlin, Heidelberg: Springer; 2013. pp. 3–17.

[24] Kochetova O., Osipov A., Novikova Y. *Future Attack Scenarios Against Authentication Systems, Communicating with ATMs.* Kaspersky Lab Technical Report, 2016.

[25] Raghavendra R., Raja K., Busch C. 'Detecting Morphed Facial Images' *Proceedings of 8th IEEE International Conference on Biometrics: Theory, Applications and Systems (BTAS-2016)*; Niagra Falls, USA, September 2016. Washington, D.C., USA: IEEE; 2016. pp. 6–9.

[26] Kraetzer C., Makrushin A., Neubert T., Hildebrandt M., Dittmann J. 'Modeling Attacks on Photo-ID Documents and Applying Media Forensics for the Detection of Facial Morphing'. Accepted for *Proceedings of the ACM Information Hiding and Multimedia Security Workshop* (IH&MMSec '17), Philadelphia, PA, USA, June, 2017, ACM, 2017.

[27] Li Y., Xu K., Yan Q., Li Y., Deng R.H. 'Understanding OSN-Based Facial Disclosure Against Face Authentication Systems' *Proceedings 9th ACM Symposium on Information, Computer and Communications Security*. Kyoto, Japan, June 2014. New York, NY, USA: ACM; 2014.

[28] Vilnai-Yavetz I., Tifferet S. 'A Picture Is Worth a Thousand Words: Segmenting Consumers by Facebook Profile Images'. *Journal of Interactive Marketing*. 2015;32; pp. 53–69.

[29] Sun Z., Meng L., Ariyaeeinia A., Duan X., Tan Z.H. 'Privacy Protection Performance of De-Identified Face Images With and Without Background'. *Proceedings of 39th International Convention on Information and Communication Technology, Electronics and Microelectronics (MIPRO)*; Opatija, Croatia, 2016. IEEE; pp. 1354–1359.

[30] Pfitzmann A., Hansen M. *Anonymity, Unobservability, Pseudonymity, and Identity Management – A Proposal for Terminology*. Technical Report University of Dresden, Germany, 2004.

[31] International Organization for Standardization (ISO). *ISO/IEC 19566 (JPEG Systems) – WG1: Call for Proposals 'JPEG Privacy and Security'*. ISO, 2017.

[32] Dittmann J., Hildebrandt M. 'Context Analysis Of Artificial Sweat Printed Fingerprint Forgeries: Assessment of Properties for Forgery Detection'. *Proceedings Second International Workshop on Biometrics and Forensics (IWBF2014)*; Valletta, Malta, March 2014. Washington, D.C., USA: IEEE; 2014. pp. 1–6.

[33] Wolf A. *Portrait Quality (Reference Facial Images for MRTD) Version: 0.06*. ICAO Report, Published by authority of the Secretary General, 2016.

[34] International Organization for Standardization (ISO). *ISO/IEC 30107-1:2016 – Information Technology – Biometric Presentation Attack Detection*. ISO, 2016.

[35] Makrushin A., Neubert T., Dittmann J. 'Automatic Generation and Detection of Visually Faultless Facial Morphs'. *Proceedings of the 13th International Conference on Computer Vision Theory and Applications (VISAPP)*, 2017.

[36] Goljan M., Fridrich J., Filler T. 'Large Scale Test of Sensor Fingerprint Camera Identification'. *Proceedings SPIE, Electronic Imaging, Media Forensics and Security XI*; Volume 7254, January 2009. SPIE; 2009.

[37] Bureau of Consular Affairs, U.S. Department of State. *Photo Requirements* [online]. 2016. Available from https://travel.state.gov/content/visas/en/general/photos.html [Accessed 25 April 2016].

[38] REWIND Consortium. *REWIND Deliverable D3.1 State-of-the-Art on Multimedia Footprint Detection*. Technical Report of the REWIND Project (EU Grant Agreement No. 268478), 2011.

[39] Neubert T., Hildebrandt M., Dittmann J. 'Image Pre-processing Detection – Evaluation of Benford's Law, Spatial and Frequency Domain Feature Performance'. *2016 First International Workshop on Sensing, Processing and Learning for Intelligent Machines (SPLINE)*. New York: IEEE; DOI: http://dx.doi.org/10.1109/SPLIM.2016.7.

Part II

Privacy and security of biometrics within general security systems

Chapter 4

Physical layer security: biometrics vs. physical objects

Svyatoslav Voloshynovskiy*,[1], Taras Holotyak*,[1],
and Maurits Diephuis*,[1]

This chapter compares and describes the biometrics and physical object security fields, based on physical unclonable functions. Both lay at the foundation of authentication and identification architectures based on the assumption that the used primitives are both nonclonable and unique.

First, it will cover the physical phenomena that form the basis for both biometrics and physical object security next to highlighting the specificities of the used verification schemes. Second, it will cover relevant properties and requirements such as the security principles, feature extraction, the effect of leaks, possible attack vectors and the practical technological requirements needed for the implementation.

4.1 Introduction

In this chapter, we bridge human security with physical object security and investigate the techniques developed in both domains. Besides of remarkable similarity in the basic assumptions and approaches, these domains have developed in two different communities with little interaction between each other. However, a careful analysis of modern techniques developed in both applications might be very interesting for enriching each domain and the development of new protection methods.

From a general point of view, both human and physical entities are objects from the real world that have their specific features that uniquely characterize them. In the case of *human security*, these specific features are known as *biometrics*. Figure 4.1 schematically summarizes the main classes of biometrics used in modern authentication and identification applications. We refer the reader to other chapters of this book for a detailed classification and analysis of different biometrics.

While the object of protection based on biometrics is a human, *physical object security* covers a broad variety of objects that range from items of everyday usage

*Department of Computer Science, University of Geneva, Switzerland
[1]The authors are thankful to the Swiss National Science Foundation, their industrial partners in the CTI project, and the contributions of Dr. P. Bas, CNRS, Lille, France.

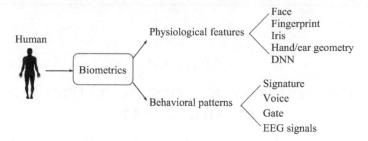

Figure 4.1   Examples of human biometrics as security features

to unique master pieces. The goal of physical object protection is to distinguish and protect authentic physical objects from fakes using various anticounterfeiting technologies. Similar to human security, physical object security solves this problem based on a unique set of features that are easy to verify but difficult to clone or copy. There exist two types of features: *extrinsic* and *intrinsic*. For physical object protection, the *extrinsic features* refer to devices, labels or materials that are added to objects to be protected to make it difficult or economically not attractive to clone, copy or reproduce. The extrinsic features include holograms, special IR/UV/magnetic materials or inks, microparticles added to the surface of object or reproduce on it, graphical codes, radio-frequency identification (RFID), etc. The *intrinsic features* refer to the unique properties of materials from which the objects are produced that are inherent to this material and particular object. Intrinsic features are elicited and then measured using a *challenge*. In the general case, these properties are difficult to characterize, model or describe analytically due to their inherently random nature. Figure 4.2 shows several examples of physical objects that would benefit greatly from anticounterfeiting measures, using both extrinsic and intrinsic features.

Summarizing, physical unclonable functions (PUF) lay at the basis of both bio-metrics and physical object security. Both share conceptually similar enrollment and verification architectures. Major differences are found when looking at the origin of the PUF, the number of potentially enrolled items or the security threat model.

4.2   Fundamentals of physical layer security based on unclonable functions

The security of both human and physical objects strongly relies on unique and unclonable features that unambiguously characterize each instance in the real world.

Therefore, in this section we will link biometrics and physical objects using the notion of PUF.

4.2.1   Randomness as a source of unclonability

It is commonly assumed that biometric features are unique to a certain degree for each person [1]. Similarly, the properties of physical materials are known to be unique

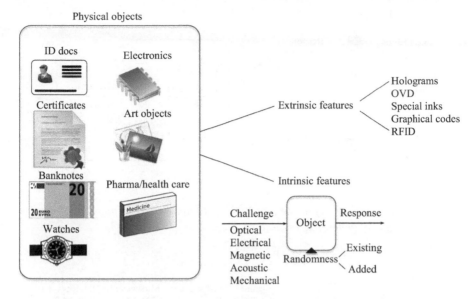

Figure 4.2   Physical objects of interest for anticounterfeiting protection using extrinsic and intrinsic features

for each object [2]. Both biometric features and properties of physical materials can be acquired using different techniques that include optical (in a wide range of wavelengths), acoustic and even electric or magnetic measurements. The physical item under investigation is subject to the excitation by one or several modalities and the measured outcome is registered as a security feature of this item. We will refer to the particular excitation as a *challenge* and to the outcome as a *response*.

A PUF is a function that maps a set of challenges to a set of responses based on an intractably complex physical phenomenon or randomness of physical object. The function can only be evaluated with a physical system and is unique for each physical instance. PUFs provide high physical security by extracting secrets from complex physical systems. Therefore, one prefers in practice complex biometrics and physical materials that look almost 'random' to the observer that tries to repeat or characterize them. Without loss of generality, one can present PUF systems for humans and physical objects in terms of the challenge–response pairs as shown in Figure 4.3.

A person w is subject to a challenge $\mathbf{a}_i$ that results into a response $\mathbf{r}_i(w)$ according to the person's biometrics. Similarly, a physical object with index w provides a response $\mathbf{r}_i(w)$ to a challenge $\mathbf{a}_i$. Both systems can be generalized as shown in Figure 4.3(c).

We will assume that some PUF system, i.e., either a person or physical object, is assigned an index $w \in \{1, \ldots, M\}$ and provides a response:

$$\mathbf{r}_i(w) = \mathrm{PUF}_w(\mathbf{a}_i), \tag{4.1}$$

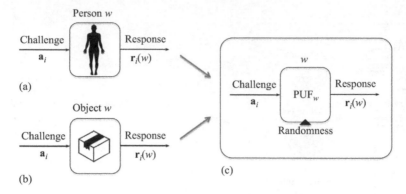

Figure 4.3   Generalized deployment of PUF: (a) biometrics, (b) physical objects and (c) generalized notations

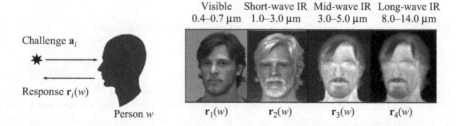

Figure 4.4   Example of challenge–response pairs for humans [3]. The same person is observed in four different bands which are considered as four challenge–response pairs

$\mathbf{r}_i(w) \in \mathbb{R}^N$, to the challenge $\mathbf{a}_i \in \mathbb{R}^l$, with $1 \leq i \leq K_a$. The number of latent parameters that are responsible for this response is very high, and there exists an *exponential* number of possible combinations of these parameters. Therefore, it is assumed that to clone a set of these parameters or to predict a possible response to a given challenge is technically unfeasible. We use a sign ▲ in Figure 4.3 to denote the embedded PUF randomness. Once the challenge is presented, it is very easy to verify the system response by comparing it with the registered enrolled response.

We show several examples of challenge–response for humans in Figure 4.4, where the same person w is 'challenged' by four challenges $\mathbf{a}_1, \ldots \mathbf{a}_4$ ranging from optical to long-range IR excitation, and for physical objects in Figure 4.5, where some object w can be excited and observed under different wavelength λ_i, incident light angle θ_i, observation angle β_i and time of excitation τ_i. The resulted responses are denoted as $\mathbf{r}_1(w)$ and $\mathbf{r}_2(w)$.

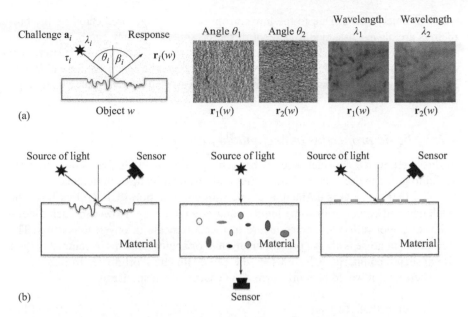

Figure 4.5   *Example of challenge–response pairs for physical objects: (a) examples of paper response to the light directed under two different angles and in two different wavelengths and (b) examples of optical PUF acquisition for the reflected (left) and transparent (middle) acquisitions with the existing randomness and reflected acquisition with the added (artificial) randomness (right)*

In theory, one can keep all possible challenge–response pairs for a given item w in secret. The final decision about the items authenticity can be based on the fusion of all correspondences between the real and enrolled responses for a specified set of challenges. However, in practice, to reduce the cost and time of verification only a limited number of challenges or even sometimes only one challenge is used which obviously reduces the security of system.

Although conceptually simple, biometrics and physical object PUFs have one significant difference in the level of randomness and thus in security and unclonability. The biometric PUFs for different w_s and identical challenge $\mathbf{a}_i$ can be considered as variations $\widetilde{\mathbf{r}}_w(\mathbf{a}_i)$ from some average response $\bar{\mathbf{r}}(\mathbf{a}_i)$, i.e., $\mathbf{r}_w(\mathbf{a}_i) = \bar{\mathbf{r}}(\mathbf{a}_i) + \widetilde{\mathbf{r}}_w(\mathbf{a}_i)$. For example, the shape of a face, finger, eye, ear, etc., are evolutionary developed to be very similar for all humans, i.e., $\bar{\mathbf{r}}(\mathbf{a}_i)$ dominates the $\widetilde{\mathbf{r}}_w(\mathbf{a}_i)$ term. In contrast, the surface microstructures of physical objects taken at lower scale are very different, i.e., $\widetilde{\mathbf{r}}_w(\mathbf{a}_i)$ is significantly larger than $\bar{\mathbf{r}}(\mathbf{a}_i)$.

Translating these observations into objective metrics, one can conclude that the entropy of PUFs for physical objects is significantly higher than those of biometrics. As a result, it is one of the most significant differences in usage of biometric and physical object PUFs. Due to the low entropy causing high predictability, the biometrics are always assumed to be kept in secret; otherwise, they can be reproduced with a descent accuracy by modern reproduction means. In contrast, the physical object PUFs are very complex and even an attacker, who has a genuine object in his/her disposal, might face a hard time to reproduce or clone it with sufficiently high accuracy.

4.2.2   Basic properties of unclonable functions

In this section, we consider a set of basic properties of PUFs that should be very useful to guide the selection of biometrics and physical features in practice.

The PUF in (4.1) is defined as a deterministic function $\mathrm{PUF}_w(\cdot)$ that links the challenge and response in some highly nonpredictable way. However, such a definition does not reflect the noisy nature of measurements in physical systems. The measurement noise is always present in physical or analog systems in contrast to their digital secure counterparts from where the name 'function' originates from.

Therefore, it would be more correct to describe the response as

$$\mathbf{r}_i(w) = \mathrm{PUF}_w(\mathbf{a}_i) + \mathbf{z}, \tag{4.2}$$

where $\mathbf{z}$ is a noise component that follows some statistical distribution $\mathbf{Z} \sim p(\mathbf{z})$. Hence, challenging a given PUF_w with some challenge $\mathbf{a}_i$, one will observe different realizations of $\mathbf{r}_i(w)$ that will be concentrated around the same mean $\mathrm{PUF}_w(\mathbf{a}_i)$.

Consequentially, the presence of noise considerably differs the design, analysis, performance and security of cryptographic (noiseless) systems from physical world (noisy) systems. In this respect, the biometrics and physical object security systems share a common problem related to the necessity to operate in noisy environments in contrast to crypto primitives. Therefore, one can use the practices developed in these domains to apply the cryptographic primitives to noisy data using techniques such as *fuzzy commitment* [4] or *helper data* [5,6].

Summarizing the earlier considerations, one can formulate the main requirements to PUF systems as shown in Figure 4.6. For simplicity, we will assume for a moment that response $\mathbf{r}_i(w)$ of PUF_w is binary to link the considered properties with the cryptographic analogs.

We briefly list the main properties of PUF systems:

- *uniqueness*: given the same challenge $\mathbf{a}_i$, different PUFs with $\mathrm{PUF}_w \neq \mathrm{PUF}_k$ should generate different responses such that $\mathbf{r}_i(w) \neq \mathbf{r}_i(k)$;
- *robustness*: given the same challenge $\mathbf{a}_i$, the PUF_w should produce the same response in multiple interaction sessions;
- *unclonability*: given either a physical object with PUF_w or leaked pair challenge–response $\mathbf{a}_i - \mathbf{r}_i(w)$, it should be unfeasible to reproduce or clone the underlying physical object. Practically, it means that the response of a faked PUF'_w from the genuine sample PUF_w should be different from the genuine one, i.e., $\mathbf{r}_i(w) \neq \mathbf{r}_i(w')$ for any challenge $\mathbf{a}_i$.

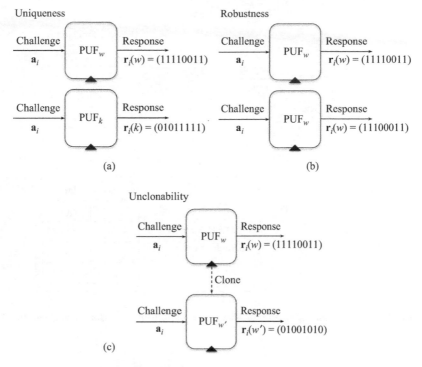

Figure 4.6   Main properties of PUFs: (a) uniqueness; (b) robustness;
(c) unclonability

4.2.3   Basic enrollment-verification architecture of PUF systems

The protection of humans and physical objects based on PUF consists of two stages:
(a) enrollment and (b) verification.

At the enrollment stage shown in Figure 4.7, given a human or physical object
with index w, a set of challenges $\{a_i\}$, $1 \leq i \leq K_a$, is generated and the set of responses
$r_i(w)$ is acquired and stored in the secured database. This enrollment is performed on
a secure terminal.

At the verification stage, the verifier presents a challenge to the object with
the claimed index w and expects to obtain the response according to the registered
database as shown in Figure 4.7.

The actual PUF index of the object is unknown. That is why its PUF is denoted
as $PUF_?$. The verification should confirm whether the claimed index w indeed corresponds to the actual one. The enrollment is performed on a secure terminal while the
challenge–response session might be executed on a possibly insecure terminal.

At the same time, as described earlier, the PUF system is based on some analogue
phenomena and is characterized by the inherent presence of noise. Therefore, it is
important to ensure a stable repeatability of the response to a given challenge and a

Enrollment (secure terminal)

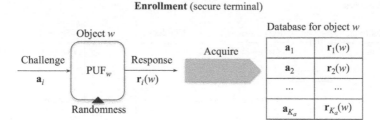

Verification at (possibly) unsecure terminal

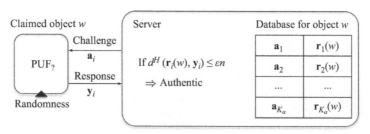

Figure 4.7   PUF general challenge–response principle, enrollment and verification

reliable decision. In its basic version, the server can compute the distance between the received response $\mathbf{y}_i$ and the expected one $\mathbf{r}_i(w)$ according to the Hamming distance $d^H(\mathbf{y}_i, \mathbf{r}_i(w))$. If this distance does not exceed some expected number of errors defined by the threshold εn, the claimed object is accepted to be w. Otherwise, its authenticity is not confirmed. The threshold ε ($0 \leq \varepsilon \leq 1$) is chosen to satisfy the trade-off between the probability of false acceptance and probability of miss that will be considered below.

The above-described multichallenge principle corresponds to both biometrics, i.e., multibiometrics or multimodal biometrics, when, for example, face and finger-print are used jointly, and physical systems, when either several wavelengths or angles or several different areas of object surface are used. However, in practice, mostly, a single challenge–response pair is used due to various issues that mostly deal with the overall complexity of data acquisition and management including memory storage, complexity, etc. Moreover, most of the multimedia security systems used in smart-phones, tables or security access systems are designed to be user-friendly in terms of speed of verification. Since optical sensors offer such a possibility most verification systems are based on biometrics that can be characterized as *image*-PUFs (Figure 4.8). A similar situation is also observed on the market of physical object security, where there is a trend to move the verification from special devices and trained professionals to end users, who perform the verification on their smartphones. Thus, most of modern physical object security systems perform the authentication and identification based on microstructure images taken directly with built-in cameras with or without

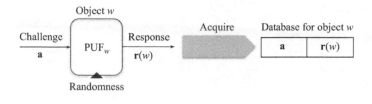

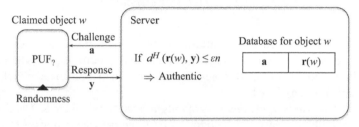

Figure 4.8   Image-PUF enrollment and verification

light-emitting diode (LED). In this respect, this form of verification refers also to the image-PUFs. Therefore, in following, when we refer to PUF we will mostly assume image-PUF, unless indicated specially.

4.3   Image-PUFs: fundamentals of security with noisy data

In this section, we consider a practical design of an image-PUF system that is common for both biometrics and physical object security. The real-world applications of image-PUF systems face two main practical challenges related to the *degradations* during the acquisition process and *scalability* due to a phenomena commonly referred to as a Big Data.

The generalized diagram of practical image-PUF system is shown in Figure 4.9. The biometric or physical object protection system provides a response $\mathbf{r}(w)$ to a challenge $\mathbf{a}$. The response of PUF_w is registered by an acquisition system in the form of digital image $\mathbf{x}(w) \in \mathscr{X}^{N_1 \times N_2}$, where $N_1 \times N_2$ is the size of image in pixels and $\mathscr{X}$ is a set of values in a chosen digital representation.

The *first practical challenge* is related to the unavoidable degradations that occur during the image acquisition. These degradations manifest themselves in differences between several samples taken from the same PUF_w and even the same challenge. The source of these degradations is related to a number of factors among which the most critical are: noise of optical charge-coupled device (CCD) and complementary metal-oxide-semiconductor (CMOS) sensors, demosaicing, defocusing especially for hand-hold devices, lens distortions (generally of a nonlinear character from the center of a lens to its borders), lossy image compression, geometrical desynchronization

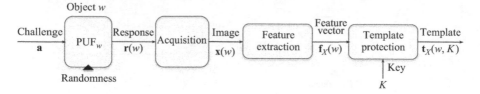

Figure 4.9   Practical image-PUF system comprising acquisition of PUF response,
feature extraction and template protection

between the images taken at the enrollment and verification stages, difference in enrollment and verification camera resolutions, light conditions, etc. Obviously, all this variability should be properly taken into account when robustness of PUF systems (Figure 4.6(b)) is considered.

The *second practical challenge* is mainly related to two factors. The first factor corresponds to the increase of resolution of imaging devices that leads to a considerable increase in image dimensions and as a consequence data size. This factor is compensated by advanced image compression tools in most multimedia applications, where compression distortions might be tolerated. However, it is not the case for PUF imaging, where the microscopic details at micron-level play a crucial role for PUF uniqueness and security. In addition, the increase of resolution of optical sensors has led to a decrease of physical pixel size causing the corresponding increasing of the noise level in acquired images. The second factor is related to the growing increase of biometric applications and devices as well as the increasing number of people using them on an everyday basis. Here, we talk about millions of people per country and hundreds of millions of devices involved in biometric protocols. Similarly, one can imagine a situation when hundred millions objects of the same type are distributed all over the world. Therefore, both large-size data from imaging sensors and big volumes of used biometrics, devices and physical objects create a Big Data issue. Although, the storage, processing and communication technologies advance with every year, one should not forget about the security and/or privacy concerns related to the Big Data. Without going into a detailed analysis of all possible misusage scenarios based on Big Data, we only point out those open problems that are of importance for the physical clonability or reproduction of genuine biometrics and object microstructures. Having the access to the high-resolution scans of image-PUFs one can try to reproduce them with high accuracy. This is especially true for the biometrics, where the entropy of original data is not very high. In addition, storing the 'open' biometrics raises privacy concerns as analyzed in this book. Finally, in many cases, the authentication/identification services are provided by the third parties, where the distribution of original data is not desirable or even prohibited.

4.3.1   Feature extraction

To cope with both types of degradation and Big Data issues and as a consequence the associated security/privacy challenges, the state-of-the-art biometrics and physical object security systems use recent achievements in computer vision and machine learning. The main conceptual idea behind the design of these systems is to *convert* a big image $\mathbf{x}(w) \in \mathbb{R}^{N_1 \times N_2}$ into a feature vector $\mathbf{f}_X(w) \in \mathbb{R}^N$ of dimension N (Figure 4.9). The feature vector should provide a *robust*, *compact* and *discriminative* representation of $\mathbf{x}(w)$. The robust representation refers to both image degradations, geometrical variations and acquisition conditions variability. The compact representation assumes that the dimension of feature vector N is considerably smaller than the original image size $N_1 \times N_2$. As a consequence, it should relax the memory storage and verification complexity issues. In turn, since dimensionality reduction in general leads to loss of informative content, one should ensure that enough information is preserved in the feature vector $\mathbf{f}_X(w)$ to discriminate different instances of $\mathbf{x}(w)$, $1 \le w \le M$, and distinguish them from fakes. At the same time, a plausible effect of dimensionality reduction is the enhancement of the security level in a way that the attacker might face certain difficulties in reverting the original $\mathbf{x}(w)$ from their low-dimensional counterpart $\mathbf{f}_X(w)$.

To generalize the analysis of biometrics and physical object feature extraction techniques, that follow the same design principles, we introduce a mapping:

$$\mathbb{F} : \mathbf{x}(w) \in \mathbb{R}^{N_1 \times N_2} \longmapsto \mathbf{f}_X(w) \in \mathbb{R}^N, \tag{4.3}$$

and consider two main state-of-the-art approaches to feature extraction that are based on a so-called *hand-crafted* design and *machine learning* principles. However, without loss of generality, both approaches can be presented as a generalized mapping $\mathbb{W} \in \mathbb{R}^{N \times (N_1 \times N_2)}$ with an element-wise nonlinearity:

$$\mathbf{f}_X(w) = \sigma(\mathbb{W}\mathbf{x}(w)), \tag{4.4}$$

where each instance of $\mathbb{W}$ will be considered below, and $\sigma(\cdot)$ represents a selection rule that has some nonlinear character.

Hand-crafted feature design is schematically shown in Figure 4.10(a). Given an acquired image $\mathbf{x}(w)$, *local descriptors* such as scale-invariant feature transform (SIFT) [7], speeded up robust features (SURF) [8], orientated fast and rotated brief (ORB) [9], KAZE [10], etc. or *global descriptors* such as GIST [11], fuzzy color and texture histogram (FCTH) [12], etc. are computed followed by feature aggregation such as bag of words (BOW) [13], vector of locally aggregated descriptors (VLAD) [14], residual vectors [15], triangulated embedding [16], generalized max pooling [17], etc., that produces a resulting vector of defined length N. In addition, given M' training samples $\mathbf{x}(i)$, $1 \le i \le M'$ and their features $\mathbf{f}_X(i)$, one can organize these features as a matrix $\mathbb{F}_X = \{\mathbf{f}_X(1), \ldots, \mathbf{f}_X(M')\} \in \mathbb{R}^{N \times M'}$ and apply a singular value decomposition (SVD) [18] decomposition $\mathbb{F}_X = \mathbb{U}\Sigma\mathbb{V}^T$ to extract the most important common components based on the analysis of significant singular values in the matrix $\Sigma \in \mathbb{R}^{N \times M'}$, where $\mathbb{U} \in \mathbb{R}^{N \times N}$ is an orthogonal matrix containing the eigenvectors of $\mathbb{F}_X\mathbb{F}_X{}^T$, and $\mathbb{V}$ is an orthogonal matrix containing the eigenvectors of $\mathbb{F}_X{}^T\mathbb{F}_X$. Typically, the dimension of the resulting feature vector can vary in range 1,000–10,000. The name 'hand-crafted'

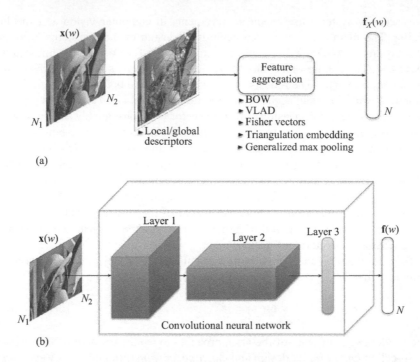

Figure 4.10   *Two main strategies to feature extraction: (a) hand-crafted approach*
and (b) deep learning approach as a representative of machine
learning approach

refers to the nature of feature construction that is solely based on the experience and
expertise of the system designer.

In addition, local descriptors can be stored together with their coordinates within
the image. In this case, one can explore the geometrical matching procedures between
the enrolled descriptors and those of image to be verified. Such a matching is typically
applied to a list of similar images returned based on aggregated descriptor and it is
referred to as *geometrical reranking* [19].

Machine learning feature design is shown in Figure 4.10(b) and refers to
automatic methods of feature extraction that might originate from unsupervised or
supervised (with the available training examples and labels) machine learning. In the
most general case, one can formulate the problem as a learning of mapping $\mathbb{W}$ (4.4)
given a set of training input images organized into a *data matrix* $\mathbb{X} = \{\mathbf{x}(1), \ldots, \mathbf{x}(M')\}$
and predefined desirable features $\mathbb{F}_X = \{\mathbf{f}_X(1), \ldots, \mathbf{f}_X(M')\}$:

$$\hat{\mathbb{W}} = \arg\min_{\mathbb{W}} \|\mathbb{W}\mathbb{X} - \mathbb{F}_X\|_F^2 + \lambda\Omega_W(\mathbb{W}), \tag{4.5}$$

where $\| \cdot \|_F$ denotes the Frobenius norm, and the term $\Omega_W(\mathbb{W})$ integrates all con-
strains on the desired properties of feature extraction transform, λ denotes the
Lagrangian multiplier.

In many practical cases, the set of features is not predefined in advance but instead only some desirable properties of features such as sparsity, discrimination, etc., are integrated into the optimization problem (4.5) via a constraint $\Omega_F(\mathbb{F}_X)$, which can include ℓ_1-norm, ℓ_2-norm or their combination (elastic nets), thus leading to:

$$(\hat{\mathbb{W}}, \mathbb{F}) = \underset{\mathbb{W}, \mathbb{F}_X}{\arg\min} \, \|\mathbb{W}\mathbb{X} - \mathbb{F}_X\|_F^2 + \lambda\Omega_W(\mathbb{W}) + \alpha\Omega_F(\mathbb{F}_X), \tag{4.6}$$

where α denotes the Lagrangian multiplier.

In addition, a solution of (4.6) is based on an iterated two-step procedure. Assuming some transform estimate $\hat{\mathbb{W}}$ from random initialization or SVD, one may find the feature vector in closed form as

$$\hat{\mathbb{F}}_X = \sigma(\hat{\mathbb{W}}\mathbb{X}), \tag{4.7}$$

where the nonlinear mapping $\sigma(\,\cdot\,)$ is determined by the constraint $\Omega_F(\,\cdot\,)$. For example, for the ℓ_2-norm constraint $\sigma(\,\cdot\,)$ is a linear function, for the ℓ_0-norm it is a hard thresholding function and for the ℓ_1-norm, it is a soft-thresholding function.

Given the estimate $\hat{\mathbb{F}}_X$, the transform estimation reduces to the solution of equation (4.5).

In addition, in the general case, a practical solution of (4.6) is unfeasible for the entire image of size $N_1 \times N_2$. Therefore, images are considered locally in small neighborhoods a.k.a. patch. In addition, to simplify the optimization the patches are treated at different resolution levels thus leading to a *multilayer* or so-called *deep learning structure*. Mathematically, it means that the matrix $\mathbb{W}$ is factorized by L matrices $\mathbb{W}_l$, $1 \leq l \leq L$, as:

$$\mathbb{F}_X(w) = G(\mathbb{W}\mathbf{x}(w)) = \sigma_L(\mathbb{W}_L \ldots \sigma_1(\mathbb{W}_1\mathbf{x}(w) + b_1) + \cdots + b_L), \tag{4.8}$$

where $\sigma_L(\cdot)$ are element-wise nonlinearities and $b_1, \ldots, b_L$ are biases at each level. Although, there is generally no theoretical guarantee that (4.6) with the factorization (4.8) can achieve a global minimum w.r.t. searched $\mathbb{W}$, there are many empirical results reported in [20], about the success of these architectures in different applications. Moreover, several recent findings of [21] and [22] propose interesting *unfolding* methods for efficient minimization of (4.6) with the matrix $\mathbb{W}$ in the form (4.8).

4.3.2 Template protection

Although the feature vector $\mathbf{f}_X(w)$ represents a reduced version of $\mathbf{x}(w)$, it contains a considerable amount of information that might lead to the reconstruction of $\hat{\mathbf{x}}(w)$ from $\mathbf{f}_X(w)$ with the following reproduction of faked items. Therefore, one should envision a set of measures for feature vector protection a.k.a. *template protection* (Figure 4.9). It should be pointed out that a direct application of cryptographic techniques to the considered problem is not feasible due to the noise presence in the final feature vectors, i.e., feature vector available at the verification stage will be a noisy version of enrolled feature vector.

Nowadays, the arsenal of developed template protection techniques for both biometrics and physical objects is quite rich and includes methods of fuzzy commitment [4], helper data [5,6], bit reliability [23], etc. However, without loss of generality, one can represent the template protection as a mapping:

$$\mathbb{T}_k := (\mathbf{f}_X(w), k) \rightarrow \mathbf{t}_X(w, k), \tag{4.9}$$

where the feature vector $\mathbf{f}_X(w)$ is mapped into a protected template $\mathbf{t}_X(w, k)$ based on given key k.

4.3.3 Performance analysis

At the *authentication stage*, the degraded version $\mathbf{y}$ of $\mathbf{x}$ is observed along with the provided claimed identity w that will be considered as a hypothesis $\mathcal{H}_w$. In some cases, the identity w should be also protected in the authentication system. The authentication system retrieves the corresponding template $\mathbf{f}_X(w)$ or the protected template $\mathbf{t}_X(w, k)$, based on the claimed identity w and takes the decision whether the template corresponds to the wth acceptance region $\mathcal{D}_w$, i.e., if $\mathbf{t}_Y \in \mathcal{D}_{(w,k)}$ or not. The decision region $\mathcal{D}_{(w,k)}$ is defined with respect to the template $\mathbf{t}_X(w, k)$.

Similar to the *false accept rate* and *false rejection rate* performance measures used in biometrics, authentication systems may be formally characterized by the *probability of miss* P_M, and the *probability of successful attack* P_{SA} as follows:

$$P_M = \Pr\left[\mathbf{T}_Y \notin \mathcal{D}_{(w,k)} | \mathcal{H}_w\right], \tag{4.10}$$

where $\mathbf{T}_Y$ corresponds to the noisy version of the extracted template from the original item with index w.

Observing the unprotected feature vector $\mathbf{f}_X(w)$ or its protected counterpart $\mathbf{t}_X(w, k)$, the attacker tries to reproduce the best estimate $\hat{\mathbf{x}}$ of $\mathbf{x}$ to physically reproduce it as $\tilde{\mathbf{x}}$ and to present it to the verification system as an authentic one. To introduce a measure of attacker successes, we define a *probability of successful attack* or probability of false accept as

$$P_{SA} = \Pr\left[\mathbf{T}_Y \in \mathcal{D}_{(w,k)} | \mathcal{H}_0\right], \tag{4.11}$$

where $\mathbf{T}_Y$ represents the template extracted from the noisy observation of the reproduced fake biometrics or product PUFs $\tilde{\mathbf{x}}$ under the hypothesis $\mathcal{H}_0$ and the acceptance region $\mathcal{D}_{(w,k)}$ of the decision device for the index w and key k. It is important to note that the fake $\tilde{\mathbf{x}}$ for the index w can be produced based on all available information about $\mathbf{x}(w)$ leaked from the feature vector $\mathbf{f}_X(w)$ or protected template $\mathbf{t}_X(w, k)$. In this case, the attacker is considered to be *informed*. In case, when the fake $\tilde{\mathbf{x}}$ is produced without $\mathbf{f}_X(w)$ or $\mathbf{t}_X(w, k)$, the attacker is considered to be *blind* or *uninformed* [24].

The *goal of the attacker* is to maximize the probability of its successful attack P_{SA} based on the earlier information. It is important to remark that in the case of a physical object PUF, the attacker might even have the access to the original object. In any case, the defender selects the algorithm $\mathbb{F}$ (4.3), the protection method $\mathbb{T}_k$ (4.9) and the decision region $\mathcal{D}_{(w,k)}$ to minimize the probability of miss P_M (4.10) for the

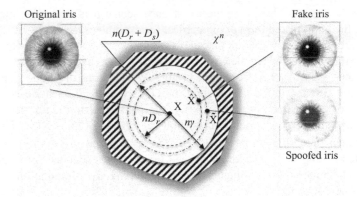

Figure 4.11   Acceptance region of an authentication system in the original image domain. The security of the system should ensure that all observations produced from the original object or authentic human should be within the acceptance region while ensuring that all fakes remain outside

claimed identity w and acceptable level of P_{SA} that naturally leads to a max–min game between these parties.

Figure 4.11 shows the desired space constellations between the original and faked data. Practically, the decision is taken in the feature or template domain, where the geometry of the acceptance region $\mathcal{D}_{(w,k)}$ might be quite complex. We assume that the attacker first estimates image $\hat{\mathbf{x}}$ from feature vector $f_{\mathbf{x}}(w)$ or its template $t_{\mathbf{x}}(w,k)$ and then reproduces it. The image acquired from such a clone is denoted as $\tilde{\mathbf{x}}$. The attacker will target to create such a fake whose image $\tilde{\mathbf{x}}$ will be as close as possible to $\mathbf{x}(w)$. Obviously, if the attacker knows less about $\mathbf{x}(w)$, it is better for the security of the system. At the same time, limited physical reproduction accuracy is a serious obstacle on the way to accurate counterfeiting. However, the limited resolution of the acquisition module of the verification device might be of help for the attacker. Therefore, the mathematical max–min game is often reduced to a game between the accuracy of counterfeiting facilities and the accuracy of acquisition devices that will be considered below.

At *the identification stage*, the identity w is unknown and the identification system should establish it based on the presented biometrics or physical object PUFs $\mathbf{y}$ by establishing an index of region $\mathcal{D}_w$, $w \in \{1, \ldots, M\}$, to which it belongs to for a given key k. The lack of knowledge about w on the side of the defender creates an advantage for the attacker to produce faked (spoofed) biometrics or physical object PUFs $\tilde{\mathbf{x}}$ such that it is at any of the M acceptance regions. That is why P_{SA} is drastically increased in the case of identification in comparison to authentication case.

The most dangerous scenario for the defender is when the attacker has access to the original object PUFs or biometrics and can scan them with very high precision and reproduce them. In the case of biometrics, it is generally assumed that the original biometric is private. Therefore, the worst case scenario is when the attacker has access to the unprotected form of biometrics, for example, scan a copy of fingerprint left on certain surface in public place. If the attacker has only access to a feature vector $\mathbf{f}_X(w)$ or protected template $\mathbf{t}_X(w, k)$, the success of a spoofing attack critically depends on a *reconstruction* from these vectors. We will consider modern approaches to the reconstruction in the next section. Obviously, since the feature vector $\mathbf{f}_X(w)$ represents a 'compressed' version of $\mathbf{x}(w)$, the reconstruction is not perfect and is characterized by a reconstruction distortion $D_r = \frac{1}{N_1 \cdot N_2} \mathbb{E}[\|\mathbf{X} - \hat{\mathbf{X}}\|_2^2]$. Finally, the reconstructed biometrics or object PUFs $\hat{\mathbf{x}}$ is reproduced in physical form either by simple printing or using more advanced technologies such as, for example, 3D printing. The image $\tilde{\mathbf{x}}$ taken from this fake is characterized by the additional distortion $D_s = \frac{1}{N_1 \cdot N_2} \mathbb{E}[\|\hat{\mathbf{X}} - \tilde{\mathbf{X}}\|_2^2]$.

The defender has the final word in the game and can accept or reject the presented PUF or biometrics depending on the distance between the templates. Considering an original high-dimensional image space, the attacker will succeed with high probability, if a spoof object PUF or biometrics satisfies a proximity constraint $d(\mathbf{x}, \tilde{\mathbf{x}}) \leq N_1 N_2 \gamma$ for some nonnegative γ.

The goal of the defender is thus to design such a template protection method that the probability P_{SA} is minimized that consequentially should lead to the maximization of the reconstruction distortion D_r. In this case, the fake object PUF or biometrics will most likely be outside the acceptance region.

4.4   Attacks against biometrics and physical object protection

In this section, we consider the main classes of attacks against biometrics and physical object protection systems. We will consider the entire chain from the challenge to the decision making and analyze attacking strategies on each element of this chain.

To provide a uniform consideration, we will cover the biometrics and physical object protection systems using a common architecture for the enrollment and verification and indicate all differences along our analysis.

4.4.1   Attacks against biometric systems

A schematic explanation of attacks against biometrics and physical object verification systems is shown in Figure 4.12.

We will start the analysis with the biometric systems. The biometrics verification system consists of two main parts: *enrollment* and *verification*. It is important to emphasize that the attacker does not have direct access to physical biometrics of a human with some assigned identifier $w \in \{1, \ldots, M\}$. The attacker is trying to do the best to get a desired outcome $\mathcal{H}_w$ ⑪ at the output of the decision device ⑩.

Following a standard classification of attacks accepted in the biometric community [25], one can distinguish *direct attacks* and *indirect attacks*.

The *direct attacks* target the generation and reproduction of 'synthetic' biometric samples. The fake samples ⑫ are presented to the input of the verification system.

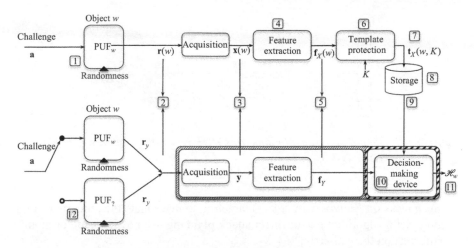

Figure 4.12   Attacks against biometrics and physical object protection systems

The reproduction of synthetic samples is based on information the attacker has in his/her disposal. Generally, one can distinguish three cases when such an information can be extracted:

* Directly from the *available biometrics*.
* From the *'left' traces* like left or stolen fingerprint traces on some surfaces or images available in social networks when face, iris, etc. can be estimated from one or multiple instances.
* From system knowledge that includes knowledge of the feature extraction, template protection and decision-making algorithms and leakages of features, templates or decision regions.

Direct attacks are performed in the *analog domain*. If successful, the fake biometric presented to the sensor of the acquisition system will be accepted as genuine. This will result in a positive hypothesis at authentication or the generation of the hypothesis $\mathcal{H}_w$ corresponding to any $w \in \{1, \ldots, M\}$, i.e., false acceptance of any enrolled instance, at identification.

Two further chapters in this book are devoted to these type of attacks to biometric system: Korshunov and Marcel discuss *presentation* attacks and their detection for voice biometrics. Furthermore, Marta Gomez for example, discusses attacks based on synthetical biometric data.

The class of *indirect attacks* is broader and includes:

* the attacks on weaknesses in the communication channels ②, ③, ⑤, ⑦, ⑨, ⑪ targeting the extraction, addition and modification of information to achieve a false acceptance;

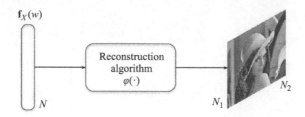

Figure 4.13   General diagram of biometrics/PUF reconstruction based on feature vector $\mathbf{f}_X(w)$

- the attacks on *Feature Extractor* ④ targeting to investigate the properties of the underlying algorithm for the direct attack planning or to extract, add or modify information there directly;
- the attacks on *Template Protection* ⑥ also aiming to learn secret key k;
- the attacks on *Storage* ⑧ targeting manipulation of database records to gain access, extracting a key k or extracting any privacy related information about the relationship between the database entries;
- the attacks on *Decision device* ⑩ targeting to investigate the 'decision'/'acceptance' regions for the direct attack planning.

The above attacks assume access to the system elements via some intrusion. In many cases, one can assume that the enrollment modules are protected and secured by the enrolling party. Therefore, the most vulnerable element for these attacks is a verification device that might be in the disposal of an attacker. However, one can assume that the decision device should be well protected in this case.

While one can easily imagine direct attacks targeting the reproduction on modern equipment including printing on 3D-printers, the estimation of synthetic biometrics generated from feature vectors or from protected templates is not a trivial task. For this reason we will briefly overview modern approaches to a reconstruction that can be also useful for PUFs-based systems.

4.4.2   Overview of reconstruction techniques

The goal of reconstruction techniques is to get an estimate $\hat{\mathbf{x}}(w)$ from its feature $f_x(w)$ of original biometric image $\mathbf{x}(w)$ using some reconstruction method $\varphi(\cdot)$ as shown in Figure 4.13.

The reconstruction problem can be formulated as a regression:

$$\|\mathbf{x}(w) - \varphi(\mathbf{f}_X(w))\|_2^2, \tag{4.12}$$

where the function $\varphi(\cdot)$ produces an estimate $\hat{\mathbf{x}}(w) = \varphi(\mathbf{f}_X(w))$.

For a long time, it was believed that this problem was ill-posed, and it was not possible to recover the original image from its low-dimensional representation $\mathbf{f}_X(w)$ due to the lossy nature of the feature extraction operation.

However, the recent progress in the solution of inverse problems based on sparsity-regularized methods and impressive achievements of machine learning techniques supported my massive training datasets, clearly demonstrate that it is possible to reconstruct high-quality images even from low-dimensional and low-level features [26]. Therefore, the risks associated to the biometrics reconstruction from the templates are not negligible.

Without pretending to give an exhaustive overview of all existing approaches, we will briefly summarize the main approaches to image reconstruction depending on available data and priors. Along this way, we consider two approaches that will be called a *signal processing reconstruction* and *machine learning reconstruction*.

4.4.2.1 Signal processing reconstruction

The signal processing reconstruction combines a broad family of methods that assume that a feature vector $\mathbf{f}_X(w)$ is available due to any attack considered above as well as one can deduce some knowledge about the feature extraction algorithm. We will consider a model with feature vector extraction (4.12):

$$\mathbf{f}_X(w) = \sigma(\mathbb{W}\mathbf{x}(w)), \tag{4.13}$$

where $\sigma(\cdot)$ and $\mathbb{W}$ are assumed to be known and defined as in Section 4.3.1.

Finding $\mathbf{x}(w)$ based on $\mathbf{f}_X(w)$ is a mathematically ill-posed problem that requires some additional information about $\mathbf{x}(w)$ for a proper regularization:

$$\hat{\mathbf{x}}(w) = \underset{\mathbf{x} \in \mathbb{R}^{N_1 \times N_2}}{\arg\min} \|\mathbf{f}_X(w) - \sigma(\mathbb{W}\mathbf{x}(w))\|_2^2 + \lambda\Omega(\mathbf{x}(w)), \tag{4.14}$$

where $\Omega(\cdot)$ denotes a regularization operator.

The main difficulty in the solution of (4.14) consists in a proper definition of regularization term $\Omega(\mathbf{x}(w))$, which is consistent with data and leads to a low-complexity and tractable solution.

Therefore, the main diversity of modern reconstruction techniques is related to the definition of $\Omega(\mathbf{x}(w))$. In the recent years, *sparsity-based* methods have become very popular due to numerous properties in a solution of inverse problems. The main idea behind these approaches is to use a model of representation of $\mathbf{x}(w)$ in some codebook with the activation of a few number of its elements. Obviously, such a codebook should be quite 'representative' to ensure that any $\mathbf{x}(w)$ can be presented as a linear combination of several codebook elements. To reduce the complexity of codebook and its learning, the entire image is split into patches which are reconstructed independently and then aggregated to the whole image. Besides of a broad number of existing models, we will focus on two representations that we will refer to as the *synthesis model* and *transform learning model*. To avoid any confusion related to the terminology, we will define a synthesis model as

$$\mathbf{x}(w) = \Phi\mathbf{a}(w) + \mathbf{e}, \tag{4.15}$$

where $\Phi \in \mathbb{R}^{(N_1 \times N_2) \times L}$ is an overcomplete dictionary, i.e., $L > N_1 \times N_2$, $\mathbf{a}(w)$ is a sparse vector indicating which elements of dictionary are used for the approximation $\mathbf{x}(w)$ and $\mathbf{e} \in \mathbb{R}^{N_1 \times N_2}$ denotes the approximation error.

Under the model (4.15), the reconstruction problem (4.14) is reduced to

$$\hat{\mathbf{x}}(w) = \arg\min_{\mathbf{x}\in\mathbb{R}^{N_1\times N_2}} \|\mathbf{f}_X(w) - \sigma(\mathbb{W}\mathbf{x}(w))\|_2^2 + \lambda\|\mathbf{x}(w) - \Phi\mathbf{a}(w)\|_2^2 + \beta\Omega(\mathbf{a}), \qquad (4.16)$$

where it is assumed that the transform Φ is learned and the approximation error is to be i.i.d. Gaussian thus leading to the l_2-norm in the prior term. The term $\Omega(\mathbf{a})$ reflects a sparsity nature of $\mathbf{a}$ that can be modeled by ℓ_0- or ℓ_1-norms. The scalars λ and β denote Lagrangian multipliers.

In many practical cases to simplify the solution of (4.16), it is assumed that the approximation error $\mathbf{e} = \mathbf{0}$ and thus $\mathbf{x}(w) = \Phi\mathbf{a}(w)$ that leads to the disappearance of the second term in (4.16), and in turns, it gives two formulations of reconstruction:

$$\hat{\mathbf{a}}(w) = \arg\min_{\mathbf{a}} \|\mathbf{f}_X(w) - \sigma(\mathbb{W}\Phi\mathbf{a}(w))\|_2^2 + \beta\Omega(\mathbf{a}), \qquad (4.17)$$

where $\hat{\mathbf{x}}(w)$ is found from $\hat{\mathbf{a}}(w)$ as $\hat{\mathbf{x}}(w) = \Phi\hat{\mathbf{a}}(w)$ and directly

$$\hat{\mathbf{x}}(w) = \arg\min_{\mathbf{x}\in\mathbb{R}^{N_1\times N_2}} \|\mathbf{f}_X(w) - \sigma(\mathbb{W}\mathbf{x}(w))\|_2^2 + \beta\Omega(\Phi^+\mathbf{x}), \qquad (4.18)$$

where Φ^+ denotes a pseudoinverse of Φ.

The transform learning model is based on

$$P\mathbf{x}(w) = \mathbf{a}(w) + \xi, \qquad (4.19)$$

where $P \in \mathbb{R}^{L\times(N_1\times N_2)}$ is a sparsifying transform that maps the input image $\mathbf{x}(w) \in \mathbb{R}^{N_1\times N_2}$ into its sparse representation $\mathbf{a}(w)$ and $\xi(w) \in \mathbb{R}^L$ denotes the residual noise in the transform domain. The inspiration for this model comes from a traditional family of signal/image independent transforms such as discrete cosine transform (DCT), discrete wavelet transform (DWT), discrete Fourier transform (DFT), etc., whereas the model (4.19) is signal/image adaptive. In contrast to the solution of (4.17) that in general is a NP-hard problem, the sparse approximation based on (4.19) leads to a close form solution $\mathbf{a}(w) = \sigma(P\mathbf{x}(w))$, where $\sigma(\cdot)$ is a nonlinear element-wise operator associated with the desired model of sparsity in $\mathbf{a}(w)$. From this point of view, the residual noise ξ is an approximation error between $P\mathbf{x}(w)$ and $\sigma(P\mathbf{x}(w))$ i.e., $\xi = P\mathbf{x}(w) - \mathbf{a}(w) = P\mathbf{x}(w) - \sigma(P\mathbf{x}(w))$.

The resulting problem of reconstruction under the transform learning model therefore reduces to

$$\hat{\mathbf{x}}(w) = \arg\min_{\mathbf{x}} \|\mathbf{f}_X(w) - \sigma(\mathbb{W}\mathbf{x}(w))\|_2^2 + \lambda\|P\mathbf{x}(w) - \mathbf{a}(w)\|_2^2 + \beta\Omega(\mathbf{a}(w)) \qquad (4.20)$$

that can be efficiently solved using alternative minimization of finding a closed form solution of $\hat{\mathbf{a}}(w)$ and iteratively finding $\hat{\mathbf{x}}(w)$ for a fixed $\hat{\mathbf{a}}(w)$ [27].

4.4.2.2   Reconstruction based on machine-learning approach

The reconstruction approach considered above only assumes the presence of a single feature vector $\mathbf{f}_X(w)$ with the knowledge of feature extraction algorithm (4.13) that might not be always in the disposal of attacker as well as certain assumptions about the model of data $\mathbf{x}(w)$ expressed via some sparsification strategy.

In contrast, the machine-learning approach is based on an assumption that the attacker might have access to multiple pairs of images and their feature vectors $\mathbf{x}(1) \leftrightarrow \mathbf{f}_X(1), \ldots, \mathbf{x}(M') \leftrightarrow \mathbf{f}_X(M')$ generated by some feature extraction algorithm, which might be even unknown for the attacker.

The goal of attacker is to use this training data as a prior to train a mapper $\varphi(\cdot)$ (4.12) as:

$$\sum\nolimits_{j=1}^{M'} \|\mathbf{x}(j) - \varphi(\mathbf{f}_X(j))\|_2^2. \tag{4.21}$$

using a strategy similar in spirit to the considered deep net factorization (4.8).

The solution to this problem seems to be difficult but the first results reported for image recovery from many local descriptors and the last layers of deep nets are very encouraging. For example, in [26], it was demonstrated that SIFT, histogram of oriented gradients (HOG) and convolutional neural network (CNN) last layer features can be efficiently mapped back to images. Therefore, the leakage of features or their weak protection might be very dangerous in view of recent capabilities of machine learning methods.

Further chapters of this book review a variety of prominent practical reconstruction approaches: Gomez and Galabry discuss actual approaches based on synthesizing and reconstruction of spoof data, whereas Rathgeb and Busch review the state-of-the art work on biometric template protection to avoid such kind of attacks. Jassim presents a new concept based on random projections.

4.4.3 Attacks against PUFs-based systems

The family of attacks against the PUFs-based systems is broader to those of biometrics systems. The main source of attacks is related to the fact that physical objects might be in direct disposal of the attackers in contrast to biometrics systems. Therefore, the attacker has broader possibilities for studying security features and trying to reproduce them with the necessary accuracy.

In addition, the attacker might improve the counterfeiting success rate by submitting the produced fakes to the verification service via public devices. This option is generally quite limited in biometrics systems, where the service might be denied for some time after several sequential unsuccessful attempts. Such a 'sensitivity' attack was discovered for the first time in digital watermarking applications [28] making it possible to investigate the geometry of acceptance regions in the decision device. Therefore, proper countermeasures should be taken against such kind of attacks.

Obviously, the investigation of extracted features might be also of interest for counterfeiters to investigate what regions, parameters and accuracy of cloning are needed to ensure the closeness of fakes to their authentic counterparts.

Finally, if the attacker has access only to the extracted features instead of originals, the earlier reconstruction techniques can be used to produce an estimate of the original image for the following reproduction.

4.5   Main similarities and differences of biometrics and PUFs-based security

The analysis performed in the previous sections makes it clear that the biometrics and PUFs-based systems possess a lot of similarities:

Main similarities

Uniqueness

- Both biometrics and PUF systems assume that each person or physical object possesses a set of unique features that can be used for reliable and nonambiguous verification

Addressed classes of problems

- Authentication
- Identification

Verification system architectures

- Both systems share similar architectures with the same building blocks consisting of acquisition, feature extraction, template protection and decision making. These architectures in both systems can be used interchangeably without significant modifications thus making both approaches quite unified

Verification equipment

- In both the cases, the dominating majority of verification modalities are based on visual information that can be acquired by on-shelf or specially designed imaging devices. Although the class of PUF's is quite broad, we do not address various electronic, magnetic, etc. PUFs systems here and focus our analysis on optical systems only

---

Besides of remarkable similarities, the biometrics and PUFs verification systems possess a number of significant differences that should be taken into account in practical considerations. We list the main differences.

Security principles

Biometric systems:

- Keep original biometrics secret
- Protect templates to avoid reversibility attack
- Once disclosed, it can be cloned with sufficiently high accuracy

PUF systems:

- Physical objects might be in the direct disposal of attackers to measure PUFs

- No special care is needed to protect templates
- Even if PUFs are measured directly from a physical object or recovered from a template, it is assumed that it is difficult or economically unattractive to reproduce fakes of sufficient accuracy

---

Disclosure

Biometric systems:

- Any public disclosure due to occasional leaks or targeted attacks are highly undesirably due to the relative ease with which they can be cloned

PUF systems:

- Physical objects are in circulation in the public domain and it is highly likely that the attackers have a direct access to them

---

Entropy and needed cloning accuracy

Biometric systems:

- Entropy of most types of biometrics is relatively low
- May be reproduced or cloned with sufficient accuracy to pass as authentic
- To increase the entropy either multimodal biometrics or extra tests for liveness are applied

PUF systems:

- Entropy of PUFs is relatively high even for the same class of objects
- To reproduce fakes from authentic objects or from estimates recovered from features or protected templates is generally a very difficult technical task when high accuracy of reproduction is needed
- In many practical cases, the randomness (entropy) of PUFs can be increased by extra modulation a.k.a. *active content fingerprinting* [29]

Possibility to add extra security features

Biometric systems:

- Generally, there is no possibility to modify human biometrics and the system deals with the existing ones (noninvasive principle)

PUF systems:

- One can introduce various modifications to physical objects applying either random or specially encoded modulation

Technical requirements for enrollment and verification equipment

Biometric systems:

- Generally low-resolution acquisition

PUF systems:

- High-resolution acquisition equipment is required

Maximum number of identifiable items

Biometric systems:

- Due to low entropy and low quality of acquisition, the number of people that can be uniquely recognized based on a single biometric modality is generally limited by millions

PUF systems:

- Due to the relatively high entropy of PUFs and stringent requirements to the acquisition equipment, the envisioned number of uniquely recognizable items is very high. The desired number ranges between hundreds of millions to billions

Impact of aging factors and variability in acquisition conditions

Biometric systems:

- Generally very good robustness to variability in acquisition conditions (pose, lightning, etc.)
- Some variability in biometrics is possible due to aging during the 'life' cycle

PUF systems:

- Relatively high sensitivity to variable acquisition conditions especially those that lead to the different appearance of PUFs, i.e., lightning, but good robustness to geometrical variability
- Relatively low variability due to aging during the life cycle, but some surface damage is possible during circulation or delivery

Requirements for enrollment

Biometric systems:

- Biometrics can be enrolled at any moment of the life cycle
- Enrollment is generally performed by trusted parties or state authorities producing ID documents

PUF systems:

- PUFs are mostly enrolled during the manufacturing cycle or before delivery/distribution to control the corresponding chains
- Enrollment is mostly performed by manufacturers

──────────────── **Requirements for verification** ────────────────

Biometric systems:

- Verification can be performed by third parties (border control, shopping, gadgets, etc.) but on trusted equipment with protected sensors, feature extraction and verification devices

PUF systems:

- Verification can be performed with both special equipments or publicly available generic devices, equipped with a corresponding sensor
- Verification can be local (if application is installed on a device) or on a server when the connected device sends either the whole image or the extracted features

4.5.1   Summary and conclusions

Summarizing, in the analysis of biometrics and PUFs systems, one can conclude:

- Since the entropy and required acquisition accuracy of PUFs systems is higher than those of biometrics, the security and number of distinguishable items of PUFs outperform biometrics in this respect.
- The main differences in the security assumptions for biometrics and PUFs recall the necessity to keep biometrics secret while PUFs can be and are revealed in the public domain and assumed to be difficult to clone with needed accuracy.
- Up to which point such a protection will be still valid? Nowadays, the accuracy of acquisition (imaging) is considerably higher to the accuracy of reproduction. Therefore, at least theoretically as soon as high quality (resolution) imaging is provided in targeted applications, the defender will have an advantage over the attacker.
- To increase the randomness (entropy) and to complicate reproducibility, the defender can add or introduce additional modifications to physical items while the biometrics systems are generally 'noninvasive'.

References

[1]   Anil K. Jain, Arun Ross, and Sharath Pankanti, "Biometrics: a tool for information security," *IEEE Transactions on Information Forensics and Security*, vol. 1, no. 2, pp. 125–143, 2006.

[2]   Pim Tuyls, Boris Skoric, Tom Kevenaar, *Security with noisy data: on private biometrics, secure key storage and anti-counterfeiting*, London: Springer Science & Business Media, 2007.

[3]   EQUINOX Corpotation, "Equinox face database," http://www.equinox sensors.com/products/HID.html [accessed on April 25, 2012].

[4]   Ari Juels and Martin Wattenberg, "A fuzzy commitment scheme," in *Proceedings of the 6th ACM Conference on Computer and Communications Security*. ACM, 1999, pp. 28–36.

[5]   Joep de Groot and Jean-Paul Linnartz, "Optimized helper data scheme for biometric verification under zero leakage constraint," in *Proceedings of the 33rd WIC Symposium on Information Theory in the Benelux*, 2012, pp. 108–116.

[6]   Tanya Ignatenko and Frans Willems, "Biometric systems: privacy and secrecy aspects," *IEEE Transactions on Information Forensics and Security*, vol. 4, no. 4, pp. 956–973, 2009.

[7]   David G. Lowe, "Object recognition from local scale-invariant features," in *Proceedings of the IEEE International Conference on Computer Vision*, 1999, vol. 2, pp. 1150–1157.

[8]   Herbert Bay, Andreas Ess, Tinne Tuytelaars, and Luc Van Gool, "SURF: speeded up robust features," *Computer Vision and Image Understanding*, vol. 110, no. 3, pp. 346–359, 2008.

[9]   Ethan Rublee, Vincent Rabaud, Kurt Konolige, and Gary Bradski, "ORB: an efficient alternative to SIFT or SURF," in *Proceedings of the International Conference on Computer Vision*, 2011, pp. 2564–2571.

[10]   Pablo Fernández Alcantarilla, Adrien Bartoli, and Andrew J. Davison, "KAZE features," in *Proceedings of the 12th European Conference on Computer Vision, Part VI*, 2012, vol. 7577, pp. 214–227.

[11]   Aude Oliva and Antonio Torralba, "Modeling the shape of the scene: a holistic representation of the spatial envelope," *International Journal of Computer Vision*, vol. 42, no. 3, pp. 145–175, 2001.

[12]   Savvas A. Chatzichristofis and Yiannis S. Boutalis, "FCTH: fuzzy color and texture histogram – a low level feature for accurate image retrieval," in *Proceedings of the 9th International Workshop on Image Analysis for Multimedia Interactive Services*, 2008, pp. 191–196.

[13]   Gabriella Csurka, Christopher Dance, Lixin Fan, Jutta Willamowski, and Cédric Bray, "Visual categorization with bags of keypoints," in *Proceedings of the Workshop on Statistical Learning in Computer Vision, ECCV*, 2004, vol. 1, pp. 1–22.

[14]   Hervé Jégou, Matthijs Douze, Cordelia Schmid, and Patrick Pérez, "Aggregating local descriptors into a compact image representation," in *Proceedings of the IEEE Conference on Computer Vision and Pattern Recognition*. IEEE, 2010, pp. 3304–3311.

[15]   Yongjian Chen, Tao Guan, and Cheng Wang, "Approximate nearest neighbor search by residual vector quantization," *Sensors*, vol. 10, no. 12, pp. 11259–11273, 2010.

[16] Hervé Jégou and Andrew Zisserman, "Triangulation embedding and democratic aggregation for image search," in *Proceedings of the IEEE Conference on Computer Vision and Pattern Recognition*, 2014, pp. 3310–3317.

[17] Naila Murray and Florent Perronnin, "Generalized max pooling," in *Proceedings of the IEEE Conference on Computer Vision and Pattern Recognition*, 2014, pp. 2473–2480.

[18] Gene H. Golub and Charles F. Van Loan, *Matrix computations (3rd Ed.)*, Baltimore and London: Johns Hopkins University Press, 1996.

[19] Hervé Jégou, Matthijs Douze, and Cordelia Schmid, "Product quantization for nearest neighbor search," *IEEE Transactions on Pattern Analysis and Machine Intelligence*, vol. 33, no. 1, pp. 117–128, 2011.

[20] Ian Goodfellow, Yoshua Bengio, and Aaron Courville, *Deep learning*, MIT Press, 2016, http://www.deeplearningbook.org.

[21] Benjamin Haeffele and René Vidal, "Global optimality in tensor factorization, deep learning, and beyond," *arXiv preprint arXiv:1506.07540*, 2015.

[22] Miguel Carreira-Perpiñán and Mehdi Alizadeh, "ParMAC: distributed optimisation of nested functions, with application to learning binary autoencoders," *arXiv preprint arXiv:1605.09114*, 2016.

[23] Svyatoslav Voloshynovskiy, Oleksiy Koval, Taras Holotyak, Fokko Beekhof, and Farzad Farhadzadeh, "Privacy amplification of content identification based on fingerprint bit reliability," in *Proceedings of the IEEE International Workshop on Information Forensics and Security*, 2010.

[24] Fokko Beekhof, Svyatoslav Voloshynovskiy, and Farzad Farhadzadeh, "Content authentication and identification under informed attacks," in *Proceedings of the IEEE International Workshop on Information Forensics and Security*, 2012.

[25] Javier Galbally, Julián Fiérrez, and Javier Ortega-García, "Vulnerabilities in biometric systems: attacks and recent advances in liveness detection," *Database*, vol. 1, no. 3, 2007.

[26] Alexey Dosovitskiy, Jost Tobias Springenberg, and Thomas Brox, "Learning to generate chairs with convolutional neural networks," in *Proceedings of the IEEE Conference on Computer Vision and Pattern Recognition*, 2015, pp. 1538–1546.

[27] Saiprasad Ravishankar and Yoram Bresler, "Learning sparsifying transforms," *IEEE Transactions on Signal Processing*, vol. 61, no. 5, pp. 1072–1086, 2013.

[28] Jean-Paul Linnartz and Marten van Dijk, "Analysis of the sensitivity attack against electronic watermarks in images," in *Proceedings of the Second International Workshop on Information Hiding*, 1998, pp. 258–272.

[29] Svyatoslav Voloshynovskiy, Farzad Farhadzadeh, Oleksiy Koval, and Taras Holotyak, "Active content fingerprinting: a marriage of digital watermarking and content fingerprinting," in *Proceedings of the IEEE International Workshop on Information Forensics and Security*, 2012.

Chapter 5

Biometric systems in unsupervised environments and smart cards: conceptual advances on privacy and security

*Raul Sanchez-Reillo**

Biometric systems can be implemented following different schemas and also deployed in a huge variety of scenarios. Each of these combinations should be studied individually as to guarantee both, a proper functionality and a high level of preservation of the users' privacy. Obviously, covering all different possibilities in a single book chapter is impossible, so this chapter will first create a taxonomy of the possibilities, in order to choose a group of them to be studied in detail. As the chapter title shows, this chapter will be focussed on those implementations working in an unsupervised environment and where identification tokens, such as smart cards, can play an important role.

A taxonomy should consider both parameters, implementation schemas and deployment scenarios. Regarding the schemas, biometric systems are typically divided between systems working in identification mode, i.e. comparing the acquired biometric sample with the whole set of previously enrolled biometric references, and in authentication (also known as verification) mode, i.e. the presented biometric sample is only compared with the biometric reference[1] corresponding to the identity previously claimed [2].

But in addition to this classification, there is another important parameter to consider, regarding the implementation schema. Such parameter is the location of the enrolled biometric references. Biometric references can be stored in a centralized way (i.e. a central database holding the references of all users) or in a distributed way, where each reference (or a small subset of them) is stored in its own location, not connected to the rest of the references [3].

From a security and privacy point of view, centralized systems are typically considered as worse, as a vulnerability in the server containing the database can compromise the biometric references of all users in the system. But it is also true

*Carlos III University of Madrid – Electronics Technology Department, University Group for Identification Technologies, Spain
[1]The terminology used in this book chapter is based on the standardized biometric vocabulary [1]

that having the problem centralized is easier to focus all security mechanisms on such server.

Considering now the deployment scenario, we can consider two main alternatives for the scope of this chapter. The first one is to consider that the environment will be supervised, i.e. someone is controlling the correct operation of the biometric system. The other one is considering an unsupervised environment, where nobody is in charge of controlling how the system is working. These two main scenarios can be further classified following the definitions stated in ISO/IEC 30107-2 [4]. For example, in case of a supervised environment, the following three cases can be found:

- Controlled scenario: where an operator physically controls the subject to acquire biometric sample.
- Assisted scenario: where there is a person available to provide assistance to the subject submitting the biometric sample.
- Observed scenario: where there is a person present to observe the operation of the device but provides no assistance (e.g. using a CCTV installation).

In the case of an unsupervised scenario, named in [4] as unattended, there are two main possibilities:

- Visible scenario: where the subject is using the system with the possibility of some other person being able to see how the biometric data has been presented.
- Private scenario: where the subject is in its own environment, out of the sight of any person who could see how the subject interacts with the biometric system.

Table 5.1 shows the taxonomy presented and illustrates each of the combinations with a potential application example that applies. It also reflects the combinations that are in the scope of this chapter, although the main focus will be on the unsupervised distributed authentication systems.

Therefore, the next section will provide an overview of the main challenges that an unsupervised environment presents. Then, as the use of ID tokens is recommended in authentication schemas, the case of smart cards will be presented in Section 5.2. After explaining how smart cards work, Section 5.3 will show how smart cards can be used in order to improve security and privacy in biometric systems. The chapter will end with a set of conclusions and recommendations.

5.1   Challenges of an unsupervised scenario

The obvious challenge that an unsupervised scenario brings is the possibility of the user attacking the system, e.g. presenting spoofed samples. But this is not the only challenge to face. This section will show three major challenges to consider when implementing and deploying a biometric system in an unsupervised scenario: probability of suffering attacks, usability of the system, and how data is handled as to analyse the existence of privacy risks.

Table 5.1   Taxonomy with application examples. The shadowed areas are the ones in the scope of this chapter

		Implementation schema		
		Identification	Authentication	
Deployment scenario		Centralized	Centralized	Distributed
Supervised	Controlled	Forensic investigation	Border control without ePassport but with previous enrolment	Border control with ePassport
	Assisted	VIP entrance without using a ID token	VIP entrance using a ID token	Automatic border control
	Observed	Physical access control with on-line connection and a security guard around	Point of selling with a non-biometric ID token	Point of selling with biometric ID token
Unsupervised	Visible	Unattended physical access control with on-line connection	ATM with a non-biometric ID token	ATM with biometric ID token
	Private	Access control to a mobile device where several users can be enrolled	Internet banking with a non-biometric ID token	Internet banking with biometric ID token

5.1.1   Presentation attacks

A presentation attack is the hostile action of a user based on interacting with the biometric data capture subsystem with the goal of interfering with the operation of the biometric system [5]. This kind of attacks can be implemented through a number of methods, e.g. artefact, mutilations, replay, etc. The goal targeted by the attacker may be, mainly, either to impersonate (i.e. to be recognized by the system as another user) or just to not being recognized (e.g. not being detected as a deported traveller when entering a country).

Presentation attacks are not unique for unsupervised scenarios. They also exist in supervised ones, with a certain level of success in certain cases. For example, it is famous the case of a Korean woman that entered Japan after having been deported. She passed the biometric check by placing tape in their fingertips not being able to be located into the deported people list [6]. As the border control system is a controlled system, this could have been avoided by providing border guards with a proper control policy, such as asking the traveller to show the fingertips to the officer and requesting the officer to assure that nothing is placed over the surface of the fingers.

It is true that several attacks may not be viable in supervised scenarios, while being perfectly suitable for unsupervised ones. For example, using a face mask [7] in a border control may lead to the security guard to detect such a mask and, therefore, being the attacker detained. That same attack can be difficult to apply in the case of a visible scenario, but easily applicable in a private scenario. The success of such attack will depend on the robustness of the whole biometric system to that kind of attacks.

Each biometric modality (e.g. fingerprint, face, voice, hand, signature, etc.) has a huge variety of potential attacks. Although many of them are already documented, some of them may not be applicable to supervised scenarios, while being available in private scenarios. For example, in [8], there is a description of how faked fingerprints can be obtained and presented. In this paper, all attacks considered are based in creating an artificial fingerprint that can be presented to the sensor. Sometimes, the fake can be done as a very thin film, so as to be placed over the fingerprint of the attacker. In all cases, in a controlled scenario, the officer visually inspects the hands and fingertips of the user and detects this attack, while in an unsupervised scenario, this may become a viable attack.

Considering facial recognition, there are attacks [7] such as printing a photograph of the impersonated user in a piece of paper or on a T-shirt, showing a video to the camera, showing a 3D image with a mobile device or even manufacturing a mask with different materials. All these attacks may be easily detected by the supervisor of the system.

In the case of behavioural biometrics, such as handwritten signature, the spoofing consists typically in how the attacker is able to imitate the behaviour of the real user (i.e. to forge his signature). In [9], it can be seen how the success of the forgery changes considering the level of knowledge of the signature by the forger. Within the 11 levels defined, the first seven levels allow the attacker to use specific tools, but in levels 8–11, the attacker shall only use his own memory, which will be the case in any supervised scenario. Therefore, the problem in unsupervised scenarios is related to the fact that the attacker may be still using the specific tools, as nobody is watching his act.

Consequently, in the case of unsupervised scenarios, it is extremely important to provide the whole system with the presentation attack detection (PAD) mechanisms available, as to be able to reject this kind of attacks. Therefore, the whole system should be evaluated for its robustness, as it is being defined in ISO/IEC 30107-3 [10]. If this kind of measures is not taken, a serious problem about the privacy of the impersonated citizen will appear, being able for the attacker to achieve an identity theft.

5.1.2   Usability

Another important challenge in unsupervised environments is the need of the system to be operated by all users without human assistance. Therefore, the system shall be self-explained and convenient to use, so that users are not tempted to either not use it or to misuse it. For the service provider deploying the system, it is of major importance that users will interact with the system easily, because if not, they are a customer lost, and then, a loss in the benefits.

But for the scope of this chapter, it is much more important the fact that the user, in order to make the system more convenient, starts misusing it. These kinds of actions typically reduce the security of the system, compromising unconsciously the privacy of the user. Let's show an example outside of the biometric world. A typical recommendation is to have a different PIN code or a different password for each application. Following that recommendation will make it impossible for a regular citizen to remember them. Therefore, the user starts making the system more convenient by many different fashions: grouping passwords by application type (e.g. all banking systems have the same password), creating an easy rule for building the password for a certain site or just even having the same password for all applications (ranging from e-mail and social networking, till banking servers). But one of the worst misuses of this policy is to write down on paper (or in the mobile phone) all passwords. As it can be easily deducted, if someone obtains that list, or the rule to create the passwords or the password for that kind of application, the result will be that the user would have unintentionally compromised all his information and, therefore, his privacy.

This also applies to biometrics. It is always said that biometrics is much more convenient, as the user won't have to remember a complex password, but only which finger (in the case of a fingerprint system) he or she has to use. The best proof for this assessment is the success of integrating fingerprint recognition in smartphones [11–13], where most users of those smartphones have changed the unblocking method from PIN or pattern to fingerprint recognition [14]. But this success case cannot be extrapolated to other deployments. The author of this chapter has been notified of cases using physical access control systems with fingerprint recognition which have been substituted by placing a chair as to keep the door open, due to the lack of convenience (or acceptance by users), wasting all the money invested.

Therefore, it is essential to work around this challenge as to design the system in the most convenient way. In the case of a biometric system, the solution has to come in two different ways: ergonomics of the capture device and a proper design and implementation of the system functionality. Regarding ergonomics, it is important to design the capture device in a way that all users know how to interact with the system. For example, in the case of a fingerprint system, the capture device shall either allow the positioning of the finger in any potential orientation (which is not recommended in many cases) or to firmly guide the user to place the finger in a specific position and orientation. In the case of an iris system, it may be better to have a capture device that is able to adapt itself to the position of the user, than to force the user to adapt himself to the camera.

Considering the functionality, the system has to be designed to follow two main objectives. The first one is to adapt the thresholds (quality, decision, etc.) to the application to serve, i.e. act not too restrictive so that the user will get tired of being rejected, but not too relaxed so that the probability of an attacker to be accepted is low enough. The second objective is to design the feedback mechanisms that the user will need in order to better use the system.

In order to completely overcome this challenge, it is extremely important to start the design and development of the system following a user-centred design [15] and to evaluate the usability of the system [16] before deploying it. Sometimes, it is recommended to first build a mock-up version of the system, with the user interface included, in order to perform an initial usability evaluation as soon as possible. With such results, the final design could be modified early enough as not to compromise development resources.

5.1.3   Personal data handling

It has to be highlighted that all personal data shall be highly protected, as much as possible, as to preserve user's privacy. Consequently, both, the location and the access to the information, shall be considered. In other words, to the above-mentioned challenges, there are other challenges such as how data is handled and the vulnerabilities that can appear. There are two main issues to consider: (a) access to data stored and (b) access to data exchanged. Both issues have a different impact depending on whether the system is centralized or distributed.

If either the user biometric reference or the comparison algorithm is centralized, major vulnerabilities are related to the exchange of information. In case of having the reference stored in a central database, either the server has to send the reference to the point of service (PoS) or the PoS has to send the actual biometric sample to server. In the first case, the comparison and the decision are taken at PoS, so the PoS has to be highly secured and, what is even more important, the communication of the biometric reference from the server has to be ciphered and authenticated with powerful algorithms. This last issue is extremely important, because revealing the biometric reference of a user may lead to invalidate such biometric trait for that user. In addition, if the most sensitive piece in the chain is the PoS, and it is deployed in an unsupervised scenario, an attacker can freely tamper with the device and execute a great variety of attacks. Due to all the above-mentioned sensitive issues, this implementation is not recommended.

In case of a centralized system, a better option is to have the comparison algorithm and the decision taken at the server. In such a case, the biometric reference is never exchanged, only the biometric sample. In such a case, the items to be protected are, in addition to the presentation attacks, both the communication of the biometric sample to the server and the communication of the decision taken by the server. Tampering with each of those parts of the communication may allow an attacker to gain the identity of another user. Obviously, in the case of unsupervised scenario, the PoS shall contain the mechanisms to deny any tampering of a user with the device, either to commute the decision taken or to inject a previously captured biometric sample.

Considering now a distributed system, everything is handled at the PoS, from the acquisition till the decision making. In this case, the vulnerabilities are strictly located at the PoS, so it has to include the mechanisms to protect:

- The user's biometric reference
- The presentation of the biometric trait
- The decision taken and
- The firmware of the PoS so as not to be able to change the algorithm, the thresholds, etc.

In few words, the PoS shall be a tamper-proof device and integrating a tamper-proof operating system [17]. It should also be evaluated independently as to be able to check if the device is really conformant with the latest available security requirements, such as the ones defined by common criteria [18].

All the above-mentioned vulnerabilities makes it quite difficult to consider that the PoS is based on a generic biometric sensor, connected to a personal computer, unless the system is built as a secure element or contain a secure element such as a smart card.

5.2 Smart cards as secure elements

Invented and developed in the early 1970s, mainly in Japan and France, smart cards have been during all these years one of the most secure components that can be integrated in a system, in order to provide security to information and processes. Nowadays, they can be found in most credit and debit cards, as the subscriber identification module in a mobile phone (i.e. SIM card) or as the security module of a PoS [i.e. a security aid module (SAM) module]. The technological evolution in the last years has made some of the characteristics of the smart cards (e.g. communication speed, storage capability or processing capability) a bit limited as compared to the advances in other computational devices. But the philosophy behind smart cards is still valid. Therefore, as it will be seen at the end of this section, there are new implementations where the term 'card' may end up losing its meaning in the term 'smart card'. Before reaching that point, the main features of smart cards will be described in the following subsections.

5.2.1 Internal architecture

The concept of a smart card comes from the invention of the integrated circuit cards (ICCs) [19]. ICCs were developed in order to improve the capacity and security of magnetic stripe cards. When implementing the idea of ICCs, two main approaches appeared. The first one was to implement the integrated circuit as a memory with a simple access logic. This kind of cards were lately known as memory cards and used mainly for prepaid telephony or fidelity applications. Nowadays, they have lost most of its business case, as there are other personal devices which provide better storage capacity and much faster communication.

The other implementation of the initial idea of an ICC was to create such integrated circuit as a microcontroller (i.e. a microprocessor with a memory module and some peripherals embedded) with a dedicated operating system to control all data flow, including mechanisms to access information and taking decisions. This implementation, due to the capability of taking its own decision, leads to the name of smart cards, which nowadays are deployed in many applications, being the most important SIM cards, SAM cards and EMV (Europay, Mastercard and VISA) debit and credit cards.

Although there are many implementations of smart cards, with quite different product qualities, all of them have the following generic architecture, as it can be

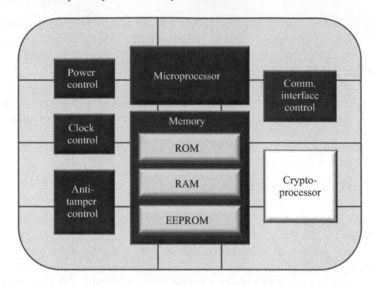

Figure 5.1   Block diagram of the internal architecture of a smart card

seen in Figure 5.1. As said, the central point of the smart card is a microprocessor, which is accompanied by a set of memories and some peripherals.

Regarding memories, a smart card has three kind of memories. It has a ROM memory (i.e. non-volatile, non-writable) for hosting the smart card operating system (SCOS), which will be described in Section 5.2.2. Another block of memory, as in any microprocessor-based system, is a piece of RAM memory, for the exclusive use of the SCOS. Both of those memories are not accessible from the external world. The only piece of memory that is accessible to the external world is the EEPROM (i.e. the non-volatile, but writable memory). This is the memory where all the information related to the application and to the cardholder is stored and processed.

In addition to the microprocessor and the memory, a smart card has a set of modules which are in control of several parameters, as to avoid bad uses or attacks. At least a smart card has a module for controlling the communication interface with the external world, a module to control the power supply levels and a module for controlling the clock levels, frequency and duty cycles. A high end product also has a set of modules to detect and handle different possibilities of attacks, as it will be briefly described in Section 5.2.4.

Figure 5.1 shows another module, painted in a different colour, as this module is not present in many smart cards. There are smart cards that also have the possibility of processing public key cryptography [20]. The mathematical calculations needed for this functionality are extremely complex, so it is better to implement them in a hardware module embedded in the microcontroller, as to be able to speed them.

5.2.2   Smart card operating system

From all the features of a smart card, the crucial one is the existence of an operating system. The SCOS is the firmware stored in the microcontroller which controls everything within the smart card. In summary, the SCOS is in charge of

- Managing the storage and retrieval of information in the EEPROM. In most smart cards, this is done using a folder-file schema, such as the one used in personal computers, although with some limitations in the number of files and the number of levels for the folders. Folders are called dedicated files (DF), while files are called elementary files (EF). Other smart cards are also able to handle data objects (DO) independently if they are implemented as EFs in DFs.
- Executing and controlling the security architecture implemented in the smart card. This will be detailed in Section 5.3.4, but as an introduction, this means handling PINs and passwords, cryptographic algorithms, access control rules, etc.
- Analysing the commands received and executing them after a robust checking of all their parameters. The main rule is that no command is executed unless it is one of the previously defined ones (i.e. submission of random commands shall not allow gaining access neither a malfunctioning).

These are the rules that all SCOS integrate, but it is important to note that there are two different kinds of SCOS. There are closed and proprietary SCOS, design and developed by smart card manufacturers. These SCOS are general purpose and, as they are closed, there is no possibility to perform any adaptation specific to an application. The quality of the SCOS, and therefore the fulfilment of all its characteristics and security mechanisms, depends in a large percentage on the smart card manufacturer. The rest of the responsibility belongs to the application designer, who should design the personalization of the cards in the proper way, setting the access rules and security parameters properly.

The other kind of operating systems is called open operating systems, which are manufactured with a minimum set of the SCOS, being the rest of the SCOS programmed by the application designer, including the command set, the data structure and the security mechanisms. The good thing about this kind of SCOS, which its most known version is Java Card [21], is that the application designer can adapt the smart card to the specific application. The major drawback of this technology is that the performance and security of the SCOS depend exclusively on the application developer. Therefore, if the application developer is not security skilled, having a detailed knowledge about smart card technology, the cards provided to the users may be full of security leaks. In order to avoid future privacy concerns, the specific implementations of SCOS, using open operating systems, should go through detailed evaluation, covering both functionality and security.

5.2.3   Functionality

From the functional point of view, smart cards are mainly thought as to be a personal identification token, being applicable to any application where earlier magnetic stripe cards are used. But due to its technology, they can go even further. Smart cards are

Case 1: No data transfer between the IFD and the ICC

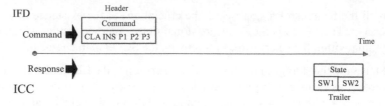

Case 2: Data transfer between the IFD and the ICC

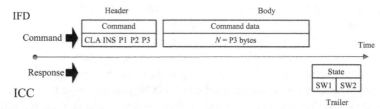

Case 3: Data transfer between the ICC and the IFD

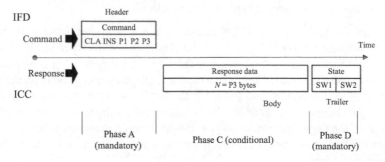

Figure 5.2   Illustration of the command–response pair in IFD–ICC communication

not passive storage devices, but active ones. The terminal (also known as inter face device – IFD), to which the smart card is connected (i.e. either inserted or at a close distance), shall send commands to the smart card and wait for its answer, following the schema of the command–response pair shown in Figure 5.2.

Once the smart card receives a command through its serial interface, the SCOS analyses such command and if everything is correct, it executes it. After the execution, the smart card sends a response to the terminal including, always, a status word of 2 bytes summarizing the result of the operation (i.e. if everything has gone ok or if some kind of error has happened).

As it can be seen in Figure 5.2, there are mainly three possibilities for such a command–response pair. There first case is where no data is exchanged, but only a command (5 bytes) is sent from the IFD to the ICC, and the status word is responded by the latter. The second case is when the terminal sends not only the command, but also some additional data, replying the card with the status word. Finally, the third case is when the terminal is requesting data from the card, so it sends only the command (5 bytes), and then receives the data demanded (if there has not been any error), plus the status word. In all the three cases, the card checks the command and the data, and it is in the hands of the SCOS to detect any kind of error. If an error occurs, the smart card does not exchange data and send the status word.

The additional case of sending and receiving data in the same command–response pair is not considered in the most populated communication protocol. In case this is needed, it is done by concatenating two commands, the first one sending the data and the second one requesting the response.

Further details in this communication protocol can be found in ISO/IEC 7816 parts 3 and 4 [22,23]. But it is important to note the general way this communication is implemented. From the very beginning, the communication between the smart card and the terminal was considered to be an asynchronous serial communication, through a single wire (i.e. bidirectional and half-duplex). This is what is usually called smart cards with contacts. Obviously, this is in contrast to an alternative communication mode that is implementing a contactless interface. There are three different standards for the contactless communication: (a) close-coupled devices, (b) proximity cards and (c) vicinity cards.

From those three, the most populated solution is the one based on proximity cards, which are the ones being able to exchange information at a distance from 0 to 10 cm [24]. This limitation in distance is intentional in many applications, as it requires the action from the user to approach the card to the terminal, validating therefore his/her intention to perform the information exchange. This is typically known as contactless smart cards or, just simply, contactless cards. This interface has the following advantages to the contact one: faster transaction as the act of inserting the card is removed, much more comfortable for the user, and the maintenance of the terminal is reduced as there is no friction between the card and the contacts of the terminal. This is why they have become so popular in many applications, starting with public transport titles, electronic passports or, even nowadays, debit and credit cards. The major disadvantage may come from the fact of being able to intercept wirelessly the communication, what shall be protected by using a whole set of logical security mechanisms, creating a secure channel, as it is usually done with contact smart cards. But regardless of whether the card is contact or contactless, the communication protocol is the same, as well as the security mechanisms available.

5.2.4 Security mechanisms

As it has been mentioned earlier, one of the most interesting features of smart cards is the security they provide. The amount of security mechanisms could be extremely huge and go from hardware security from logical security. Regarding hardware mechanisms, not many details are available, as they are all part of the industrial property of each manufacturer. Obviously including the latest mechanisms mean a more expensive product, and that's the reason why there are smart cards in the market which do not provide all these mechanisms, ending up in vulnerable devices. But there are others products in the market that include many more security mechanisms than the ones here mentioned, obtaining certificates of extremely high level according to Common Criteria [25].

The hardware mechanisms are extremely diverse. Some of them include encapsulating the microcontroller as a Faraday jail, distributing the internal memory in a pseudo aleatory manner, adding sensors to discover when the outer world is tampering with the encapsulation or ciphering all the information stored in the smart card. An illustration about the mechanisms and how some attacks could be successful can be seen in a presentation from Tarnovsky in [26].

Talking now about logical security mechanisms, there are a lot of them that are not only known, but also standardized. The one best known is the use of PINs and passwords in order to gain access to parts of the smart card. There are two main features in this security mechanism. The first one is that each single EF in the card can have each of its operations involved, protected by one of the many PIN codes and passwords available in the card. For example, an EF (being the basic data storage unit on a smart card, as introduced in Section 3.2) can be protected by PIN1 for reading, and by PASSWORD4 for writing. The second feature is the one related to how PINs and passwords are handled by the SCOS. Typically, all PINs and passwords have a maximum number of wrong presentations defined, and once that number is exceeded, that PIN or password will be blocked, and therefore that access condition won't be able to be satisfied, and all operations in all files, where that PIN or password is used, will be forbidden. A well-known example is the PIN code of a SIM card, where, in order to access the subscriber number, that PIN will have to be presented correctly. After three consecutive wrong presentations, that PIN is blocked, and then the mobile phone won't be able to be used, as the subscriber number won't be accessed. If the application designer wishes, an unblocking PIN or password can be defined, as to be able to unblock a certain PIN or password. If that unblocking PIN is presented correctly, then the relevant blocked PIN may be initialized with a new value given. The best example of this is the case of the PUK code in a SIM card.

But the most important feature in terms of logical security mechanisms is the possibility of combining some advanced ones in order to create a *secure channel*, through which the exchange of information may be considered secure. From the different mechanisms, the basic ones required are the following three: (a) authentication of both ends of the communication; (b) generation of session keys and (c) securing the message exchanged. All the three mechanisms are explained in the following paragraphs.

In order to authenticate both ends in the communication, there are some mechanisms, but the one explained here is using symmetric cryptography [20]. Figure 5.3 shows how the terminal can authenticate the card, known as *internal authentication*. As it is based on symmetric cryptography, where it is commonly referred to as shared key authentication, both parts (i.e. the terminal and the smart card) share the knowledge of a secret key (K). As it is the terminal who wants to check if the card is reliable, it sends a random number (R) to the card, so as to ask the card to generate the results coming out of a certain algorithm that involves K and R. The card responds with the resulting number, and the terminal repeats the same process, applying the same

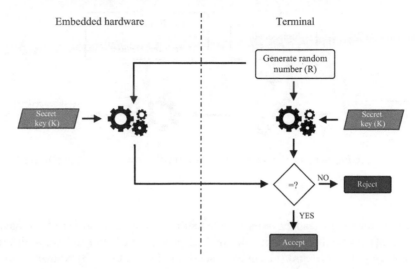

Figure 5.3   Block diagram for the internal authentication

algorithm with R and K. If both results match, then the terminal can determine that the card belongs to the system and, therefore, it is authenticated.

The same process can be done exchanging the terminal and the card, where the one generating the random number and the one checking if both results match, is the card. In that case, the process is called ***external authentication***, as it is the card which is authenticating the external world. Some cards provide joint process merging internal and external authentication, by the generation of two different random numbers, one from the card and one from the terminal, and both parts checking the result. In that case, the process is called ***mutual authentication***.

Although this process achieves the authentication of both ends in the communication, it may suffer from a serious vulnerability: if K is the same for all cards and terminals, then either the disclosure of K or its discovery through cryptanalysis will compromise the security of the whole system and the privacy of all users. Therefore, it is necessary to have a different key for each card. But it is not viable to have a random K for each card, as they will have to be stored somewhere for the terminal to access them. Obviously, this solution creates more vulnerabilities that the one it solved. But there is a very simple way to solve this. All keys within an application or a system can be different by deriving them from a common ***seed*** using a hashing algorithm and some unique identification for each card. This way K will be different for each card, minimizing the possibility of a cryptanalysis success, as it may not have enough data to obtain K from the results given by the card. This process is called ***diversification***. However, in this case the problem remains that, if K is compromised one way or another, an attacker can construct the seeding himself, and this could happen if the attacker can have access to a large number of cards belonging to the same application.

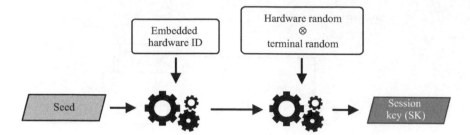

Figure 5.4   An implementation of the generation of a session key (seed is kept as a secret)

Going one step ahead, in order to avoid still the minimum possibility of having the diversified key of a card being discovered, it would be better to have a different key in each use of the card (i.e. in each session). This is typically known as using *session keys (SKs)*, and the process to generate them is exemplified in Figure 5.4. The SK is obtained by applying a hashing algorithm (or a set of them) combining the seed (which is kept secret), some unique identifier from the smart card, and random numbers generated by the card and the terminal. Once generated the SK, this key will be used for the cryptographic algorithms during the session using the card, e.g. for performing the mutual authentication.

By combining these two processes (authentication and SK-based encryption for the communication channel), the removal of vulnerabilities derived from cryptanalysis is achieved, as well as the possibility to authenticate both ends in the communication. But it will be necessary to protect the messages exchanged between the card and the terminal. In few words, this means

- to authenticate the sender of the message, as to reject those messages not coming from the proper terminal,
- to be able to check the integrity of the message, as to deny those messages modified along the communication and
- to provide confidentiality by ciphering the content of the message. This may not be needed when exchanging some kind of messages.

Using symmetric cryptography, there are several ways to solve this need, ISO/IEC 7816-4 [23] standardizes a process called *secure messaging (SM)*. An implementation of the SM process is illustrated in Figure 5.5. As it can be seen, both the command and the data are processed with SK and a hash of the result, plus the data is cyphered again using SK. Then, the message sent from the terminal to the card is the command plus the cyphered data, which contains also the result of the hash (which is also considered to be the integrity check data), constituting the integrity check block. So the card can perform the reverse operation using SK, and finally obtain both the data and the integrity check block. The card calculates also the integrity check block and if it matches the one in the message, the command to be executed. If SK is not the same as the one from the terminal, the integrity check block calculate by the card

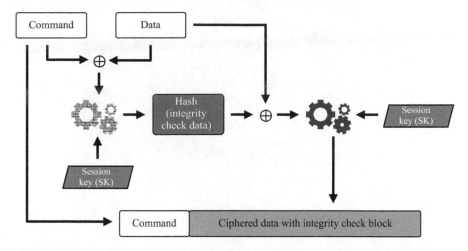

Figure 5.5   An implementation of secure messaging

will not be the same as the one in the message, and therefore, the command won't be executed.

Being this a process that can be implemented in any communication, the importance of a smart card is that it is the card which requires and enforces using it for certain operations and/or access to certain data. If an operation required by the card is not followed precisely and using the same secret key data, an error will be shown, and the session will be considered closed, having to start again the session, and therefore using a new SK.

Smart cards also have further security mechanisms. For example, keys can also have a certain number of maximum consecutive wrong uses, blocking them and the information related to them if the maximum number is exceeded. Also, keys, PINs and password are stored in the card in files or DO which forbid the possibility of being read by the terminal. Also, changing the value of a key, PIN or password requires the highest level of security the card implements.

In addition, the above-mentioned processes can also be implemented by using a public key infrastructure (PKI), e.g. using public key certificates for both authenticating the card and authenticating the cardholder. A practical application of this is implemented in the EMV payment system, both with the static and the dynamic authentication [27].

All these security mechanisms plus many others are what makes a smart card to be considered as a secure element and sometimes being the security engine of a system. In such a case, the smart card is usually called SAM.

5.2.5   Other implementations

As it has been said, smart cards were invented to substitute and improve magnetic stripe cards. Therefore, its size has always been the one of a credit card (also called

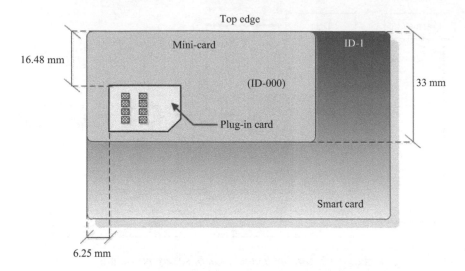

Figure 5.6   Standardized card form factors

ID-1 form factor, as defined in ISO/IEC 7810 [28]). But as applicability of smart cards increased, and technology started to create devices smaller and smaller, new form factors were defined. From those alternatives, the one that stated being a success is the so-called ***plug-in card***, thought to be used for a semi-permanent insertion in other electronic devices. The two main applications where this format has been used is the SIM module in mobile telephones or as a SAM in a PoS. Figure 5.6 shows the relationship about the standardized sizes of smart cards.

The plug-in card, being really small, has come to the need to even reduce it much more in size, in particular for being used in new smartphone generations. This way two smaller versions of the plug-in card have been designed, known as micro-SIM or the even smaller nano-SIM. The only difference in all these formats is the amount of plastic in the card, but the IC in the smart card is always the same and is located in the centre of the contacts.

Further, new applications have discovered some inconveniences of the technology of smart cards. The first one has been the need of faster communications, so ISO/IEC JTC1/SC17, the standardization body for this technology, decided to create an USB communication for contact smart cards [29] and also faster communication for proximity cards [24]. The USB interface also gave the idea of creating USB tokens that follow the philosophy of smart cards.

Another inconvenience for some applications is the use of PIN codes, trying to be substituted or complemented by biometric authentication such as fingerprint recognition. This was easily added to USB tokens using ether flat or swipe fingerprint sensors. But USB tokens, being very useful for their use with computers, are not so convenient in other applications. Therefore, a couple of family of standards have been developed recently to allow embedding other devices in the form factor of a

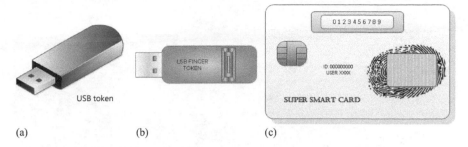

(a)   (b)   (c)

*Figure 5.7   Examples of other formats that can implement the smart card
              functionality: (a) USB token; (b) USB token with fingerprint sensor
              embedded; (c) smart card with embedded devices*

smart card. The generic one is ISO/IEC 18328 [30], which talks about all kind of
devices, from keyboards to displays. Related to biometrics, there is another standard
that talks about embedding biometric capture devices in smart cards (ISO/IEC 17839)
[31]. This will be further explained in the following section. Figure 5.7 shows some
alternative designs for the concept of smart cards.

It can be observed that the always-evolving market of mobile devices is demanding two contradictory things nowadays. On the one hand, they are trying to reduce
the number of external peripherals, such as the SIM card. On the other, they have
the need of improving the protection of all the personal data that users store in their
smartphones. In order to try to solve this situation, the semiconductor industry is
manufacturing microprocessors and microcontrollers with a security module embedded, which is usually called *secure element*. Unfortunately, these secure elements are
not disposable, so they cannot become blocked as a user would not accept that his
smartphone gets fully blocked losing the invested money and the information stored.
Therefore, for this kind of applications, smart cards are still the best solution, may be
integrating the smart card philosophy with one of the most commonly used external
devices in smartphones, such as an SD card.

5.3   Smart cards to improve security and privacy in unsupervised biometric systems

In a biometric system, smart cards can play two different roles. It can be used simply
as a SAM in each capture device, so as to secure the communication of biometric data
to the other end of the communication. In this role, smart cards could be used not only
in authentication systems, but also in identification. The second role is the typical
one for an authentication system, using the card as the token to claim the identity
of the user who wants to be authenticated. This is one of the best implementation of
authentication systems, in particular because it allows to store securely the biometric
reference of the user and as we will see, even performing more operations. Also,

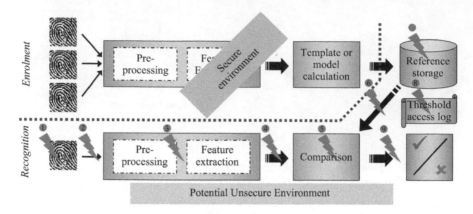

Figure 5.8   Potential vulnerable points of a biometric authentication system. PVPs are shown by using a flash symbol with the PVP number

this approach can be considered positive with regards to privacy, as the biometric reference resides in the vicinity and control of the user.

The best way to understand the benefits of using smart cards in authentication biometric systems is by studying the different potential vulnerable points (PVPs) of a biometric system, as illustrated in Figure 5.8. This figure simplifies the problem by considering the enrolment as secure and only considering the potential attacks during the recognition phase. PVP1 is the one related to presentation attacks. PVP2, PVP4, PVP6 and PVP9 refer to tampering with the communication of biometric data between different parts of the biometric system (depending how the system is implemented some of them may not be present). PVP3 and PVP5 are related to malware that may change the algorithms, either for pre-processing or for comparison. PVP8 may also be part of the malware, or, if the thresholds are modifiable, then this PVP is related to tampering with the storage of all parameters and the access log. And finally, PVP7 is related to accessing the place where the biometric reference is stored.

In a biometric authentication system, many of these vulnerabilities may be solved by simply using smart cards properly. ISO/IEC 24787 [32] defines the relationship between smart cards and biometrics, explaining the different ways this relationship can be built.

The simplest relationship is just by storing securely the biometric reference in the smart card, sending it to the biometric system by request and in a protected manner. This is known as ***store-on-card*** and illustrated in Figure 5.9. The benefits come from the properties of a smart card to store the reference securely (removing PVP7) and forcing the biometric system to access that information only if a secure channel has been created (removing PVP6). So, two PVPs have been removed just by only using a well-configured and designed smart card.

The next step is to protect the comparison algorithm, the thresholds and the decision, by including it into the smart card. This is illustrated in Figure 5.10 and

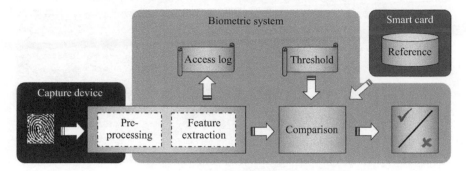

Figure 5.9   Store-on-card configuration

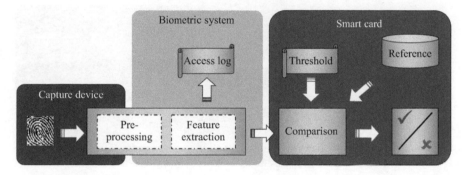

Figure 5.10   On-card biometric comparison configuration

known as *on-card biometric comparison*, although the market also calls it *match-on-card*. By using this configuration, in addition to the previous PVPs removed, also PVP4, PVP5, PVP9 and part of PVP8 are avoided. The only drawback of this configuration is that the data format of the processed biometric sample shall be compatible with the comparison algorithm and the biometric reference in the card. In the case of fingerprint, this is feasible as there is a standardized data format for the minutiae [33]. But the rest of modalities, plus the other approaches for processing fingerprints, do not have a standard related to a feature-based data format. Therefore, the algorithms in the card shall be developed ad-hoc for that biometric system. This may be possible using Java Cards, but, as it is an interpreted language, the execution time may not be viable.

Following the same idea of placing more capabilities to the smart card, we can talk about also adding processing in the card. This can be done up to different levels of integration, from just the last part of the processing, till the whole processing from the raw biometric sample. A particular implementation of this kind of solution is defined in ISO/IEC 24787, called work-sharing. If the card can include the whole processing, then the interoperability with several capture devices is possible, thanks

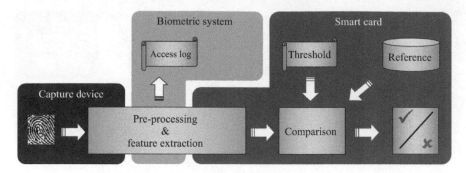

Figure 5.11   Biometric-processing-on-card configuration

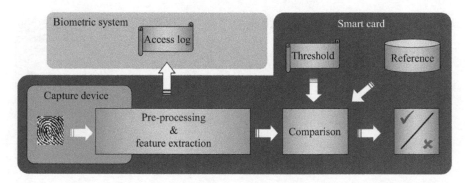

Figure 5.12   Biometric-system-on-card (BSoC) configuration

to the existence of standardized biometric data formats in raw mode, as defined in ISO/IEC 19794 [34] family of standards. Generically, this can be called biometric-processing-on-card and is illustrated in Figure 5.11.

Although there are not many initiatives related to the previous architecture, there is much more interest in going further and even including the capture device in the card. This is called ***biometric-system-on-card (BSoC)*** and is illustrated in Figure 5.12. Although it is initially defined in ISO/IEC 24787, there is a whole family of standards that define in detail this approach (ISO/IEC 17839 [31]). With this approach, all PVPs are removed, except for PVP1 and the possibility of tampering the biometric systems and attacking the log storage (part of PVP8). The challenge with this approach is how to embed the capture device in a plastic card. For some modalities embedding the capture device in the card is currently impossible (e.g. iris recognition). For some others, it may be needed a thicker version of the card which will force to use the contactless interface (e.g. some fingerprint or face recognition). But for some others, it may be viable, although the plasticity of the card may be compromised (e.g. some fingerprint approaches).

This same philosophies can be exported to other portable devices, such as smartphones, but it has to be respected the security constraints of smart cards and be able to obtain common criteria security certificates of high level, and for the whole device (not only part of it). As smartphones are multipurpose devices, this is not seen possible nowadays.

5.3.1 Practical implementations

Integrating biometrics in smart cards has been done at both, academic and industrial levels. Obviously, the schema most commonly found in the industrial world is Store-on-Card, available in many applications. One of the most interesting facts is that by 1998, the US National Security Agency had already worked on this and even published some guidelines on using biometrics in smart cards [35]. But the most important deployment is, by far, the case of ePassports [36], where the passport booklet includes a contactless smart card which stores, at least, the image of the passport holder face. Such image is read by the system, and then compared with the traveller's live captured image.

On-card biometric comparison has been addressed widely by the academics since 2000 ([37] in Spanish and a summary in English in [38]), where a first prototype was shown integrating hand geometry and iris recognition and showing the limitations for including speaker recognition. Prototypes were built using one of the first Java Card products. Also at that time, two companies were working in this kind of products, but using fingerprints: Precise Biometrics [39] and Giesecke & Devrient [40]. Precise Biometrics is still active in this field, offering different kind of products. Both cased involved native operating system cards, although later on they worked also in including the fingerprint comparison in Java Cards.

From these first works, several others were released, both by industry and academics. An interesting listing of some of this works can be found in the Encyclopaedia of Biometrics [41]. A later piece of work, which also analyses the aspects related to security and implementation, can be the PhD thesis from Barral [42], also focussed on fingerprints.

From the deployment point of view, there are several implementations that deserve special attention. The first one is the deployment of the electronic version of the Spanish National ID Card (DNIe [43]), which was initiated in 2005 and currently has issued more than 56 million cards. In this card, the fusion of smart card technology, PKI and biometrics is achieved (see Figure 5.13). Three biometric traits are used, being two of them used with the store-on-card schema (face and static handwritten signature), while the third one, which are two fingerprints, are used for on-card biometric comparison, in order to unblock the card keys or even to renew the PKI certificates.

Another important specification is the personal identity verification of federal employees and contractors, in the United States [44]. This project started its definition in 2005 by launching the first version of FIPS[2] 201 document. The extension of the project made this specification extremely interesting, although later decisions led to a reduction of its deployment from what it was initially expected. An example of it use can be seen at the US Department of Veteran Affairs [45].

---

[2]FIPS stands for Federal Information Processing Standard.

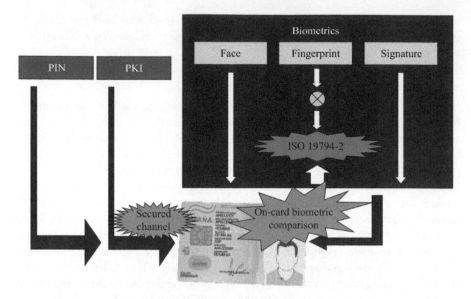

Figure 5.13   Architecture of the Spanish national ID card (DNIe)

Going beyond on-card biometric comparison, several publications have shown part of the processing inside the smart card, such as the work by Kümmel and Vielhauer [46], integrating dynamic handwritten signature biometrics using a Java Card, with not only the comparison inside but also the feature extraction.

But industry seems to have more interest in moving towards biometric system-on-card, in particular with fingerprints. There have been products with the shape of a USB memory stick, including some of them the properties of a SCOS. Unfortunately, this line has not had a great success, leading to proprietary solutions with no detailed specification about the security included in the device. Some companies have tried to bring the idea of a BSoC as a reality, but typically limiting some of the features of the smart card, in particular those related to the physical dimensions or flexibility (i.e. typically being thicker and rigid) [47,48] or only allowing swipe sensors [49]. Very recent advances have allowed creating full-size flexible fingerprint sensors [50], which will allow to boost the concept of a full-compliant BSoC product, for both, contact and contactless smart cards.

5.4 Conclusions

Throughout this chapter, the technology of smart cards and the relationship with biometric systems has been explained. As it has been seen, smart cards are the perfect complement to a biometric system, in particular when it is implemented as a distributed authentication system. The security provided by smart cards allows reducing most of the PVPs in unsupervised deployments. As the exchange of sensitive information is reduced and, when it exists, it can be forced to be cyphered, privacy is enforced.

It is important to note that the smart card functionality can be implemented as a card or as any other kind of personal token. In fact, it can even be implemented within a personal smartphone, but it is important to preserve the capability of disposing of the secure token in case it becomes blocked. This should also be accompanied by educational activities that let the users know about the importance of security and privacy, and the reasons why a secure element may become blocked, losing some of the personal information.

Even though merging biometrics with smart cards solves many of the problems, there is still one major risk, which is the possibility of getting a successful presentation attack. Therefore, there is the need of creating PAD methodology that may avoid most of the vulnerabilities.

References

[1] ISO/IEC JTC1 SC37 Biometrics. ISO/IEC 2382-37:2012 "Information technology – Vocabulary – Part 37: Biometrics". International Organization for Standardization and International Electrotechnical Commission, 2012.

[2] Jain A. K., Bolle R., Pankanti S. (eds). *Biometrics: Personal Identification in Networked Society*. Norwell, MA: Norwell, MA: Springer, 2006. p. 411. eBook ISBN 978-0-387-32659-7.

[3] Sanchez-Reillo R., Sanchez-Avila C., Lopez-Ongil C., Entrena-Arrontes L. 'Improving security in information technology using cryptographic hardware modules'. *Proceedings of the 36th International Carnahan Conference on Security Technology*, Atlantic City, US, Oct 2002. IEEE; 2002. pp. 120–123.

[4] ISO/IEC JTC1 SC37 Biometrics. ISO/IEC 30107-2:2017 "Information technology – Biometric presentation attack detection – Part 2: Data formats". International Organization for Standardization and International Electrotechnical Commission, 2017.

[5] ISO/IEC JTC1 SC37 Biometrics. ISO/IEC 30107-1:2016 "Information technology – Biometric presentation attack detection – Part 1: Framework". International Organization for Standardization and International Electrotechnical Commission, 2016.

[6] Flink Y. *Million Dollar Border Security Machines Fooled with Ten Cent Tape* [online]. 2009. Available from http://findbiometrics.com/million-dollar-border-security-machines-fooled-with-ten-cent-tape/ [Accessed 18 Aug 2016].

[7] Breithaupt R. 'Steps and stones with presentation attack detection – PAD projects @ BSI'. *Proceedings of the International Biometrics Performance Conference (IBPC 2014)*, Gaithersburg, US, Apr 2014. NIST; 2014. Available from http://biometrics.nist.gov/cs_links/ibpc2014/presentations/12_tuesday_breithaupt_2014.04.01_StepsNStones_IBPC_NIST_V07.pdf

[8]   Sousedik C., Busch C. 'Presentation attack detection methods for fingerprint recognition systems: a survey'. IET Biometrics, 2014, vol. 3 (4), pp. 219–233.

[9]   Sanchez-Reillo R., Quiros-Sandoval H. C., Liu-Jimenez J., Goicoechea-Telleria I. 'Evaluation of strengths and weaknesses of dynamic handwritten signature recognition against forgeries'. *Proceedings of the IEEE Int. Carnahan Conf. Security Technology*, 2015, Taipei, RoC, IEEE; 2015. pp. 373–378.

[10]   ISO/IEC JTC1 SC37 Biometrics. ISO/IEC 30107-3:2017 "Information technology – Biometric presentation attack detection – Part 3: Testing and reporting". International Organization for Standardization and International Electrotechnical Commission, 2017.

[11]   Wikipedia. *Touch ID* [online]. 2016. Available from https://en.wikipedia.org/wiki/Touch_ID [Accessed 18 Aug 2016].

[12]   Samsung. *Galaxy S6* [online]. 2016. Available from http://www.samsung.com/uk/consumer/mobile-devices/smartphones/galaxy-s/SM-G920FZKABTU [Accessed 18 Aug 2016].

[13]   Huawei. *Ascend Mate 7* [online]. 2016. Available from http://consumer.huawei.com/en/mobile-phones/Ascend-Mate7/index.htm [Accessed 18 Aug 2016].

[14]   Campbell M. *Average iPhone User Unlocks Device 80 Times per Day, 89% Use Touch ID, Apple Says* [online]. 2016. Available from http://appleinsider.com/articles/16/04/19/average-iphone-user-unlocks-device-80-times-per-day-89-use-touch-id-apple-says [Accessed 18 Aug 2016].

[15]   ISO/TC 159/SC 4 Ergonomics of human-system interaction. ISO 9241-210:2010 "Ergonomics of human-system interaction – Part 210: Human-centred design for interactive systems". International Organization for Standardization, 2010.

[16]   Blanco-Gonzalo R., Diaz-Fernandez L., Miguel-Hurtado O., Sanchez-Reillo R. 'Usability evaluation of biometrics in mobile environments'. *Proceedings of the 6th International Conference on Human System Interactions (HSI)*, 2013, Gdansk, Poland, June 2013. pp. 123–128.

[17]   Sanchez-Reillo R. 'Tamper-proof operating system'. *Encyclopaedia of Biometrics*, vol. 2. Springer, Berlin, 2009. pp. 1315–1321.

[18]   CommonCriteria for Information Technology Security Evaluation. The Common Criteria Portal [online]. Available at https://www.commoncriteriaportal.org [Accessed 18 Aug 2016].

[19]   Zoreda J. L., Oton J. M. *Smart Cards*. Artech House, Norwood, 1994.

[20]   Schneier B. *Applied Cryptography*. Wiley, New York, 1995.

[21]   Oracle. *Java Card Technology* [online]. 2009. Available from http://www.oracle.com/technetwork/java/embedded/javacard/overview/index-jsp-140503.html [Accessed 18 Aug 2016].

[22]   ISO/IEC JTC1 SC17 Identification cards. ISO/IEC 7816-3:2006 "Identification cards – Integrated circuit cards – Part 3: Cards with contacts – Electrical interface and transmission protocols". International Organization for Standardization and International Electrotechnical Commission, 2006.

[23] ISO/IEC JTC1 SC17 Identification cards. ISO/IEC 7816-4:2013 "Identification cards – Integrated circuit cards – Part 4: Organization, security and commands for interchange". International Organization for Standardization and International Electrotechnical Commission, 2013.

[24] ISO/IEC JTC1 SC17 Identification cards. ISO/IEC 14443 "Identification cards – Contactless integrated circuit cards – Proximity cards". International Organization for Standardization and International Electrotechnical Commission, 2016.

[25] Common Criteria Portal. *Certified Products* [online]. 2016. Available from https://www.commoncriteriaportal.org/products/ [Accessed 18 Aug 2016].

[26] Tarnovsky C. 'Attacking [the] TMP Part 2 ST19WP18'. *DEFCON 20*, 2012, Las Vegas, US, Jul 2012. Available from https://www.youtube.com/watch?v=h-hohCfo4LA.

[27] EMVco. *A Guide to EMV Chip Technology* [online]. 2017. Available from https://www.emvco.com/best_practices.aspx?id=217 [Accessed 24 May 2017].

[28] ISO/IEC JTC1 SC17 Identification cards. ISO/IEC 7810:2003 "Identification cards – Physical characteristics". International Organization for Standardization and International Electrotechnical Commission, 2003.

[29] ISO/IEC JTC1 SC17 Identification cards. ISO/IEC 7816-12:2005 "Identification cards – Integrated circuit cards – Part 12: Cards with contacts – USB electrical interface and operating procedures". International Organization for Standardization and International Electrotechnical Commission, 2005.

[30] ISO/IEC JTC1 SC17 Identification cards. ISO/IEC 18328 "Identification cards – ICC-managed devices". International Organization for Standardization and International Electrotechnical Commission, 2016.

[31] ISO/IEC JTC1 SC17 Identification cards. ISO/IEC 17839 "Information technology – Biometric system-on-card". International Organization for Standardization and International Electrotechnical Commission, 2016.

[32] ISO/IEC JTC1 SC17 Identification cards. ISO/IEC 24787:2010 "Information technology – Identification cards – On-card biometric comparison". International Organization for Standardization and International Electrotechnical Commission, 2010.

[33] ISO/IEC JTC1 SC37 Biometrics. ISO/IEC 19794-2:2011 "Information technology – Biometric data interchange formats – Part 2: Finger minutiae data". International Organization for Standardization and International Electrotechnical Commission, 2011.

[34] ISO/IEC JTC1 SC37 Biometrics. ISO/IEC 19794 "Information technology – Biometric data interchange formats". International Organization for Standardization and International Electrotechnical Commission, 2011.

[35] National Security Agency. *Guidelines for Placing Biometrics in Smartcards* [online]. 1998. Available from http://citeseerx.ist.psu.edu/viewdoc/download?doi=10.1.1.446.6459&rep=rep1&type=pdf [Accessed 24 May 2017].

[36] ICAO. 9303-9 Machine Readable Travel Documents – Part 9: Deployment of Biometric Identification and Electronic Storage of Data in eMRTDs

[online]. Available at https://www.icao.int/publications/Documents/9303_p9_cons_en.pdf [Accessed 24 May 2017].

[37]   Sanchez-Reillo R. 'Mecanismos de autenticación biométrica mediante tarjeta inteligente'. PhD Thesis, E.T.S.I. Telecomunicación (UPM). Available at http://oa.upm.es/844/ [Accessed 24 May 2017].

[38]   Sanchez-Reillo R. 'Securing information and operations in a smart card through biometrics'. *Proceedings IEEE 34th Annual 2000 International Carnahan Conference on Security Technology (Cat. No.00CH37083)*, Ottawa, ON, 2000. pp. 52–55. doi: 10.1109/CCST.2000.891166.

[39]   Precise Biometrics. *Precise Match-on-Card* [online]. 2017. Available from https://precisebiometrics.com/fingerprint-technology/match-on-card/ [Accessed 24 May 2017].

[40]   Giesecke & Devrient. *Global Website* [online]. 2017. Available from https://www.gi-de.com/en/index.jsp [Accessed 24 May 2017].

[41]   Pang C. T., Yun Y. W., Jiang X. 'On-card matching'. *Encyclopaedia of Biometrics*. Boston, MA: Springer US, 2009. pp. 1014–1021. doi: 10.1007/978-0-387-73003-5_289.

[42]   Barral C. 'Biometrics and Security: Combining Fingerprints, Smart Cards and Cryptography'. PhDThesis, École Polytechnique Fédérale de Lausanne. Available at https://infoscience.epfl.ch/record/148685/files/EPFL_TH4748.pdf.

[43]   Gobierno de España – Ministerio del Interior – Dirección General de la Policia. DNI electrónico [online]. Available at https://www.dnielectronico.es/PortalDNIe/index.jsp [Accessed 24 May 2017].

[44]   NIST. *About Personal Identity Verification (PIV) of Federal Employees and Contractors* [online]. 2016. Available from http://csrc.nist.gov/groups/SNS/piv/ [Accessed 24 May 2017].

[45]   U.S. Department of Veteran Affairs. PIV Card Project [online]. Available at https://www.va.gov/pivproject/ [Accessed 24 May 2017].

[46]   Kümmel K., Vielhauer C. "Biometric Hash Algorithm for Dynamic Handwriting Embedded on a Java Card". In: Vielhauer C., Dittmann J., Drygajlo A., Juul N.C., Fairhurst M.C. (eds). *Biometrics and ID Management*. BioID 2011. Lecture Notes in Computer Science, vol. 6583. Springer, Berlin, Heidelberg, 2011.

[47]   MoriX. MoriX Card [online]. Available at http://morix-ic.com/en/products/morix_card/ [Accessed 24 May 2017].

[48]   Samsung. *e-Smart Full Wireless Fingerprint Access Card* [online]. 2008. Available from https://www.youtube.com/watch?v=ONmb161SZzo [Accessed 24 May 2017].

[49]   Card Tech. Biometric System-on-Card [online]. Available at http://www.cardtech.it/files/pictures/prodotti/scheda_a4_eng.pdf [Accessed 24 May 2017].

[50]   NEXT Biometrics. *NB-0610-S2 Flexible Sensor Chipset* [online]. 2017. Available from http://nextbiometrics.com/products/for_smartcards/nb-0610-s2/ [Accessed 24 May 2017].

Chapter 6
Inverse biometrics and privacy

Marta Gomez-Barrero[1] and Javier Galbally[2]

In addition to an overall improvement of their performance, the widespread deployment of biometric recognition systems has also led to the disclosure of privacy and security concerns related to the use of these sensitive data. In particular, the early common belief that biometric templates are irreversible has been proven wrong. Over the last decade numerous works have studied the challenging problem of how to reconstruct synthetic samples from the stored templates, which match the original biometric samples. Such a process, known as *inverse biometrics*, poses a severe threat to the privacy offered by biometric systems: not only personal data can be derived from compromised and unprotected templates, but those synthetic samples can be as well used to launch other attacks (e.g., presentation attacks). Due to these serious implications, three different aspects of inverse biometrics have been analysed in the recent past: i. how to generate synthetic samples, ii. how to reconstruct a particular sample from its template, and iii. how to evaluate and counterfeit the aforementioned reconstruction techniques. This chapter summarises the works dealing with these three aspects in the biometric field.

6.1 Introduction

While the use of biometric information for recognition offers numerous advantages, biometric data is highly sensitive, and many people are thus currently concerned about the possible misuse of such data [1]. For instance, these data could be used to reveal medical conditions, to gather personal information, even in a covertly manner given the recent developments in biometrics at a distance [2], or to link databases. Furthermore, geographical position, movements, habits and even personal beliefs can be tracked by observing when and where the biometric characteristics of an individual are used to identify him/her [3]. In order to protect the privacy of the owner, biometric systems typically do not store raw biometric data, which may disclose sensitive information about the subjects. Rather, they store the extracted template

[1]da/sec Biometrics and Internet Security Research Group, Hochschule Darmstadt, Germany
[2]European Commission, DG-Joint Research Centre, E.3, Italy

(feature set) containing the information about the individual that is most relevant for recognition purposes.

Whereas in the past, it has been a common belief that the features extracted from biometric samples did not reveal any information about the underlying biometric characteristic and its owner [4], in 2003, Bromba explored in [5] the question of whether 'biometric raw data can be reconstructed from template data'. In that article, he reached three main conclusions, namely

- There are cases where raw data are very similar to template data by definition and therefore can hardly be distinguished.
- Often the reconstruction is possible to a degree which is sufficient for misuse.
- Even if reconstruction should not be possible in specific cases, misuse of templates remains possible.

As a consequence, in the following years, several studies have supported those facts and hence arisen serious concerns regarding the soundness of the aforementioned assumption for different characteristics [6–10]. In those works, the information stored in reference templates is exploited in order to generate synthetic samples by means of reverse engineering approaches known as *inverse biometrics*. The synthetic samples generated can be subsequently used to launch masquerade attacks (i.e. impersonating a subject), thereby decreasing the security of the system, or to derive information from its owner, thereby failing to protect the subject's privacy. Hence, there is a need to protect the biometric information stored in the database, since the feature extraction process alone does not hide our personal biometric data.

Furthermore, such synthetic biometric samples have a wide range of applications, which includes but is not limited to privacy assessment of biometric systems [11]. This has attracted the attention of the biometric community, and serious efforts have been directed to explore different synthetic samples generation techniques over the last years. Depending on the final outcome of a particular synthetic sample generation method, and the knowledge needed to carry it out, different consequences on the security and privacy of biometric systems can be expected. This chapter is hence focused on reviewing existing methods to generate synthetic samples and their implications on the subject's privacy. In Section 6.2, a classification of such methods is presented, and their possible threats on the security and privacy of biometric systems were discussed. Then, we will focus on the particular problem of inverse biometrics, reviewing the state of the art on inverse biometric methods in Section 6.3. Evaluation methodologies for those algorithms are later described in Section 6.4, and the privacy issues raised by them as well as possible countermeasures are analysed in Section 6.5.

6.2 Synthetic biometric samples generation

As already introduced, this chapter is focused on the review of inverse biometric methods for different characteristics and their privacy-related issues. In order to put inverse biometrics into context, the present section gives a general overview of the broader field of synthetic biometric samples generation. In addition, the overall taxonomy

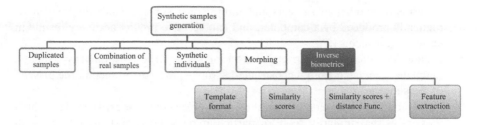

Figure 6.1 *Classification of the methods for synthetic biometric samples generation. The methods that are the main focus of the present chapter (i.e. inverse biometrics) are highlighted in gray and classified according to the knowledge required to be carried out*

considered in the chapter regarding the different techniques proposed so far to produce synthetic samples is shown in Figure 6.1.

Even though the development of synthetic images has been an active research field in other areas for a long time [12–14], with dedicated conferences such as the short for Special Interest Group on Computer GRAPHics and Interactive Techniques annual conference on computer graphics, started in 1974 [15], synthetic biometrics have not been widely used within the biometric community until very recently.

Within this community, historically, *manually* produced physical biometric traits such as fingerprints, signatures or forged handwriting have been a point of concern for experts from a forensic point of view [16,17]. More recently, such physically produced synthetic characteristics have been largely utilised for vulnerability and spoofing assessment studies in characteristics such as the iris [18] or the face [19]. However, it has not been until the fairly recent development of the biometric technology when the automatic generation of *digital* synthetic samples has been considered. The main reason behind that interest is the wide range of applications of those synthetic samples, which include but are not limited to

- Increase of biometric data: the amount of available data belonging to the enrolled subjects could be augmented with synthetic samples. This way, verification accuracy could be improved, especially for modalities like the handwritten signature, the gait and other behavioural characteristics where the intra-class variability is large [20–23].
- Parametric biometric systems testing: system robustness to specific issues can be investigated from the parametric control provided by synthetic images [24]. For instance, synthetic face images can include carefully controlled changes to pose angle, illumination, expression and obstructions [25].
- Generation of completely synthetic databases: due to the legal restrictions regarding data protection and the sharing and distribution of biometric data among different research groups, we can use synthetic data to overcome the usual shortage of large biometric databases, with no privacy concerns since data does not belong to real subjects. Furthermore, these synthetic databases have no size restrictions,

in terms of number of subjects and samples per subject, since they are automatically produced by a computer, and can serve to predict accuracy results in large-scale systems [26].

- Pseudo-identities generation: enrolled templates could be substituted by synthetic templates generated from reconstructed samples, thereby protecting the privacy of the subjects [27].
- Vulnerability and irreversibility studies: an attacker could use artificial data which maximises the similarity score of particular recognition systems [28] or with synthetically reconstructed images, which would be positively matched to the stored reference, to impersonate the enrolled subjects [7,8,29]. Furthermore, the irreversibility of the stored templates can be analysed in terms of whether it is possible to reconstruct such synthetic samples, and system robustness to specific forgeries tactics can be tested by modelling those forgery strategies [30].
- Entropy studies: from an information theory standpoint, the synthetic images may be useful to conduct experiments on the individuality of a particular biometric characteristic [31,32]. Such knowledge can be subsequently used to estimate the entropy of the corresponding templates, to develop methods that compress and reliably store the intrinsic individual-related data conveyed within the biometric pattern.

Due to that wide range of applications, over the last years, a growing interest has arisen in the scientific community in this new field of research, and different algorithms have been proposed for the synthetic generation of biometric samples for different biometric characteristics [10], such as voice [33–37], fingerprints [38], iris [8,18,39–43], handwriting [44], face [26,45,46], mouth [47,48], signature [30,49–51], mouse dynamics [52,53] or keystroke dynamics [54]. Depending on the final purpose, different types of synthetic data might be created. Therefore, synthetic sample generation methods can be broadly divided into five categories, as depicted in Figure 6.1:

- **Duplicated samples**: in these methods, the generation algorithm starts from one or more real samples of a given person and, through different transformations, produces different synthetic (or duplicated) samples corresponding to the same subject. This approach has been applied to signature [55,56], handwriting [57–59] or face synthesis [26,46,60,61].
- **Combination of different real samples**: this type of algorithms start from a pool of real units, such as *n*-phones in speech (isolated or combination of sounds) or *n*-grams in handwriting (isolated or combination of letters), and combines them, using some type of concatenation procedure, to form the synthetic samples. These techniques have been used in applications such as text to speech, typewriting to handwriting or Completely Automatic Public Turing test to tell Computers and Humans Apart, being the approach followed by most speech [34,62] and handwriting synthesisers [44]. In a similar manner, a sequence of face images can be used to reconstruct 3D shape models to improve face-recognition accuracy [63].

- **Synthetic individuals**: in this case, some kind of a priori knowledge about a certain biometric characteristic (e.g., minutiae distribution, iris structure, signature length, etc.) is used to create a model of the biometric characteristic for a population of subjects. New synthetic individuals can then be generated sampling the constructed model. In a subsequent stage of the algorithm, multiple samples of the synthetic subjects can be generated by any of the procedures for creating duplicated samples. These algorithms are therefore able to generate completely synthetic databases and constitute a very effective tool to overcome the usual shortage of biometric data without undertaking highly resource-consuming acquisition campaigns [64]. Different model-based algorithms have been presented in the literature to generate synthetic individuals for biometric characteristics such as iris [18,40,41,43], fingerprint [38,65], speech [66–68], mouth [47], handwriting [69,70], signature [50,51], mouse dynamics [52,53] or keystroke dynamics [54].
- **Morphing**: these algorithms aim at transforming the characteristics belonging to one subject (i.e. source) to characteristics belonging to a second subject (i.e. target). Numerous efforts have been directed towards this research field within the speaker recognition community [71], stemming the initial proposals from the late 1980s and the early 1990s [72,73] and leading to the very recent Voice Conversion Challenge 2016 [74]. In this challenge and other articles, it has been shown that such synthetic voice is positively identified with the target identity for different speaker-recognition systems [75], in spite of its rather unnatural sound. More recently, several methods yielding more natural speech have been proposed [76–79]. Analogously, the feasibility of generating synthetic faces which can be positively matched to both the source and the target subjects has been demonstrated in [80,81].
- **Inverse biometrics**: in these methods (see Figure 6.2), given a genuine template, the aim is the reconstruction of a synthetic biometric sample, which matches the stored biometric reference according to a particular biometric recognition system. In other words, it denotes a reverse engineering process which reconstructs synthetic samples from the information conveyed in real biometric templates, and which has already been applied to fingerprint [82,83], iris [7,8], handshape [29], face [19] or handwriting [84,85]. Most of these techniques have been used in the framework of vulnerability-related studies.

The first, second and third types of methods (i.e. duplicated samples, combination of different real samples and morphing) pose no additional privacy threats to the enrolled subjects: in order to generate synthetic samples, the potential attacker is already in possession of real biometric data belonging to the corresponding subjects. Similarly, the generation of fully synthetic individuals raises no privacy concerns, as no biometric information belonging to real subjects is derived. On the contrary, inverse biometric methods do result in a potential violation of the subject's privacy: given only a theoretically secure representation of the subject's biometric characteristics (i.e. the unprotected template), sensible biometric information can be obtained. In addition, contrary to the common belief that templates do not comprise enough information in order to reconstruct the original sample from them [4], synthetic samples can be generated and used to impersonate a particular subject, launching masquerade attacks. Since the main focus of the present chapter is privacy-related issues, we will restrict in the following to the analysis of the approaches that belong to the inverse biometrics category.

6.3   Inverse biometrics methods

As mentioned in the previous section, from all the different synthetic biometric samples generation techniques proposed in the literature, only inverse biometrics methods present a major challenge to privacy protection. Among the different methods proposed in the literature, several differences may be found, being the knowledge they require to be carried out one of the most noticeable. Therefore, Figure 6.1 shows a classification of these methods based on the knowledge requirements, where four different cases are considered, which are as follows:

- Knowledge of the template format. These methods will, for instance, inject templates into the system or extract information from particular parts of the templates.
- Knowledge of the similarity scores between one or several probes and the stored reference. These approaches include those based on some hill-climbing variant, where the impostor takes advantage of the variation of the similarity score when different probes are presented to the system.
- Knowledge of the similarity scores and the distance function used by the biometric system. In this case, the impostor also needs to know the topology of the distance function in order to extract more information regarding the similarity scores.

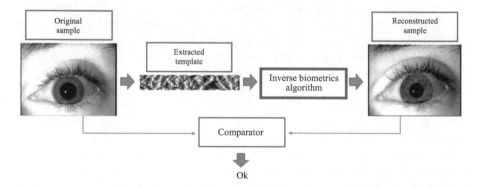

Figure 6.2   General diagram of inverse biometrics methods: starting from a template extracted from a real biometric sample, the inverse biometrics algorithm outputs a reconstructed sample. This sample is then positively matched to the original sample by the biometric comparator. Images have been extracted from [8]

Table 6.1   Summary of key inverse biometric approaches, where 'Knowledge' refers
to the type of knowledge required to carry out each method (see
Section 6.3 and Figure 6.1). 'Attack' refers to the attacking scenario
evaluated on the corresponding article (see Section 6.4), which yielded
the specified 'Success Rate' on the mentioned 'Database'. Whenever an
identification instead of a verification system was evaluated, the Attack
is denoted as 'Id'

Knowledge	Characteristic	Reference	Attack	Success rate	Database
Template format	Fingerprint	[82]	Id. - 1a	100% Rank 1	FVC2000 (110 subjects)
		[83]	2a	>90% 0.1% FMR	FVC2002-DB1 (110 subjects)
		[86]	2a	>99% 0.1% FMR	FVC2006 (140 subjects)
		[87]	Id. - 1a	>23% Rank 1	NIST-4f
Similarity scores	Face	[19]	–	–	FRS
		[6]	1a	>95% 1% FMR	NIST Mugshot (110 subjects)
	Iris	[8]	All	94% 0.01% FMR	BioSecure (210 subjects)
		[88]	1a	100% MS >0.9	CASIAv3 Interval (249 subjects)
	Handshape	[29]	All	50–90% 0.1% FMR	UST DB (564 subjects)
Distance function	Face	[89]	1a	>72% 1% FMR	FERET (1,196 subjects)
Feature extraction	Face	[90]	3	100% 0.01% FMR	BioSecure (210 subjects)
		[9]	3	99–100% 0.1% FMR	XM2VTS (295 subjects)
	Iris	[7]	1a	>96% 0.1% FMR	NIST ICE 2005 (132 subjects)
	Handwriting	[84,85]	1a	<70%	5 subjects

- Knowledge of the feature extraction method. Some algorithms require knowledge of this step in order to reverse-engineer it and reconstruct biometric samples from an optimised template.

In the following sections, we describe the inverse biometrics methods introduced in the literature in terms of the aforementioned types. A summary, including the experimental setup and success rate (SR) of the algorithms (i.e. acceptance rate of the synthetic samples, see Section 6.4 for more details), is shown in Table 6.1.

6.3.1 Template format

One of the first works that addressed the problem posed by inverse biometrics was carried out by Hill [82]. In this work, a general scheme for the reconstruction of biometric samples is proposed, consisting in four successive steps, where only access to and knowledge of the templates format stored in the database is required. In the first step, the attacker needs to gain access to one or more biometric templates. Second, he or she needs to understand the structure of the template (i.e. what information is stored, which format is used, etc.). After determining the structure of the templates, a reverse-engineering process is carried out in the third step in order to reconstruct one or more *digital* biometric samples. Finally, the eventual attacker could also reconstruct *physical* artefacts from the digital synthetic samples.

The most challenging step is the third one that is devising a method for reconstructing digital samples given only the stored templates. In [82], a particular case study on minutiae-based fingerprint templates is presented, based on three consecutive steps: (i) fingerprint shape estimation, (ii) orientation field creation and (iii) ridge pattern synthesis. In the first step, decision trees or neural networks are used, depending on the information available in the template. The orientation field creation relies on the singular points. In the final step, the ridge pattern is iteratively synthesised starting on the minutiae positions and guided by the orientation field.

A similar approach for the generation of fingerprint samples from standard minutiae-based fingerprint templates was proposed in [83]. Since the templates follow the corresponding ISO standard [91], the format is known to the attacker. This raises a new concern regarding the use of standards: on the one hand, they are necessary as they guarantee interoperability. However, on the other hand, they provide a lot of information to potential adversaries. Such a concern reinforces the need to protect biometric templates. The algorithm defined in [83] starts by estimating the fingerprint area with a greedy heuristic algorithm, based on the available minutiae positions. The Nelder–Mead simplex algorithm [92] is used to synthesise the orientation map, which is used together with the set of minutiae and a fixed frequency to estimate the ridge pattern. In contrast to Hill's approach, using different frequency values on this last step, the algorithm in [83] allows to obtain different synthetic samples from a single template. Finally, an additional rendering step is carried out, in which noise is added to the 'perfect' reconstructed image, thus yielding more realistic images. As originally suggested by [82], starting from those synthetic images, gummy fingers can be generated using a printed circuit board as demonstrated in [86] (see Figure 6.3).

A different approach is followed in [87] to reconstruct fingerprint images, in which no iterative technique is considered. Assuming that only the minutiae positions and orientations are available, the orientation field is estimated using minutiae triples. Then, based on the estimated field and the minutiae distribution, the fingerprint class is estimated. Finally, the ridge pattern is synthesised using linear integral convolution. In addition, a rendering step is also undertaken to generate more realistic fingerprints, applying a low-pass filter and histogram equalisation.

6.3.2 Similarity scores

A second set of reverse-engineering methods assumes knowledge of similarity scores between probe synthetic images and the enrolled identity, while no knowledge about the structure of the template is assumed. The method proposed in [19] for the reconstruction of face images from Eigenface-based templates relies on a hill-climbing optimisation of synthetic face images. The authors use the similarity score between the synthetic images and the stored template as feedback to improve the synthetic reconstruction. A more efficient hill-climbing technique is proposed in [6], where each quadrant of the synthetic face image is independently optimised even if only quantised scores are shared by the verification system (as recommended by the [94]).

Rathgeb and Uhl proposed in [88] a different inverse biometrics method based on a hill-climbing algorithm using synthetic samples. In spite of the high dependency of the positive verification of iris textures on the feature extraction algorithm, the authors describe a general approach for synthetic iris textures generation based on a hill-climbing algorithm, where no knowledge about the extracted features is required. This method builds upon the fact that, in spite of their differences, most iris-recognition algorithms share a common characteristic: they tend to average pixels in a block-wise manner. Taking advantage of this concept, the hill-climbing algorithm can be sped up finding the size of such blocks and applying the modification scheme in a block-wise manner.

In their work, the initial synthetic iris texture is set to either a plain texture where all pixels are set to 128, or an initial Eigeniris texture averaged out of five randomly chosen textures. The hill-climbing modification scheme updates the iris texture from the top left to the right bottom iteratively increasing or decreasing the pixel values of complete blocks by a fixed value, retaining changes only if the similarity score with respect to the stored reference template is improved.

A different scheme was proposed by Galbally *et al.* [8] to reconstruct iris patterns from their corresponding iris binary templates, using a probabilistic approach based

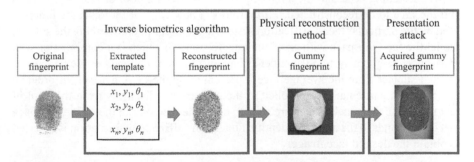

Figure 6.3   *Example of how a compromised template reconstructed through an inverse biometrics algorithm can lead to other type of stronger threats such as presentation attacks. Images extracted from [86,93]*

on genetic algorithms, which is able to reconstruct several different images from the same iris template. The approach needs to have access to a matching score which does not necessarily need to be that of the system being attacked. This way, the reconstruction approach is independent of the matcher or feature extractor being used. However, the algorithm will be most effective if the development matcher and the final test matcher use a similar feature-extraction strategy. This is usually the case, as most iris-recognition methods are based on the principles first introduced by Daugman [95]. In addition, the authors showed that the algorithm can successfully bypass black-box commercial systems with unknown feature-extraction algorithms.

Regarding the particular choice for genetic algorithms among other optimisation strategies, it is based on the unknown nature of the search space. Although previous work in [88] partially supports the assumption of smoothness/continuity of the similarity score function with respect to the iris textures, so far this fact has not been proven. Consequently, whereas the efficiency of classical stochastic gradient descent methods would be at least unclear, genetic search algorithms generally obtain better results in such scenarios: by simultaneously searching for multiple solutions in the solution space, they are more likely to avoid potential local minima or even plateaus in the search space.

Similarly, Gomez-Barrero *et al.* proposed in [29] a probabilistic inverse biometrics method based on a combination of a hand-shape images generator and an adaptation of the Nelder–Mead simplex algorithm [92], which had been previously used to recover face images in [90]. More specifically, the hand-shape generator, starting from a parameter vector, outputs a binary hand-shape image using the active shape model approach. In order to obtain realistic images, an initial pool of real images is required. Then, to impersonate a particular subject, the input of the aforementioned generator is optimised with the downhill simplex algorithm [90,92].

6.3.3   Similarity score and distance function

A different approach is followed by Mohanty *et al.*, who reconstruct face images in [89] assuming access to the similarity scores between a pool of real face images and the face to be reconstructed. Furthermore, knowledge of the distance function used by the particular face-verification system is required. In this method, the authors model the face subspace with an affine transformation, which is a combination of a PCA baseline and a system-dependent non-rigid transformation (i.e. the deviations of each subject from the average face represented by the PCA matrix). In order to reconstruct a particular face enrolled in the system, the distances from the pool of real images to the attacked face are used to compute the point in the affine subspace that corresponds to the attacked identity. Finally, the affine transformation is inverted to obtain the desired face image.

6.3.4   Feature extraction

A final set of inverse biometrics methods requires also knowledge of the feature extraction method. For instance, Venugopalan and Savvides reconstruct iris images from the iris binary templates in [7] in a two-step approach, assuming knowledge of the feature-extraction algorithm. First, using a reversed version of the Gabor function used to extract iris binary templates from iris images, and a pool of real iris images, a subject-specific iris pattern is generated. Then, this pattern is embedded into a real iris texture to make the image more realistic.

In a similar manner to the approach proposed in [6], but using a Bayesian hill-climbing algorithm to optimise the feature vectors instead of the input samples, face images are recovered from Eigenface- and GMM parts-based systems in [9]. The optimised templates are reverse-engineered to obtain the final synthetic images. Analogously, Gomez-Barrero *et al.* reconstruct face samples from Eigenface systems in [90] by means of the downhill simplex algorithm, adapted from [92]. It should be noted that the hill-climbing attack can be launched on any system so long as the adversary has access to: (i) scores and (ii) template format. However, the reconstruction process to recover the face image only works if the feature extractor uses Eigenfaces (i.e. knowledge of the feature extractor required).

Finally, Kümmel *et al.* presented in [84,85] two different approaches to reconstruct handwriting patterns from protected BioHashing templates extracted from their unprotected counterparts as proposed in [96]. Both methods exploit the information leaked by the interval matrix stored as part of the reference template, which allows a straightforward reconstruction of an unprotected feature vector. Starting from such vector, the first method, requiring longer time, is based on a combination of manual user interaction to implement part of the features with a genetic algorithm to implement the remaining features. On the other hand, the second method applies a spline interpolation function to specific features of the reconstructed unprotected vector to generate handwriting sequences.

6.3.5 Summary

The main characteristics of the methods described in this section can be summarised as follows (see Table 6.1):

- The methods proposed to reconstruct fingerprints [82,83,86,87] require knowledge of the template format.
- Different methods requiring knowledge of similarity scores have been proposed to reconstruct faces [6,9,90], iris [8,88] or handshapes [29].
- Knowledge of the distance function is required by a single algorithm based on faces [89].
- Hill-climbing approaches can be applied to reconstruct different biometrics samples [6,8,9,19,29,84,85,88–90]. In these cases, the knowledge required is: (i) template format and (ii) matching scores. In some cases, also the (iii) feature-extraction algorithm should be known by the adversary.
- In the hand-shape, face and iris reconstruction methods mentioned [6,9,19,29,90], a set of real images is used to either initialise the optimisation process or embed the identity to reconstruct into the subspace. Due to the free access to different biometric databases, this is not a strong limitation for eventual attackers [9,90].

6.4   Evaluation of inverse biometrics methods

The main goal of the evaluation of the inverse biometric methods described in Section 6.3 is to assess the threat to the security and privacy of biometric systems posed by the generated synthetic samples. In other words, to study the feasibility of reverting the feature extraction process that, in principle, should conceal our biometric data.

Whereas under normal operation conditions, the subjects try to access the system interacting with it in a straightforward manner, security and privacy evaluations are carried out under attacking scenarios where an impostor tries to access (break) the system interacting with it using some type of approach or methodology for which the application was not intended. In the particular case of inverse biometrics, the system is presented with a synthetic sample reconstructed using one of the methods presented in Section 6.3 instead of a real biometric characteristic. This way, an 'inverse biometrics' attack is carried out.

For such security and privacy evaluations, the experimental framework should be designed not only to avoid biased results but also to estimate the degree of compliance of the proposed reconstruction approaches with the next main objectives, which are as follows:

- Determine the feasibility of recovering a biometric sample from its template.
- Evaluate to what extent the reconstructed samples are able to compromise the security and privacy granted by biometric recognition systems.
- Determine whether it is possible to generate different synthetic reconstructed samples from one given template.

Keeping those goals in mind, a two-step protocol was proposed in [8,29], divided into a development and a validation stage, as depicted in Figure 6.4:

- **Development**. The purpose of this stage is 2-fold: (i) on the one hand, train any module of the reconstruction algorithm if necessary; (ii) on the other hand, generate the synthetically reconstructed data sets that will be used in the validation stage. Therefore, in this step, synthetic samples for a particular instance are reconstructed, using an inverse biometrics method. This way, a synthetic database, depicted in red in Figure 6.4, is generated, comprising at least one synthetic sample generated from each real template in the original database.
- **Validation**. The objective of this stage is to validate the proposed reconstruction scheme and to estimate its performance. For this purpose, the synthetically reconstructed samples generated in the development stage are presented to a different biometric system to determine if they are positively matched to the genuine original images, which would mean that the reconstruction approach is successful and, as a consequence, templates are reversible. In particular, three different types of attacks are carried out, reporting the SRs, defined below, for each case. If the reported SRs are high, we can conclude that traditional or unprotected biometric systems do not grant the necessary privacy to the subjects.

It should be noted that different systems might be used at the development and validations stages, as depicted in Figure 6.4. In fact, using different systems at the

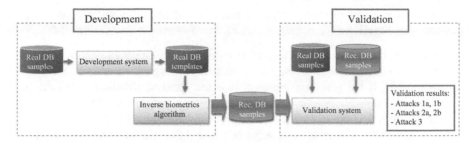

Figure 6.4   *Two-stage experimental protocol followed in the experimental evaluation: (i) in the development stage, the reconstructed database is generated using a development system and (ii) in the validation stage, the privacy threat posed by the reconstructed samples is evaluated launching attacks on validation systems, which might not necessarily be the development system. In order to obtain unbiased results, different biometric systems are used at each stage (i.e. development and validation). Finally, real databases are depicted in dark gray, and synthetic databases in light gray*

validation stage leads to a more meaningful and complete evaluation of the threat posed by the inverse biometrics method, since there is no bias in the biometric comparator which could justify the success of the reconstruction.

This way, for a given operating point of the system, defined by the false non-match rate (FNMR) at a given false match rate (FMR), the performance of the attack is measured in terms of its SR, which is the expected probability that the attack successfully reconstructs a given sample, thereby achieving the impersonation of a subject. It is computed as the ratio between successful attacks (A_s) and the total number of attacks carried out (A_T):

$$SR = A_s/A_T \times 100\ (\%) \tag{6.1}$$

This measure gives an estimation of how dangerous a particular attack is for a given biometric system: the higher the SR, the bigger the privacy threat. Or, in other words, for the case of inverse biometrics, the more reversible the templates. Since in general, the success of an attack is highly dependent on the FMR of the system, the vulnerability of the system to the attacks with the reconstructed images should be evaluated at different operating points. In [8,29], it is proposed to carry out the analysis at the points corresponding to FMR = 0.1%, FMR = 0.05% and FMR = 0.01%, which, according to [97], correspond to a low, medium and high-security application, respectively. For completeness, systems should be also tested at very high security operating points, for instance those corresponding to FMR $\ll$ 0.01%.

In particular, the key factors to compute the SR are to define: (i) what constitutes an attack and (ii) when an attack is considered to be successful. For the present analysis of the unprotected templates reversibility, three representative attacks will be taken into account in order to estimate the performance of the proposed reconstruction methods (see Figure 6.5):

1. **Attack 1**: 1 reconstructed image vs 1 real image. In this case, the attack is carried out on a 1-on-1 basis. That is, one reconstructed image is compared to one real image and, if the resulting score exceeds the fixed verification threshold, the attack is deemed to be successful. Two possible scenarios may be distinguished in this case, depending on the real image being attacked:

    (a)   **Attack 1a**. The real image being attacked is the original sample from which the synthetic image was reconstructed.

    (b)   **Attack 1b**. The real image being attacked is a different sample of the same subject present in the real database.

2. **Attack 2**: N reconstructed images vs 1 real image. In this case, N reconstructed images are computed from one real sample. Then, all reconstructed samples are compared to a real sample. The attack is successful if at least one of the synthetic images positively matches the real template. This represents the most likely attack scenario analysed in other related vulnerability studies [83], where the template of a legitimate subject in the database is compromised, and the intruder reconstructs multiple images to try and break the system. The attacker will gain access if any of the reconstructed images results in a positive score. The same two scenarios as in attack 1 can be considered here:

    (a)   **Attack 2a**. The real image being attacked is the original sample from which the synthetic images were reconstructed.

    (b)   **Attack 2b**. The real image being attacked is one of the other samples of the same subject present in the real database.

3. **Attack 3**: N reconstructed images vs average (of M real images). It is a common practice in many biometric recognition systems to compare the probe sample to several stored templates and return the average score. To emulate this scenario, each reconstructed image is compared to the M samples of the real subject available in the database. The attack is successful if the average score against the real samples of *any of the N reconstructed images* is higher than the given verification threshold. It should be noted that in this last attacking scenario, $N = 1$ would be the particular case with the lowest success chances.

The works presented in [8,29] follow the described protocol, including in the experimental evaluation all the aforementioned attacks. In addition, several biometric recognition systems were used, based on different or unknown feature-extraction techniques. This way, it was shown that the proposed inverse biometrics methods are robust to different features, achieving high SRs in all scenarios considered: in [8], at FMR = 0.1%, the SR ranges from 66.7% for Attack 1b to 96.7% for Attack 3. On the other hand, in [29], also at FMR = 0.1%, the SR ranges from 53.28% for Attack 1b to 92.58% for Attack 3. As it may be observed, a similar behaviour is found in both evaluations, where the most challenging scenario for the attacker is 1b (i.e. one real image is reconstructed, and some other image of the subject is attacked), and the most vulnerable scenario is 2 (i.e. the attacker reconstructs several synthetic images and tries to access the system with at least one of them).

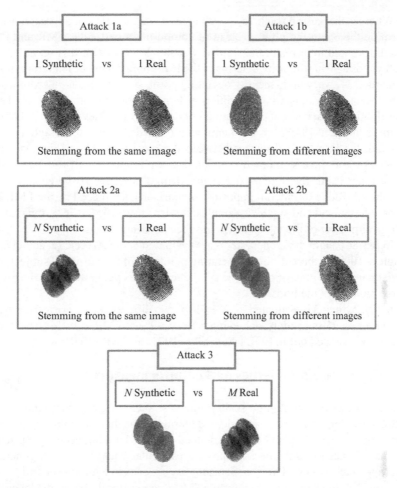

Figure 6.5   Summary of the most representative attacks for the analysis of unprotected templates reversibility

However, the remaining articles described in Section 6.3 (different from [8,29]) only consider one or a subset of the aforementioned attacks, for which the SRs are summarised in Table 6.1.

Attack 1a is evaluated in [89] for two different systems, commercial and Bayesian, obtaining SR >72%. In [7], for the same attack, several evaluation sets are considered, comprising synthetic samples produced with different training sets or parameters. The results report SRs over 96% at FMR = 0.1%. Similarly, different quantisation levels are analysed in [6] for Attack 1a, achieving a SR > 95% for FMR = 1%. Finally, the reconstruction methods proposed in [82,87] evaluate the same attack on an identification task: at rank 1, SRs of 100% and 23%, respectively, are reported.

Whereas the same **Attack 1a** is evaluated in [88], a different approach is used to determine the operating point: instead of computing a particular FMR and FNMR, a fixed threshold in terms of the similarity score (MS) is used. It is set at MS = 0.9, which yields a 90% similarity in terms of the Hamming distance. Therefore, it should be regarded as a very high-security operating point, since such a high score would be taken as genuine by most iris-recognition systems. Among the attacks defined earlier, the evaluation is carried out considering only Attack 1a, achieving SR = 100%.

In addition, in [84,85], a third approach is used to evaluate **Attack 1a**. On the one hand, the Hamming distances between the reconstructed and the real samples achieved by both inverse biometrics method, as well as a SR = 0%–70%, are shown. On the other hand, the authors report the change in terms of the equal error rate when the FMR due to random impostor comparisons is replaced by the FMR due to comparisons of real and synthetic samples: it raises from 10% to 75%, thus reflecting the high similarity of the synthetic data to the real data.

In [83,86], eight different systems are evaluated under **Attack 2a**, achieving SRs as high as 100% at three different operating points. In other words, all subjects can be impersonated with at least one of the synthetic samples generated with the proposed inverse biometrics methods.

In [9], **Attack 3** is evaluated for $N = 1$ for two different face verification systems, at different operating points, achieving a SR = 99% at FMR = 0.1%. An analogous evaluation is carried out in [90], where SR = 100% at FMR = 0.01%.

6.5 Privacy-related issues and countermeasures

Over the last years, some high-scale initiatives, such as the Indian Unique ID[1] or the SmartBorders package[2], have adopted biometrics as their recognition technology. Biometric systems are also being introduced into the banking sector [98], reaching our smartphones through specific apps for particular banks[3], through general payments apps such as ApplePay[TM] or SamsungPay[TM], or even with Mastercard[TM] 'selfie' payments[TM,4]. Furthermore, biometric ATMs[5,6] are currently being deployed. However, as mentioned in Section 6.1, in spite of the wide acceptance and deployment of biometric verification systems, some concerns have been raised about the possible misuse of biometric data [1]. Such concerns can be summarised in the following questions:

- *Fully related to the inverse biometrics issue discussed in the previous sections, an initial set of questions would be: do the stored templates reveal any information*

[1]https://uidai.gov.in/
[2]http://ec.europa.eu/dgs/home-affairs/what-we-do/policies/bordersand-visas/smart-borders/index_en.htm
[3]https://ingworld.ing.com/en/2014-4Q/7-ing-app
[4]http://www.cnet.com/news/mastercard-app-will-let-you-pay-for-things-with-a-selfie/
[5]http://www.biometricupdate.com/201301/citibank-launches-smart-atms-with-biometric-capabilities-in-asia
[6]http://www.biometricupdate.com/201508/ctbc-bank-piloting-atms-that-use-finger-vein-scanning-and-facial-recognition

about the original biometric samples? In other words, are we able to reconstruct synthetic samples whose templates are similar enough to those of the original subject? The works described in Section 6.3 have shown that, for a wide variety of biometric characteristics, it is possible to carry out such an inverse engineering process. In fact, not only one but several synthetic samples can be generated, which are positively matched to the reference template. As a consequence, an eventual attacker which manages to obtain just a template belonging to a certain subject (e.g. the iris binary template or minutiae template) is able to reconstruct the original biometric sample. The attacker can afterwards use it to illegally access the system, to steal someone's identity or to derive additional information from the obtained biometric data, thereby violating the right to privacy preservation of the subject. As a consequence, we must ensure the *irreversibility* of the templates. Section 6.4 has presented a protocol to measure their reversibility.

• *Even if templates were irreversible, are my enrolled templates in different recognition systems somehow related to each other? Can someone crossmatch those templates and track my activities?* We should not only think about protecting the stored references in order to make infeasible the inverse biometrics process. With the widespread use of biometrics in many everyday tasks, a particular subject will probably enrol in different applications, such as healthcare or online banking, with the same biometric instance (e.g., my right index finger). The right of privacy preservation also entails the right not to be tracked among those applications. If the answer to those questions is yes, we are facing an additional privacy issue: an eventual attacker who gets access to several templates enrolled in different systems could combine that information and further exploit it to gain knowledge of how many bank accounts we have or infer patterns in our regular activity. Therefore, *crossmatching* between templates used in different applications should be prevented.

• *Finally, what if someone steals a template extracted from my right index finger? Won't I be able to use that finger again to enrol into the system? Has it been permanently compromised?* Since biometric characteristics cannot be replaced, we should be able to generate multiple templates form a single biometric instance in order to discard and replace compromised templates. Furthermore, those templates should not be related to one another, in the sense that they should not be positively matched by the biometric system, to prevent the impersonation of a subject with a stolen template. Consequently, *renewability* of biometric templates is also desired. It should be noted that both crossmatching and renewability can be addressed at the same time if full *un-linkability* between templates belonging to the same subject is granted.

The relevance of these concerns and the efforts being directed to solve them within the biometric community are highlighted by some recent special issues in journals, such as the IEEE Signal Processing Magazine Special Issue on Biometrics Security and Privacy Protection [99], the development of international standards on biometric information protection, such as the ISO/IEC IS 24745 [100], specific tracks on biometric security [101,102] or privacy-enhancing technologies [103,104] at international conferences, recent publications [3,105–107] and PhD thesis [108–111] or the EU FP7 projects TURBINE on trusted revocable biometrics identities[7] and PIDaaS on private identification as a service[8]. However, it is only since very recently that those concerns have been raised, and new methods have been proposed to tackle them.

In order to address those privacy concerns, biometric data is considered *sensitive personal data* by the European Union General Data Protection Regulation [112]: biometrics is an intrinsic part of the human body and/or behaviour, which we cannot discard in case of theft. Within this directive, *personal data* is defined as *any information relating to an identified or identifiable natural person (data subject); an identifiable natural person is one who can be identified, directly or indirectly, in particular by reference to an identifier such as a name, an identification number, location data, an online identifier or to one or more factors specific to the physical, physiological, genetic, mental, economic, cultural or social identity of that natural person*. This means that *processing* of biometric data is subject to right of *privacy preservation*, where the notion of *processing* means *any operation or set of operations which is performed on personal data or on sets of personal data, whether or not by automated means, such as collection, recording, organisation, structuring, storage, adaptation or alteration, retrieval, consultation, use, disclosure by transmission, dissemination or otherwise making available, alignment or combination, restriction, erasure or destruction*.

Those definitions imply that, in order to grant the subject's privacy, biometric information should be carefully protected both in its stored form (i.e. biometric templates or references), and any time it is used for verification purposes. Therefore, with the main goal of developing secure and privacy-preserving biometric technologies, new standardisation efforts are being currently directed to prevent such information leakages. In particular, the ISO/IEC IS 24745 on biometric information protection [100] encourages the substitution of traditional biometric systems with biometric template protection schemes. In these systems, unprotected templates are neither stored in the database nor compared for verification purposes. They are substituted by protected templates so that, in case of leakage, those references disclose no biometric information about the subjects, hence protecting the privacy of the subjects. To that end, protected templates should comply with the two major requirements of *irreversibility* and *un-linkability*:

* *Irreversibility*: in order to overcome the first privacy issue described in the previous section (i.e. amount of biometric information which is leaked by the template), we require that knowledge of a protected template and corresponding auxiliary data cannot be exploited to reconstruct a biometric signal which positively matches the original biometric sample.
* *Un-linkability*: in order to overcome the second and third aforementioned drawbacks of unprotected verification systems (i.e. biometric characteristics should

---

[7]http://www.turbine-project.eu/
[8]http://www.pidaas.eu/

not be matched across systems, and they should be replaceable), given a single biometric sample, it must be feasible to generate different versions of protected templates, so that those templates cannot be linked to a single subject. This property guarantees the privacy of a subject when he/she is registered in different applications with the same biometric instance (prevents crossmatching) and also allows issuing new credentials in case a protected template is stolen.

Only fulfilling those two requirements is the privacy of the subject fully preserved. As we have seen in Section 6.3, inverse biometrics methods pose a threat to both requirements, when dealing with unprotected biometric systems. On the one hand, it has been shown for a number of characteristics, that traditional biometric templates are not irreversible. That is, we can generate a synthetic sample which matches the stored reference template. On the other hand, using the synthetic samples generated from the information stored in the templates, linking two given templates to a single or different identities, is straightforward. The need for further development of biometric template protection schemes, which fulfil the aforementioned requirements, is thus highlighted. If any of those requirements is not met, the subject's privacy cannot be guaranteed.

Nevertheless, protecting the information stored or handled by the system is not enough. Even if no information can be derived from the stored references, an eventual attacker can still create an artefact and present it to a biometric system, trying for instance to be identified as any of the identities enrolled in the system. Hence, in order to prevent further attacks with synthetic samples or artefacts, additional measures should be implemented to discriminate between real and synthetic samples. Over the last decade, serious efforts have been directed to this problem, known as *presentation attack detection* (PAD), anti-spoofing or liveness detection, for fingerprint [113], face [114] or iris [115]. In addition, an ISO/IEC international standard has been recently published on this topic [116], being the second and third parts of it under development.

6.6 Conclusions

The present chapter has shown the feasibility of recovering or reconstructing synthetic biometric samples from the information stored in unprotected reference templates. The experimental findings in most of the works described in Section 6.3 show that an attack on different biometric systems based on a wide variety of characteristics using such reconstructed images would have a very high chance of success. In addition, depending on the inverse biometrics algorithm, not only one but several synthetic samples, visually different from each other, can be generated, all matching the stored reference template. Success chances of impersonating a particular subject are hence increased.

It may be argued that the reconstruction approaches considered in this chapter can be successful only when the template stored in the database is compromised. Even if it may be difficult, it is still possible in classical biometric systems where the enrolled templates are kept in a centralised database. In this case, the attacker would have to access the database and extract the information or intercept the communication channel when the stored template is released for matching. However, the threat is increased in match-on-card (MoC) applications where an individual's biometric template is stored in a smartcard that the subject has to carry with him in order to access the system. Such applications are rapidly growing due to several appealing characteristics such as scalability and privacy [117]. This makes MoC systems potentially more vulnerable to the type reconstruction algorithms described in this chapter especially when the biometric data is stored without any type of encryption [118] or printed in the clear on plastic cards as 2D barcodes [119].

Similarly, biometric data is being stored in many official documents such as the biometric passport [120], some national ID cards[9], the US FIPS-201 Personal Identity Verification initiatives [118] or the ILO Seafarers Identity Card Program [119]. In spite of the clear advantages that these type of applications offer, templates are more likely to be compromised as it is easier for the attacker to have physical access to the storage device and, as has already been demonstrated [121], fraudulently obtain the information contained inside.

On the other hand, even if access to centralised databases is theoretically more difficult to obtain, an eventual attacker would be able to compromise not one but numerous biometric templates. Therefore, given the high SRs achieved by those algorithms, the need for including template protection schemes [122,123] in commercial biometric systems has been reinforced. In fact, the need to only store protected data, and not traditional unprotected biometric template, has been emphasised by some big database leakages which have made it to the news in the last few years. For instance, in 2013, thousands of Argentinians had their privacy rights violated when the country's electoral registration roll, which had been made available online, experienced a major leak of personal data following the presidential election[10]. In 2015, fingerprints belong to 5.6 million US federal employees were stolen[11]. And more recently, in 2016, in a government data breach, which affected 55 million voters in the Philippines, fingerprint records were disclosed[12].

At the same time, a PAD strategy [113–116] should be adopted to confirm if the biometric samples presented to the system are those of a genuine subject and not that of a digital or physical artefact. Both aforementioned measures complement each other and should be both implemented into biometric verification systems.

From a security and privacy perspective, we believe that these examples may serve as a wake-up call for vendors and developers to be aware of the potential risks of not securing biometric templates, as is the case in some operational systems already installed in sensitive areas. Even if the research community has dedicated serious efforts to the proposal and development of new biometric template protection schemes [122,123], some commercial schemes such as BioHASH[13] are being distributed, and

[9]http://www.dnielectronico.es/
[10]https://www.privacyinternational.org/node/342
[11]https://www.wired.com/2015/09/opm-now-admits-5-6m-feds-fingerprints-stolen-hackers/
[12]http://www.wired.co.uk/article/philippines-data-breach-fingerprint-data
[13]http://www.genkey.com/en/technology/biohashr-sdk

some standardisation efforts are currently being made [106,124], third-party stan-dardised evaluation of the revocable methods is still needed. To that end, two main approaches may be adopted: *security through obscurity* or *security through trans-parency* (also known as *security by design*). The security through transparency scheme follows Kerckhoffs' principle [125]: a cryptosystem should be secure even if every-thing about the system, except the key, is public knowledge. This principle can be applied to any security-related technology, in particular, biometrics: in words of the Biometric Working Group [126], in the context of biometric recognition, applying security through transparency would mean to *make public exposure of countermea-sures and vulnerabilities which will lead to a more mature and responsible attitude from the biometrics community and promote the development of more secure systems in the future* [127]. Such principle should be followed in order to thoroughly evaluate all requirements and privacy-related aspects not only of traditional biometric systems but also of biometric template protection schemes [111,128].

References

[1]   J. Bustard, "The impact of EU privacy legislation on biometric system deployment: Protecting citizens but constraining applications," *IEEE Signal Processing Magazine*, vol. 32, no. 5, pp. 101–108, 2015.

[2]   M. Tistarelli, S. Z. Li, and R. Chellappa, Eds., *Handbook of Remote Biometrics for Surveillance and Security*. London: Springer, 2009.

[3]   M. Barni, G. Droandi, and R. Lazzeretti, "Privacy protection in biometric-based recognition systems: A marriage between cryptography and signal processing," *IEEE Signal Processing Magazine*, vol. 32, no. 5, pp. 66–76, 2015.

[4]   International Biometric Group, "Generating images from templates," *White Paper*, 2002.

[5]   M. Bromba, "On the reconstruction of biometric raw data from template data," 2003. [Online]. Available: http://www.bromba.com/knowhow/temppriv.htm.

[6]   A. Adler, "Images can be regenerated from quantized biometric match score data," in *Proc. Canadian Conf. on Electrical and Computer Engineering (CCECE)*, 2004, pp. 469–472.

[7]   S. Venugopalan and M. Savvides, "How to generate spoofed irises from an iris code template," *IEEE Transactions on Information Forensics and Security*, vol. 6, no. 2, pp. 385–395, June 2011.

[8]   J. Galbally, A. Ross, M. Gomez-Barrero, J. Fierrez, and J. Ortega-Garcia, "Iris image reconstruction from binary templates: An efficient probabilis-tic approach based on genetic algorithms," *Computer Vision and Image Understanding*, vol. 117, no. 10, pp. 1512–1525, 2013.

[9]   J. Galbally, C. McCool, J. Fierrez, and S. Marcel, "On the vulnerability of face verification systems to hill-climbing attacks," *Pattern Recognition*, vol. 43, pp. 1027–1038, 2010.

[10] N. Y. Svetlana, P. S. P. Wang, M. L. Gavrilova, and S. N. Srihari, *Image Pattern Recognition: Synthesis and Analysis in Biometrics (Series in Machine*

Perception & Artificial Intelligence). Imperial College Press, London, UK, 2006.

[11]   S. N. Yanushkevich, A. Stoica, V. P. Schmerko, and D. V. Popel, *Biometric Inverse Problems.* Boca Raton, FL: Taylor & Francis, 2005.

[12]   K. Waters, "A muscle model for animation three-dimensional facial expression," ACM SIGGRAPH Computer Graphics, vol. 21, no. 4, pp. 17–24, 1987.

[13]   D. P. Greenberg, K. E. Torrance, P. Shirley, *et al.,* "A framework for realistic image synthesis," in *Proc. Conf. on Computer Graphics and Interactive Techniques (SIGGRAPH).* ACM Press/Addison-Wesley Publishing Co., 1997, pp. 477–494.

[14]   N. M. Orlans, D. J. Buettner, and J. Marques, "A survey of synthetic biometrics: Capabilities and benefits," in *Proc. Int. Conf. on Artificial Intelligence (ICAI),* 2004, pp. 499–505.

[15]   *SIGGRAPH'74 Proceedings of the 1st Annual Conference on Computer Graphics and Interactive Techniques,* Boulder, CO, July 12–14, 1974.

[16]   A. Wehde and J. N. Beffel, *Finger-Prints Can be Forged.* Michigan University: The Tremonia Publishing Co., 1924.

[17]   A. S. Osborn, *Questioned Documents.* Boyd Printing Co., Albany, NY, 1929.

[18]   J. Cui, Y. Wang, J. Huang, T. Tan, and Z. Sun, "An iris image synthesis method based on PCA and super-resolution," in *Proc. IAPR Int. Conf. on Pattern Recognition (ICPR),* 2004, pp. 471–474.

[19]   A. Adler, "Sample images can be independently restored from face recognition templates," in *Proc. Canadian Conf. on Electrical and Computer Engineering (CCECE),* vol. 2, 2003, pp. 1163–1166.

[20]   J. Fierrez and J. Ortega-Garcia, *On-line Signature Verification.* ch. Handbook of biometrics, Boston, MA: Springer, 2008, pp. 189–209.

[21]   N. V. Boulgouris, D. Hatzinakos, and K. N. Plataniotis, "Gait recognition: A challenging signal processing technology for biometric identification," *IEEE Signal Processing Magazine,* vol. 22, no. 6, pp. 78–90, 2005.

[22]   J. Galbally, J. Fierrez, M. Martinez-Diaz, and J. Ortega-Garcia, "Improving the enrollment in dynamic signature verification with synthetic samples," in *Proc. IAPR Int. Conf. on Document Analysis and Recognition (ICDAR),* 2009.

[23]   J. Galbally, M. Diaz-Cabrera, M. A. Ferrer, M. Gomez-Barrero, A. Morales, and J. Fierrez, "On-line signature recognition through the combination of real dynamic data and synthetically generated static data," *Pattern Recognition,* vol. 48, pp. 2921–2934, 2015.

[24]   R. Cappelli, "Use of synthetic data for evaluating the quality of minutia extraction algorithms," in *Proc. 2nd NIST Biometric Quality Workshop,* 2007.

[25]   J. Marques, N. M. Orlans, and A. T. Piszcz, "Effects of eye position on eigenface-based face recognition scoring," vol. 8, pp. 1–7, 2000.

[26]   K. Sumi, C. Liu, and T. Matsuyama, "Study on synthetic face database for performance evaluation," in *Proc. Int. Conf. on Biometrics (ICB),* 2006, pp. 598–604.

[27] A. Othman and A. Ross, "On mixing fingerprints," *IEEE Transactions on Information Forensics and Security*, vol. 8, no. 1, pp. 260–267, 2013.

[28] F. Alegre, R. Vipperla, N. Evans, and B. Fauve, "On the vulnerability of automatic speaker recognition to spoofing attacks with artificial signals," in *Proc. European Signal Processing Conference (EUSIPCO)*, 2012, pp. 36–40.

[29] M. Gomez-Barrero, J. Galbally, A. Morales, M. A. Ferrer, J. Fierrez, and J. Ortega-Garcia, "A novel hand reconstruction approach and its application to vulnerability assessment," *Information Sciences*, vol. 268, pp. 103–121, 2014.

[30] D. V. Popel, *Synthesis and Analysis in Biometrics*. ch. Signature analysis, verification and synthesis inpervasive environments. Hackensack, NJ: World Scientific, 2007 pp. 31–63.

[31] K. P. Hollingsworth, K. W. Bowyer, and P. J. Flynn, "The best bits in an iris code," *IEEE Transactions on Pattern Analysis and Machine Intelligence*, vol. 31, no. 6, pp. 964–973, 2009.

[32] R. M. Bolle, S. Pankanti, J. H. Connell, and N. K. Ratha, "Iris individuality: A partial iris model," in *Proc. Int. Conf. on Pattern Recognition (ICPR)*, vol. 2, 2004, pp. 927–930.

[33] J. Schroeter, Basic Principles of Speech Synthesis, In *Handbook of Speech Processing*, J. Benesty, M. Mohan Sondhi, Yiteng Arden Huang, eds. Springer Berlin Heidelberg, 2008, pp. 413–428.

[34] J. Olive, "Rule synthesis of speech from dyadic units," in *Proc. IEEE Int. Conf. on Acoustics, Speech, and Signal Processing (ICASSP)*, vol. 2, 1977, pp. 568–570.

[35] P. R. Cook, *Real Sound Synthesis for Interactive Applications*. Boca Raton, FL: CRC Press, 2002.

[36] T. Dutoit, *An Introduction to Text-to-Speech Synthesis*. Springer Netherlands, Kluwer Academic Publishers, 2001.

[37] S. Dieleman, H. Zen, K. Simonyan, *et al.*, "Wavenet: A generative model for raw audio," 2016. Accessed at https://arxiv.org/abs/1609.03499.

[38] R. Cappelli, *Handbook of Fingerprint Recognition*. ch. Synthetic Fingerprint Generation, London: Springer, 2003, pp. 203–231.

[39] S. Makthal and A. Ross, "Synthesis of iris images using Markov random fields," in *Proc. European Signal Processing Conference (EUSIPCO)*, 2005, pp. 1–4.

[40] S. Shah and A. Ross, "Generating synthetic irises by feature agglomeration," in *Proc. IEEE Int. Conf. on Image Processing (ICIP)*, 2006, pp. 317–320.

[41] J. Zuo, N. A. Schmid, and X. Chen, "On generation and analysis of synthetic iris images," *IEEE Transactions on Information Forensics and Security*, vol. 2, pp. 77–90, 2007.

[42] Z. Wei, T. Tan, and Z. Sun, "Synthesis of large realistic iris databases using patch-based sampling," in *Proc. IAPR Int. Conf. of Pattern Recognition (ICPR)*, 2008, pp. 1–4.

[43]   L. Cardoso, A. Barbosa, F. Silva, A. M. G. Pinheiro, and H. Proença, "Iris biometrics: Synthesis of degraded ocular images," *IEEE Transactions on Information Forensics and Security*, vol. 8, no. 7, pp. 1115–1125, 2013.

[44]   A. Lin and L. Wang, "Style-preserving english handwriting synthesis," *Pattern Recognition*, vol. 40, pp. 2097–2109, 2007.

[45]   V. Blanz and T. Vetter, "A morphable model for the synthesis of 3D faces," in *Proc. Conf. on Computer Graphics and Interactive Techniques (SIGGRAPH)*, 1999, pp. 187–194.

[46]   N. Poh, S. Marcel, and S. Bengio, "Improving face authentication using virtual samples," in *Proc. IEEE Int. Conf. on Acoustics, Speech and Signal Processing (ICASSP)*, 2003.

[47]   Y. Du and X. Lin, "Realistic mouth synthesis based on shape appearance dependence mapping," *Pattern Recognition Letters*, vol. 23, pp. 1875–1885, 2002.

[48]   E. Yamamoto, S. Nakamura, and K. Shikano, "Lip movement synthesis from speech based on hidden Markov models," *Speech Communication*, vol. 26, no. 1, pp. 105–115, 1998.

[49]   J. Galbally, R. Plamondon, J. Fierrez, and J. Ortega-Garcia, "Synthetic on-line signature generation. Part I: Methodology and algorithms," *Pattern Recognition*, vol. 45, pp. 2610–2621, 2012.

[50]   J. Galbally, J. Fierrez, J. Ortega-Garcia, and R. Plamondon, "Synthetic on-line signature generation. Part II: Experimental validation," *Pattern Recognition*, vol. 45, pp. 2622–2632, 2012.

[51]   M. A. Ferrer, M. Diaz-Cabrera, and A. Morales, "Static signature synthesis: A neuromotor inspired approach for biometrics," *IEEE Transactions on Pattern Analysis and Machine Intelligence*, vol. 37, no. 3, pp. 667–680, 2015.

[52]   A. Nazar, I. Traore, and A. A. E. Ahmed, "Inverse biometrics for mouse dynamics," *International Journal of Pattern Recognition and Artificial Intelligence*, vol. 22, no. 3, pp. 461–495, 2008.

[53]   A. Nazar, *Synthesis and Simulation of Mouse Dynamics: Using Inverse Biometrics to Generate Realistic Mouse Actions that Have Behavioral Properties*. Saarbrücken, Germany: VDM Verlag, 2008.

[54]   F. Rashid, "Inverse biometrics for keystroke dynamics," Master's thesis, 2009.

[55]   M. E. Munich and P. Perona, "Visual identification by signature tracking," *IEEE Transactions on Pattern Analysis and Machine Intelligence*, vol. 25, no. 2, pp. 200–217, 2003.

[56]   C. Oliveira, C. A. Kaestner, F. Bortolozzi, and R. Sabourin, "Generation of signatures by deformations," in *Proc. IAPR Int. Conf. on Advances in Document Image Analysis (ICADIA)*. Springer LNCS-1339, 1997, pp. 283–298.

[57]   M. Mori, A. Suzuki, A. Shio, and S. Ohtsuka, "Generating new samples from handwritten numerals based on point correspondence," in *Proc. IAPR Int. Workshop on Frontiers in Handwriting Recognition (IWFHR)*, 2000, pp. 281–290.

[58] J. Wang, C. Wu, Y.-Q. Xu, H.-Y. Shum, and L. Ji, "Learning-based cursive handwriting synthesis," in *Proc. IAPR Int. Workshop on Frontiers of Handwriting Recognition (IWFHR)*, 2002, pp. 157–162.

[59] H. Choi, S.-J. Cho, and J. Kim, "Generation of handwritten characters with Bayesian network based on-line handwriting recognizers," in *Proc. Int. Conf. on Document Analysis and Recognition (ICDAR)*, 2003, pp. 1–5.

[60] H. Wang and L. Zhang, "Linear generalization probe samples for face recognition," *Pattern Recognition Letters*, vol. 25, pp. 829–840, 2004.

[61] H. R. Wilson, G. Loffler, and F. Wilkinson, "Synthetic faces, face cubes, and the geometry of face space," *Vision Research*, vol. 42, no. 34, pp. 2909–2923, 2002.

[62] A. Black and N. Campbell, "Optimizing selection of units from speech database for concatenative synthesis," in *Proc. European Conf. on Speech Communication and Technology (EUROSPEECH)*, 1995, pp. 581–584.

[63] C. van Dam, R. Veldhuis, and L. Spreeuwers, "Face reconstruction from image sequences for forensic face comparison," *IET Biometrics*, vol. 5, no. 2, pp. 140–146, 2015.

[64] D. Maltoni, *Generation of Synthetic Fingerprint Image Databases*. New York, NY: Springer, 2004, pp. 361–384.

[65] M. Kücken and A. C. Newell, "A model for fingerprint formation," *Europhysics Letters*, vol. 68, no. 1, p. 141, 2004.

[66] D. H. Klatt, "Software for a cascade/parallel formant synthesizer," *Journal Acoustic Society of America*, vol. 67, pp. 971–995, 1980.

[67] N. B. Pinto, D. G. Childers, and A. L. Lalwani, "Formant speech synthesis: improving production quality," *IEEE Transactions on Acoustics, Speech, and Signal Processing*, vol. 37, pp. 1870–1887, 1989.

[68] P. L. De Leon, V. R. Apsingekar, M. Pucher, and J. Yamagishi, "Revisiting the security of speaker verification systems against imposture using synthetic speech," in *Proc. Int. Conf. on Acoustics, Speech, and Signal Processing (ICASSP)*, 2010, pp. 1798–1801.

[69] J. M. Hollerbach, "An oscillation theory of handwriting," *Biological Cybernetics*, vol. 39, no. 2, pp. 139–156, 1981.

[70] R. Plamondon and W. Guerfali, "The generation of handwriting with delta-lognormal synergies," *Biological Cybernetics*, vol. 78, no. 2, pp. 119–132, 1998.

[71] Y. Stylianou, "Voice transformation: a survey," in *Proc. IEEE Int. Conf. on Acoustics, Speech and Signal Processing (ICASSP)*, 2009, pp. 3585–3588.

[72] M. Abe, S. Nakamura, K. Shikano, and H. Kuwabara, "Voice conversion through vector quantization," in *Proc. IEEE Int. Conf. on Acoustics, Speech, and Signal Processing (ICASSP)*, 1988, pp. 655–658.

[73] H. Valbret, E. Moulines, and J.-P. Tubach, "Voice transformation using PSOLA technique," in *Proc. IEEE Int. Conf. on Acoustics, Speech, and Signal Processing (ICASSP)*, vol. 1, 1992, pp. 145–148.

[74]   T. Toda, L.-H. Chen, D. Saito, *et al.*, "The voice conversion challenge 2016," in *Proc. Int. Conf. on Biometrics: Theory, Applications, and Systems (BTAS)*, 2016, pp. 1632–1636.

[75]   T. Kinnunen, Z.-Z. Wu, K. A. Lee, F. Sedlak, E. S. Chng, and H. Li, "Vulnerability of speaker verification systems against voice conversion spoofing attacks: the case of telephone speech," in *Proc. Int. Conf. on Acoustics, Speech, and Signal Processing (ICASSP)*, 2012, pp. 4401–4404.

[76]   L.-H. Chen, Z.-H. Ling, L.-J. Liu, and L.-R. Dai, "Voice conversion using deep neural networks with layer-wise generative training," *IEEE/ACM Transactions on Audio, Speech, and Language Processing*, vol. 22, no. 12, pp. 1859–1872, 2014.

[77]   T. Toda, A. W. Black, and K. Tokuda, "Voice conversion based on maximum-likelihood estimation of spectral parameter trajectory," *IEEE Transactions on Audio, Speech, and Language Processing*, vol. 15, no. 8, pp. 2222–2235, 2007.

[78]   S. Desai, A. W. Black, B. Yegnanarayana, and K. Prahallad, "Spectral mapping using artificial neural networks for voice conversion," *IEEE Transactions on Audio, Speech, and Language Processing*, vol. 18, no. 5, pp. 954–964, 2010.

[79]   N. Xu, Y. Tang, J. Bao, A. Jiang, X. Liu, and Z. Yang, "Voice conversion based on Gaussian processes by coherent and asymmetric training with limited training data," *Speech Communication*, vol. 58, pp. 124–138, 2014.

[80]   M. Ferrara, A. Franco, and D. Maltoni, "The magic passport," in *Proc. IEEE Int. Joint Conf. on Biometrics (IJCB)*, 2014, pp. 1–7.

[81]   R. Ramachandra, K. B. Raja, and C. Busch, "Detecting morphed face images," in *Proc. Int. Conf. on Biometrics: Theory, Applications, and Systems (BTAS)*, 2016.

[82]   C. J. Hill, "Risk of masquerade arising from the storage of biometrics," Master's thesis, Australian National University, 2001.

[83]   R. Cappelli, D. Maio, A. Lumini, and D. Maltoni, "Fingerprint image reconstruction from standard templates," *IEEE Transactions on Pattern Analysis and Machine Intelligence*, vol. 29, pp. 1489–1503, September 2007.

[84]   K. Kümmel, C. Vielhauer, T. Scheidat, D. Franke, and J. Dittmann, "Hand-writing biometric hash attack: A genetic algorithm with user interaction for raw data reconstruction," in *Proc. IFIP Int. Conf. on Communications and Multimedia Security*. Linz, Austria: Springer, 2010, pp. 178–190.

[85]   K. Kümmel and C. Vielhauer, "Reverse-engineer methods on a biometric hash algorithm for dynamic handwriting," in *Proc. ACM Workshop on Multimedia and Security*, 2010, pp. 67–72.

[86]   J. Galbally, R. Cappelli, A. Lumini, *et al.*, "An evaluation of direct and indirect attacks using fake fingers generated from ISO templates," *Pattern Recognition Letters*, vol. 31, pp. 725–732, 2010.

[87]   A. Ross, J. Shah, and A. K. Jain, "From template to image: Reconstructing fingerprints from minutiae points," *IEEE Transactions on Pattern Analysis and Machine Intelligence*, vol. 29, pp. 544–560, 2007.

[88] C. Rathgeb and A. Uhl, "Attacking iris recognition: An efficient hill-climbing technique," in *Proc. Int. Conf. on Pattern Recognition (ICPR)*, 2010.

[89] P. Mohanty, S. Sarkar, and R. Kasturi, "From scores to face templates: A model-based approach," *IEEE Transactions on Pattern Analysis and Machine Intelligence*, vol. 29, pp. 2065–2078, 2007.

[90] M. Gomez-Barrero, J. Galbally, J. Fierrez, and J. Ortega-Garcia, "Face verification put to test: A hill-climbing attack based on the uphill-simplex algorithm," in *Proc. International Conference on Biometrics (ICB)*, 2012, pp. 40–45.

[91] ISO/IEC JTC1 SC 37 Biometrics, *ISO/IEC 19794-2:2011, Information Technology – Biometric Data Interchange Formats – Part 2: Finger Minutiae Data*, Geneva, Switzerland: International Organization for Standardization, 2011.

[92] J. A. Nelder and R. Mead, "A simplex method for function minimization," *Computer Journal*, vol. 7, pp. 308–313, 1965.

[93] J. Galbally, R. Cappelli, A. Lumini, D. Maltoni, and J. Fierrez, "Fake fingertip generation from a minutiae template," in *Proc. Intl. Conf. on Pattern Recognition (ICPR)*, December 2008.

[94] BioAPI Consortium, "BioAPI specification (version 1.1)," March 2001. Available: www.bioapi.org/Downloads/BioAPI.

[95] J. Daugman, "How iris recognition works," in *Proc. IEEE Int. Conf. on Image Processing (ICIP)*, 2002, pp. I.33–I.36.

[96] C. Vielhauer, *Biometric User Authentication for IT Security: From Fundamentals to Handwriting*. New York, NY: Springer Science & Business Media, 2006, vol. 18.

[97] ANSI-NIST, "ANSI x9.84-2001, Biometric Information Management and Security," 2001.

[98] European Association for Biometrics (EAB), *Proc. 5th Seminar on Biometrics in Banking and Payments*, London, 2016.

[99] N. Evans, S. Marcel, A. Ross, and A. B. J. Teoh, *IEEE Signal Processing Magazine, Special Issue on Biometric Security and Privacy*, Vol. 32, n. 5, 2015.

[100] ISO/IEC JTC1 SC27 IT Security Techniques, *ISO/IEC 24745:2011. Information Technology – Security Techniques – Biometric Information Protection*, Geneva, Switzerland: International Organization for Standardization, 2011.

[101] K. Bowyer, A. Ross, R. Beveridge, P. Flynn, and M. Pantic, Eds., *Proc. of 7th IEEE Int. Conf. on Biometrics: Theory, Advances and Systems (BTAS)*, 2015.

[102] F. Alonso-Fernandez and J. Bigun, Eds., *Proc. of the 8th International Conference on Biometrics (ICB)*, 2016.

[103] M. Locasto, V. Shmatikov, and U. Erlingsson, Eds., *Proc. of the 37th IEEE Symposium on Security and Privacy*, 2016.

[104] B. D. Decker, S. Katzenbeisser, and J.-H. Hoepman, Eds., *Proc. of the 31st International Information Security and Privacy Conference (IFIP SEC)*, 2016.

[105] K. Nandakumar and A. K. Jain, "Biometric template protection: Bridging the performance gap between theory and practice," *IEEE Signal Processing Magazine*, vol. 32, no. 5, pp. 88–100, 2015.

[106]   S. Rane, "Standardization of biometric template protection," *IEEE Multimedia*, vol. 21, no. 4, pp. 94–99, 2014.

[107]   M. Ferrara, D. Maltoni, and R. Cappelli, "A two-factor protection scheme for MCC fingerprint templates," in *Proc. Int. Conf. of the Biometrics Special Interest Group (BIOSIG)*, 2014, pp. 1–8.

[108]   Y. Sutcu, "Template security in biometric systems," Ph.D. dissertation, New York University, 2009.

[109]   A. Nagar, "Biometric template security," Ph.D. dissertation, Michigan State University, 2012.

[110]   X. Zhou, "Privacy and security assessment of biometric template protection," Ph.D. dissertation, Technische Universität, Darmstadt, 2012.

[111]   M. Gomez-Barrero, "Improving security and privacy in biometric systems," Ph.D. dissertation, Universidad Autonoma de Madrid (UAM), Spain, 2016.

[112]   European Parliament, "EU Regulation 2016/679 on the protection of natural persons with regard to the processing of personal data and on the free movement of such data (General Data Protection Regulation)," 2016. [Online]. Available: http://europa.eu/rapid/press-release_IP-15-6321_en.htm.

[113]   C. Sousedik and C. Busch, "Presentation attack detection methods for fingerprint recognition systems: A survey," *IET Biometrics*, vol. 3, no. 4, pp. 219–233, 2014.

[114]   J. Galbally, S. Marcel, and J. Fierrez, "Biometric antispoofing methods: A survey in face recognition," *IEEE Access*, vol. 2, pp. 1530–1552, 2014.

[115]   J. Galbally and M. Gomez-Barrero, "A review of iris anti-spoofing," in *Int. Workshop on Biometrics and Forensics (IWBF)*, 2016, pp. 1–6.

[116]   ISO/IEC JTC1 SC37 Biometrics, *ISO/IEC 30107:2016, Information Technology – Biometric Presentation Attack Detection*, International Organization for Standardisation, Geneva, Switzerland.

[117]   C. Bergman, *Advances in Biometrics: Sensors, Algorithms and Systems*. ch. Match-on-card for secure and scalable biometric authentication. London: Springer, 2008, pp. 407–422.

[118]   NIST, NIST Special Publication 800-76, "Biometric Data Specification for Personal Identity Verification," February 2005.

[119]   ILO, 2006, ILO SID-0002, "Finger Minutiae-Based Biometric Profile for Seafarers' Identity Documents," Int'l Labour Organization.

[120]   ICAO, ICAO Document 9303, Part 1, Volume 2: Machine Readable Passports – Specifications for Electronically Enabled Passports with Biometric Identification Capability, International Civil Aviation Organization, Montréal, Quebec, Canada, 2006. Published in separate English, Arabic, Chinese, French, Russian and Spanish editions by the INTERNATIONAL CIVIL AVIATION ORGANIZATION 999 Robert-Bourassa Boulevard, Montréal, Quebec, Canada H3C 5H7.

[121]   J. van Beek, "ePassports reloaded," in *Black Hat USA Briefings*, 2008.

[122]   C. Rathgeb and A. Uhl, "A survey on biometric cryptosystems and cancelable biometrics," *EURASIP Journal on Information Security*, vol. 2011, no. 3, 2011.

[123]   V. M. Patel, N. Ratha, and R. Chellappa, "Cancelable biometrics: A review," *IEEE Signal Processing Magazine*, vol. 32, no. 5, pp. 54–65, 2015.

[124]   ISO/IEC JTC1 SC37 Biometrics, *ISO/IEC WD 30136, Performance Testing of Biometric Template Protection Schemes*, Geneva, Switzerland: International Organization for Standardization, 2015.

[125]   A. Kerckhoffs, "La cryptographie militaire," *Journal des sciences militaires*, vol. 9, pp. 5–83, 1883. Available: http://www.petitcolas.net/fabien/kerckhoffs/#english.

[126]   BWG, "Communications-electronics security group (CESG) – biometric working group (BWG) (UK government)," 2009. Available: http://www.cesg.gov.uk/policy_technologies/biometrics/index.shtml.

[127]   BWG, "Biometric security concerns, v1.0," CESG, UK Government, Tech. Rep., 2003.

[128]   M. Gomez-Barrero, C. Rathgeb, J. Galbally, C. Busch, and J. Fierrez, "Unlinkable and irreversible biometric template protection based on bloom filters," *Information Sciences*, vol. 370–371, pp. 18–32, 2016.

Chapter 7

Double-layer secret-sharing system involving privacy preserving biometric authentication

*Quang Nhat Tran[1], Song Wang[2], Ruchong Ou[3]
and Jiankun Hu[1]*

Introduction

The advent of cloud computing has made sharing unlimited. People now store their data on the cloud and enjoy the convenience of immediate access with authentication ensured. This raises an imminent issue: privacy protection. How to protect the identity of a user when using the online services is becoming vitally important. The traditional methods are password and tokens. However, they cannot authenticate genuine users since authentication based on these methods is dependent on knowledge of the password and possession of the token. Biometrics such as fingerprint have been used effectively in addressing this issue [1]. From a biometric security point of view, in order to protect the privacy of a user's identity, the template stored in the database for matching should be protected. Little work has been done in regards to the application that both private biometrics templates and shared secret images are stored in the cloud environment. We believe that such application will have a great potential via following observations: (i) There are already many images in the cloud and the rate of the image storage in the cloud is increasing every day. This will provide a good platform to embed secret among these images. (ii) Biometrics can be used to enhance the strength of privacy. Yet, it needs to be protected as well when it is stored in the cloud. The steganography technology can provide a win–win solution. This chapter puts secret image sharing in a new angle of view by proposing a novel method in which the privacy of the user is protected by biometrics.

In combination with secret sharing, biometric and steganography can provide inter-protection for privacy. The topic of this chapter is to apply biometrics to steganography in secret sharing in order to protect the privacy of a user when sharing a secret image. This new approach is a double-layer secret-sharing system (DLSSS) as both steganography and biometric data are utilized to provide a two-layer security system.

[1]School of Engineering and Information Technology, University of New South Wales at ADFA, Australia
[2]College of Science, Health, and Engineering, School of Engineering and Mathematical Sciences, Department of Engineering, La Trobe University, Australia
[3]Alfred Hospital, Monash University, Australia

The chapter is organized as follows: Section 7.1 presents some concepts to be applied in the proposed system. Related works are reviewed in Section 7.2. In Section 7.3, we detail out the structure of the new system. Section 7.4 presents and discusses experimental results and Section 7.5 concludes the chapter.

7.1   Conceptual clarifications

7.1.1   Steganography

Steganography is defined as the science of hiding information in a transmission medium. The term is derived from the combination of two ancient Greek words 'stegos' and 'grafia', which mean 'covered' and 'writing', respectively. Unlike encryption, whose objective is to transform data into unreadable information, steganography aims at creating a medium with undetectability, robustness, and capacity. Undetectability refers to human visual system's (HVS's) inability to recognize hidden information in the medium. Robustness means the amount of modification that the medium can withstand before an adversary destroys hidden information. Lastly, capacity indicates the amount of data that the medium can hide [2].

Steganography has been used for a long time: Histaiacus shaved a slave's head and tattooed the secret message on it so that when the hair grew back, he could dispatch with the secret. On the other hand, some ancient Arabic manuscripts on secret writings are believed to have been 1,200 years old [3]. With the bloom of computer technology, steganography has been digitalized as well. It can be conceptually applied to various kinds of data object since the techniques and methods for each format are similar [4]. With digital images, steganography has a wide range of application, some of which are: Copyright Protection, Feature Tagging, and Secret Communications [5].

There is more and more research done on steganography as technology has been developing drastically thanks to its wide applicability. Copyright Protection applies steganography in terms of watermarking, which is considered to be a sub-category of steganography [6]. On the other hand, illegal activities also utilize steganography. It is believed that Al-Qaeda used steganography to transmit their messages over social media for the attack on September 11, 2001 [7,8]. Therefore, researchers have been working on new methods of steganography and advanced systems that can detect steganograms in order to not only improve the security of information but also prevent the world from potential threats.

Steganography methods have been developed to utilize the power of computer, categorized as: image and transform domain techniques [6]. Sometimes, these two domains are also referred to spatial and frequency.

7.1.2   (k, n)-Threshold secret sharing

Shamir *et al.* [9] developed a secret-sharing scheme method that separates and distributes the secret into n parts. In order to recover the secret, a threshold of k parts must be present [9]. This section describes Shamir's secret-sharing scheme and the theory behind it.

Shamir's method to share secret integer y uses a $(k-1)$-degree polynomial as a function of x:

$$F(x) = y + m_1 \times x + m_2 \times x^2 + \cdots + m_{k-1} \times x^{k-1} \tag{7.1}$$

Shares distributing process: The steps to generate n shares for sharing are detailed as follows:

Step 1: Choose an arbitrary number n, where n is a positive integer and $n \geq 2$

Step 2: Choose an arbitrary integer number k, where $k \leq n$

Step 3: Choose randomly $k-1$ integer values to be the values for coefficients $m_1, m_2, \ldots, m_{k-1}$

Step 4: Choose randomly n distinctive values of x_i, where $i = 1, 2, \ldots, n$

Step 5: For each value x_i, calculate the corresponding $F(x_i)$

Step 6: Distribute each pair $(x_i, F(x_i))$ as a share to a participant. The values of the coefficients m become insignificant after all the shares have been distributed as those values can be recovered along with the secret y to retrieve the secret in the recovery process.

Recovery process:

Step 1: Collect at least k shares from the n distributed shares to form a system of equations:

$$F(x_1) = y + m_1 \times x_1 + m_2 \times x_1^2 + \cdots + m_{k-1} \times x_1^{k-1} \tag{7.2}$$

$$F(x_2) = y + m_1 \times x_2 + m_2 \times x_2^2 + \cdots + m_{k-1} \times x_2^{k-1} \tag{7.3}$$

$$\cdots \tag{7.4}$$

$$F(x_k) = y + m_1 \times x_k + m_2 \times x_k^2 + \cdots + m_{k-1} \times x_k^{k-1} \tag{7.5}$$

Step 2: Solve the equation system from Step 1 to acquire coefficient ms using a polynomial interpolation technique, such as Lagrange interpolation as follows:

$$F(x) = F(x_1)\frac{(x-x_2)(x-x_3)\ldots(x-x_k)}{(x_1-x_2)(x_1-x_3)\ldots(x_1-x_k)}$$

$$+ F(x_2)\frac{(x-x_1)(x-x_3)\ldots(x-x_k)}{(x_2-x_1)(x_2-x_3)\ldots(x_2-x_k)} + \cdots$$

$$+ F(x_k)\frac{(x-x_1)(x-x_2)\ldots(x-x_{k-1})}{(x_k-x_1)(x_k-x_2)\ldots(x_k-x_{k-1})} \tag{7.6}$$

Step 3: Calculate the value of y, i.e. recover the secret.

We can draw from the equation system that if less than k values of x are collected, the equation is impossible to be solved. In other words, if less than k shares are gathered, the secret cannot be recovered.

A more detailed version of the scheme that includes proof is available in [9].

7.1.3   Biometric template protection

Biometric authentication is a leap in protecting a user's privacy. An individual's biometric data is associated with the identity of the user. However, if a biometric identifier is compromised, there exist two threats: First, the identifier is considered compromised and lost forever. Second, if the same identifier is used in multiple applications, those are exposed to be exploited as well. Therefore, Ratha *et al.* [10] proposed a method of template protection to prevent retrieving biometric template. Biometric template protection can be broadly classified into two categories: bio-cryptosystems and cancellable biometrics [11].

Bio-cryptosystems either secure the cryptographic key using biometric features or generate the cryptographic key directly from biometrics. The former gives rise to the fuzzy scheme, motivated by the fuzzy vault concept [12]. It creates a secret string from a set of biometric data and compares the secret strings when matching two sets of biometric data. As long as the two data sets' difference is less than a pre-defined threshold, the secret string can be reproduced. Biometric key generation uses the biometric data to derive a key. This key is stored as template so that when matching is performed, the query biometric data is used to generate a key. The actual matching is performed between the keys. Apart from access control and identity authentication, bio-cryptosystems have also been applied to network security [13].

The idea of cancellable biometrics is to use non-invertible transformation to convert the original biometric data into intentionally distorted information. If a cancellable template is compromised, the transformation parameters can be changed so that a new template is obtained to replace the compromised template. Biohashing [14,15] is an early technique on cancellable biometrics. This technique uses random information from the user to increase the entropy of the biometric template. Raw biometric data, in combination with the user's random information, is fed into a hashing function to generate a hashed template, which is then stored in the database.

In this book, there is a chapter dedicated to reviewing of biometric template protection schemes, namely 'Biometric template Protection: State-of-the-Art, Issues, and Challenges'.

7.2   Related work

This section reviews some of the related work in the fields of privacy preserving with biometrics and secret sharing with steganography.

7.2.1   Authentication with biometric template protection

Motivated by the idea that a local structure needs to possess the ability to tolerate non-linear distortion, Yang *et al.* [16] proposed a biometric template protection method by extracting minutiae features from a Delaunay quadrangle-based structure. Moreover, they used the topology code to not only quantize the features so that structural changes are mitigated but also to enhance the security of the template. In the end, matching is performed using PinSketch [17]. Due to the local stability of Delaunay triangles, Yang *et al.* [18] designed cancellable fingerprint templates based on Delaunay triangle-based structures.

Yang *et al.* in [19] utilized Voronoi neighbour structures (VNSs) to generate a bio-cryptosystem. Specifically, the VNSs were constructed from a fingerprint image, each of which contains a set of minutiae. The feature vector of each minutia is then used to generate a fixed-length bit string for each VNS. These strings are then used to form a secret key that is used for matching. The key is protected by PinSketch to ensure the security of the template.

Wang *et al.* [20–23] focused on the study of non-invertible transformations in the design of cancellable biometrics. A common feature of the cancellable templates generated by these methods is that the binary string obtained from the pair-minutiae vectors is protected by a non-invertible transformation, either through infinite-to-one mapping or the application of digital signal-processing techniques. The resultant cancellable templates meet the requirements of revocability, diversity, non-invertibility and performance.

There has been considerable research in secret sharing using secure biometric authentication, e.g. [24,25]. When it comes to using steganography and watermarking techniques, there have been various works that aim at protecting biometric data. In 2003, Jain *et al.* [26] proposed to use steganography to hide minutiae data in casual data that is not related to fingerprint. In 2011, Ross and Othman [27] proposed to protect the raw biometric data by decomposing the biometric image into two images. The raw image can only be recovered given that both sub-images are available. These sub-images are stored in different databases. Recently, Whitelam *et al.* [28] proposed a framework that has multiple layers to ensure the security of multimodal biometric data. Specifically, the features from a face image are hidden into fingerprint images using a watermarking technique. Afterward, steganography is applied to embed the fingerprint images in casual images. In both phases, Whitelam used a key to determine the location into which the features were going to be embedded.

7.2.2 *Secret sharing with steganography*

Different approaches have been conducted in secret image sharing. But the most commonly used technique is the (k, n)-threshold secret sharing. Lin and Tsai [29] proposed a secret image-sharing scheme using the (k, n)-threshold secret sharing. The secret image's pixel values are embedded into n camouflage images along with a parity signal bit to be used for authentication. This scheme is simple to implement and is capable of utilizing the power of steganography in combination with authentication. However, the authentication feature in the system might be compromised easily in case of having dishonest participants and this scheme is still dependent on the transmission of a secret key, which is exploitable. Yang proposed an improved version of this scheme by enhancing the authentication feature for each stego-image [30]. However, the newer version has not been able to solve the cryptographic-dependent problem. In 2010, Lin and Chan [31] proposed an invertible image sharing scheme that is capable of losslessly recovering the secret and the stego-images.

7.3   Double-layer secret-sharing system

In this section, we propose a new secret scheme whose power lies in the combination of biometric-based authentication with steganography. At first, the fingerprint images are transformed using a cancellable template method to protect the privacy of the fingerprint data. Here, we used the method of Delaunay triangle-based local structure proposed by Yang *et al.* [18]. The secret image to be shared is then embedded in the transformed fingerprint images of authenticated users to create bio-stego-images. These images are embedded in camouflage images and stored in the cloud with dummy images. This scheme involves the use of a trusted third party (TTP). We assume that this TTP is an organization such as a bank where the communication channel between the bank and the participants are secure.

Let us assume that Alice uses DLSSS to share an image with Bob. She creates an account and registers with the TTP by submitting her biometric data and a set of images, including the one she wants to share with Bob. Bob also registers by submitting his biometric data.

Alice sends a request via a traditional secure channel to the TTP. The request consists of Bob's ID and the image's ID. TTP asks Alice to re-scan her biometric in order to authenticate the request. Alice replies with the newly scanned version of biometric data. TTP confirms and retrieves the image as well as Bob's biometric data to embed the secret image in and produces bio-stego-images. Each of the bio-stego-images is embedded into some camouflage images to produce stego-images. These images are stored in the cloud with other images.

TTP sends to Bob a notification that Alice has shared with him an image. Bob logs into the system and sends a request to download the stego-images. TTP requires Bob to re-scan his biometric data to verify his identification. Bob re-scans his biometric. TTP performs a matching between his newly scanned biometric data and the version in bio-stego-images, which has already been recovered by TTP when Bob successfully logs into the system. It confirms Bob's identification and recovers the secret image for Bob.

In the scheme, two users do not make any direct communication though possess a secure channel to utilize the power of steganography. More details are presented in Section 7.3.3.

Figure 7.1 shows the registration step for the system. Figures 7.2 and 7.3 present the embedding and recovery process of the scheme, respectively.

The proposed secret-sharing system has double-layer protection: layer one: generation of bio-stego-images and layer two: generation of camouflage-stego-images. The detail of this scheme is described in the remaining part of this section.

7.3.1   Embedding process

Any secret-sharing scheme that employs steganographic method can be applied to this double-layer scheme. In this chapter, we tested the system with Lin and Tsai's

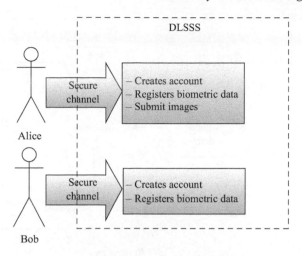

Figure 7.1 Registration

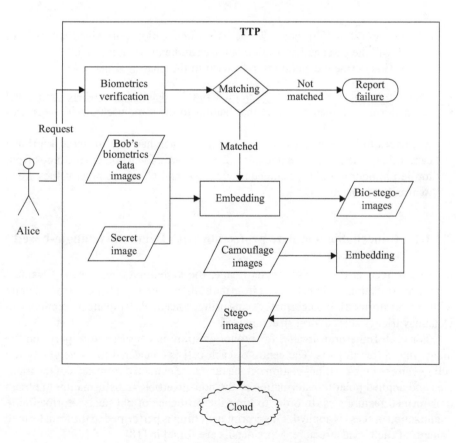

Figure 7.2 Embedding process

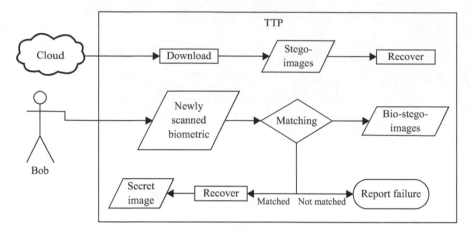

Figure 7.3   Recovery

embedding algorithm [29] but left out its authentication part since we are not concerned with the pixel authentication but user authentication instead.

Let us first define the notations to be used in this sharing scheme:

- Secret image S whose size is $m \times m$. Each pixel is denoted as s_i where $i = [1, m^2]$
- n transformed biometric images that belong to the same user. Each has size of $2m \times 2m$
- n^2 images to be used as camouflage images. Each has size of $4m \times 4m$. Each camouflage image is segmented into n 2×2 blocks. Each pixel in the block from top-left to bottom-right is denoted as X, W, V and U, whose pixel values are x, w, v and u, respectively.

7.3.1.1   Cancellable template protection with Delaunay triangle-based local structure

In order to recover the original secret image, the system requires a newly scanned transformed version of biometric data from the authenticated user to perform matching with the transformed bio-stego-images data using cancellable template protection with Delaunay triangle-based local structure [18].

For each fingerprint image, A Voronoi diagram is constructed to partition the image into different parts. The centre of each cell is connected to its neighbouring cell's centre to create a triangulation net. Features are extracted from each of the triangles and applied polar transformation as the non-invertible transformation to obtain transformed feature sets. In order to limit the influence of physical deformation, a quantization process is applied. Fingerprint matching is performed in the transformed domain of each local structure. More details are found in [18].

X_i'	W_i'
$x_i' = (x_{i1}, x_{i2}, \ldots, x_{i7}, x_{i8})$	$w_i' = (w_{i1}, w_{i2}, \ldots, w_{i6}, F_{i1}, F_{i2})$
V_i'	U_i'
$v_i' = (v_{i1}, v_{i2}, \ldots, v_{i5}, F_{i3}, F_{i4}, F_{i5})$	$u_i' = (u_{i1}, u_{i2}, \ldots, u_{i5}, F_{i6}, F_{i7}, F_{i8})$

Figure 7.4   4-pixel block after embedding

7.3.1.2   Layer one: bio-stego-images

Algorithm 1: *Basic block steganography*

Input: Secret image pixel value s;
        n distinct 2×2 camouflage image blocks;
Output: n manipulated camouflage image blocks with n shares and signal bits embedded

Step 1: Take the value x_i of pixel X_i as the value of x in the following polynomial (Figure 7.4).

$$F(x) = (y + m_1 \times x + m_2 \times x^2 + \cdots + m_{k-1} \times x^{k-1}) \bmod 251 \qquad (7.7)$$

Step 2: Take s to be y in (7.7).
Step 3: Choose arbitrarily $k - 1$ integers as m_i's values in (7.7)
Step 4: Calculate $F(x_i)$ for each x_i
Step 5: Hide the bits of $F(x_i)$ in W_i, U_i and V_i with the following LSB. The resultant block is illustrated in Figure 7.4 technique:

- Partition the eight bits of $F(x_i)$ into three parts, each of which consists of 2, 3 and 3 bits, respectively.
- Replace the last 2 bits of W_i with the first 2 bits of $F(x_i)$. The new byte is W_i' with value w_i'
- Replace the last 3 bits of V_i with the next 3 bits of $F(x_i)$. The new byte is V_i' with value v_i'
- Replace the last 3 bits of U_i with the last 3 bits of $F(x_i)$. The new byte is U_i' with value u_i'

Notes:

- The number of bits in $F(x_i)$ might outnumber the capacity of the three pixels W_i, V_i and U_i. Therefore, in order to limit the magnitude of $F(x_i)$, a modular operation *mod q* where $q = 255$ is performed. However, here comes another issue: the result of the modular operation *mod q* might not be unique [9] unless q is a prime. On the other hand, the image pixel's range of value is $[0, 255]$. Therefore, setting $q = 251$ which is the greatest prime number in the range satisfies this problem.
- In order to perform the modular operation with $q = 251$, all the values of x_i, y and the coefficients m_i must fall into the range $[0, 250]$. This means that pixels that

possess grey-value greater than 250 must be truncated to 250 in a pre-processing procedure. This might cause little distortion in the stego-image comparing with the original one. However, the HVS is not sensitive enough to recognize the difference. This truncation may allow a design of stego-detector that scans the abnormally in pixel values of the transmitting image. However, the hacker will have to overcome another layer of biometric authentication in order to recover the secret image.

Assume that the secret image is an $m \times m$ grey-scale image. There is a database of camouflage images, each of which has the size of $4m \times 4m$. In order to ensure the steganography, the camouflage images should be commonly seen pictures.

As we expand Algorithm 1 to the whole image, the complete algorithm is described as follows:

Algorithm 2: *Detailed process for secret image sharing*

Input: An $m \times m$ secret image S to be shared by n participants;
        A database of more than n $2m \times 2m$ camouflage images
Output: n stego-images in which the secret image pixels and signal bits are embedded.

Step 1: Select n distinct images $I_1, I_2, \ldots, I_n$ from the camouflage database.
Step 2: Divide the secret image into m^2 individual pixels S_i where $i = 1, 2, \ldots, m^2$.
Step 3: Divide each of the camouflage images $I_j (j = 1, 2, 3, \ldots, n)$ into m^2 blocks $B_j i$, each is a 2×2 block. Denote the four pixels in each block as: X_{ji}, W_{ji}, V_{ji}, and U_{ji}.
Step 4: For each secret image pixel S_i, implement Algorithm 1, we have a sequence of 2×2 stego-image blocks $B'_{1i}, B'_{2i}, \ldots, B'_{ni}$.
Step 5: Compose the stego-images by grouping together 2×2 blocks $B'_{1i}, B'_{2i}, \ldots, B'_{ni}$. In the end, n $2m \times 2m$ stego-images are distributed to n secret keepers.

We can see that Algorithm 2 extends Algorithm 1 by applying Algorithm 1 to all pixels of the secret image and all blocks of the camouflage image.

After phase one, we have a set of bio-stego-images.

7.3.1.3   Camouflage-stego-images

Layer two applies the same algorithm as layer one on each of the bio-stego-images to produce a set of n^2 camouflage-stego-images.

7.3.2   Recovery process

The recovery process with authentication feature is divided in two steps: Step One recovers the bio-stego-images; step Two recovers the secret image.

Algorithm 3: Bio-stego-images recovery

Input: Set of n^2 camouflage-stego-images
   User log-in session
Output: n bio-stego-images or report of failure in authentication.

Step 1: Divide each of the collected stego-image I'_j into m_2 2×2 blocks B'_{ji} where $(i = 1, 2, \ldots, m^2)$. Each of the four pixels in the block is denoted from top, left to right: $X'_{ji}, W'_{ji}, V'_{ji}, U'_{ji}$.
Step 2: For each bio-stego pixel image, the following steps are taken:

- For each $j = 1, 2, \ldots, k$, take value x'_{ji} of each block B'_{ji} as x and reconstruct $F(x_{ji})$ from two bits embedded in W'_{ji}, three bits embedded in U'_{ji}, and three bits embedded in V'_{ji} to use as $F(x)$ in the polynomial:

$$y = (-1)^{k-1}\left[F(x_1)\frac{x_2 x_3 \ldots x_k}{(x_1 - x_2)(x_1 - x_3)\ldots(x_1 - x_k)}\right.$$
$$+ F(x_2)\frac{x_1 x_3 \ldots x_k}{(x_2 - x_1)(x_2 - x_3)\ldots(x_2 - x_k)} + \cdots$$
$$\left. + F(x_k)\frac{x_1 x_2 \ldots x_{k-1}}{(x_k - x_1)(x_k - x_2)\ldots(x_k - x_{k-1})}\right] \bmod 251 \qquad (7.8)$$

- Compute y to get the pixel value of bio-stego-image.

Step 3: Compose all the pixels to get the secret bio-stego-image.

After being retrieved, the bio-stego-images are used as part of input to recover the original secret image.

7.3.2.1 Recover secret image
After matching the new version of fingerprint data image and the bio-stego-images, the process of recovering the secret image is detailed as follows:

Input: n^2 bio-stego-images $P_1, P_2, \ldots, P_n$
   Newly scanned version of biometric images from user
Output: The secret image S or report of failure in recovering.

Step 1: Perform matching between the bio-stego-images and the newly scanned version of biometric data. If the number of matched pairs is more than k, perform the next step. If not, stop the algorithm and report failure.
Step 2: Divide each of the collected bio-stego-image into m^2 2×2 blocks B'_{ji} $(i = 1, 2, \ldots, m^2)$. Each of the four pixels in the block is denoted from top, left to right: $X'_{ji}, W'_{ji}, V'_{ji}, U'_{ji}$.

Step 3: For each secret pixel image S_i, where $i = 1, 2, \ldots, m^2$, the following steps are taken:

- For each $j = 1, 2, \ldots, k$, take value x'_{ji} of each block B'_{ji} as x and reconstruct $F(x_{ji})$ from two bits embedded in W'_{ji}, three bits embedded in U'_{ji}, and three bits embedded in V'_{ji} to use as $F(x)$ in the polynomial:

$$y = (-1)^{k-1}\left[F(x_1)\frac{x_2 x_3 \ldots x_k}{(x_1 - x_2)(x_1 - x_3) \ldots (x_1 - x_k)}.\right.$$
$$+ F(x_2)\frac{x_1 x_3 \ldots x_k}{(x_2 - x_1)(x_2 - x_3) \ldots (x_2 - x_k)} + \cdots$$
$$\left.+ F(x_k)\frac{x_1 x_2 \ldots x_{k-1}}{(x_k - x_1)(x_k - x_2) \ldots (x_k - x_{k-1})}\right] \bmod 251 \qquad (7.9)$$

- Compute y to get the pixel value s_i of the secret image pixel S_i.

Step 4: Compose all the pixels to get the secret image.

7.3.3 Security analysis

The security of this scheme is provided by the double-layer steganography method and the biometric template protection algorithm. After being registered to TTP, the biometric data are kept in a secure fashion so that unintended users cannot recover them for any purpose. In other words, we protect the biometric data templates of the registered users in such a way that an adversary cannot retrieve the raw (or original) biometric data from a compromised template, even when an attack on the database possibly leads to exposing the templates to the attacker. This is because what is stored in the database is the transformed version of raw biometric data. The transformed template reveals no information whatsoever about the original biometric data.

We choose to implement cancellable biometric [18] for biometric template protection in the proposed system. Security offered by this cancellable template design method is twofold. First, the many-to-one mapping of the non-invertible transformation safely protects the original biometric data because no information about them can be obtained from the transformed templates that are stored in the database. Even in the worst case scenario where both a transformed template and its associated parameter key are compromised, the original biometric data cannot be retrieved, so its security is guaranteed. Second, when a stored (transformed) template is compromised, it can be revoked and a new template can be generated to replace it. This can be readily done by changing the parameter key so that a new, non-invertible transformation is produced to generate a new template which is totally different and unrelated to the old one. The parameter key is randomly generated, which means that there is no restriction on the number of new templates that can be built.

A TTP is defined as a trusted domain by all the users [32]. In this system, TTP performs all the communication and transmission between two users. On the other hand, the two users though possess a secure channel, do not make any direct communication to transmit the secret image because the power of steganography lies on the concealing of information in a transmission medium. Making use of steganography, when transmitting secret over a TTP, the users can rely on a twofold security. The secret is fragmented and outsourced to a cloud service. Let us assume that an attacker knows of a secret is hidden in the images. He uses a stego-detector that inspects all pixels of the images being transmitted and finds out that the intensity values are all under or equal to 250. However, in order to recover the secret image, the attacker needs to: (i) Know the order that the eight bits of the secret pixel values were hidden. (ii) Know the colour channel that the pixels are hidden into. (iii) Know the biometric information of the authorised user. This means that he needs to be able to invert the one-way transformation used in cancellable template. Moreover, he also needs to catch the exact all n^2 camouflage-stego-images and possess the transformation that the biometric data is applied. This proves the mathematical difficulty of breaking this scheme without compromising the biometric authentication. Even if all the parameters of the system are public, the attacker needs to bypass the biometric authentication.

7.4 Experimental results

7.4.1 Settings and parameters

This section presents some experimental results to examine the feasibility of the proposed scheme. We implemented the scheme with (3, 4)-threshold case. The secret image has the size of 187×187. Therefore, the each of the bio-stego-images is 374×374 and each of the stego-images is 748×748. The camouflage images are provided by [33]. Moreover, the secret image and the biometric images are grey scale; meanwhile, the camouflage images are full colour. The pixels from each of the bio-stego-images are embedded in the green channel of each of the 16 stego-images.

For template transformation, we used Yang *et al.*'s result [18] for matching evaluation between fingerprint images in the public database FVC2002 [34]. This means that the biometric template is stored in the system using cancellable template and that after retrieving the bio-stego-images, the system applies cancellable template on the images to perform matching.

7.4.2 Results

What we will measure here is the PSNR values for the images.

All 16 camouflage images are shown in Figure 7.5. The corresponding stego-images are shown in Figure 7.6 for visual comparison. The PSNR values of the 16 stego-images compared to the originals are presented in the table:

Table 7.1 PSNR values

	a	b	c	d	e	f	g	h	i	j	k	l	m	n	o	p
PSNR	33.82	33.53	33.91	37.47	37.42	33.28	33.27	32.47	30.62	33.17	32.53	36.66	37.36	34.71	33.80	32.72

Figure 7.5   Camouflage images. Taken from [33].

We can see from the table that the stego-images give a fair result, which suggests it may be visually acceptable. In comparison with Lin and Tsai's results [29] of 39.21, 39.16, and 39.16, the result of this system is decent. This can be explained as the consequence of mapping the pixel values that are greater than 250 down to 250 each time the embedding process is performed. Additionally, the limitation of polynomial $F(x)$ magnitude also contributes to downgrade the image quality. Since $F(x)$ performs a modulus 251 operation, the result of the calculation is limited in the range of [0, 250]. To prevent this, we can use Galois Field GF (2^8) as suggested in [30] to process the full range of pixel intensity. Also, one way to fully limit the noise in stego-image and achieve better PSNR values is to employ a better algorithm that produces lossless image such as the one proposed in [31]. Another trivial factor that influences the quality of the images is that the JPEG is a lossy compression that aims at reducing the size of the images. For better results, an input and output of PNG format is recommended.

A biometric image from authenticated user is necessary in order to recover a secret. Therefore, the performance of biometric matching module plays a crucial rule

Figure 7.6   Camouflage-stego-images

in the system. As stated earlier, the decrease in the quality of the bio-stego-images is not visible to HVS and evidently does not influence the process of fingerprint matching. The EER result came out for the system is 5.93% as tested in FVC2002DB1 (Figure 7.7) [18].

In order to improve the security when embedding in Layer 2, we can also partition the bio-stego-images and embed the pixels in different colour channels using a cryptographic key generated in the TTP. In this case, the key is not transmitted. Therefore, the system's security is still ensured and cryptographically independent.

7.5   Conclusion and future works

In this chapter, we proposed a new way to preserve the privacy of user by the combination of steganography and biometrics template protection. The secret image to be shared is protected by double layers of steganography: embedding it into the transformed fingerprint images, which are to be hidden into casual images to prevent unauthorised access when stored online. In order for authorised users to recover the

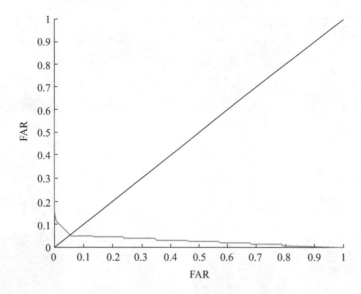

Figure 7.7   Matching performance. Graph adopted from [18].

secret image, a matching between their fingerprint and the version into which the secret image is embedded is performed in the transformed domain of the cancellable template, which is used to protect the raw biometric information. With this new framework, the problem of dishonest participant that Yang *et al.* mentioned [30] can be fully resolved since the stego-images are stored in a cloud server. Moreover, this system is no longer dependent on cryptographic key.

The main contribution of this chapter is to present a novel scheme that utilizes the power of not only steganography but also biometrics to protect the identity of the user. Although the results produced in experiments are fairly accepted, it is believed that certain factors can be improved to increase the quality of the outcome images, such as: using a wider field to process full range of the pixel values and generating output of PNG format to prevent the degradation caused by lossy compression. In the near future, looking for an appropriate algorithm and developing new techniques to apply in this system are promising research topics. In addition, in order to better evaluate the performance of the system, a new metric that shows the relationship of biometric data matching and steganographic embedding should be devised.

The book presents more insights in biometrics template protection mechanisms by reviewing some significant innovations in the field or discussing alternative concepts.

References

[1]   Li, C. and Hu, J. "A security-enhanced alignment-free fuzzy vault-based fingerprint cryptosystem using pair-polar minutiae structures". In: *IEEE Transactions on Information Forensics and Security* 11.3 (2016), pp. 543–555.

[2]   Chen, B. and Wornell, G. W. "Quantization index modulation: a class of provably good methods for digital watermarking and information embedding". In: *IEEE Transactions on Information Theory* 47.4 (2001), pp. 1423–1443.

[3] Sadkhan, S. B. "Cryptography: current status and future trends". In: *Information and Communication Technologies: From Theory to Applications, 2004. Proceedings. 2004 International Conference on.* IEEE. 2004, pp. 417–418.

[4] Provos, N. and Honeyman, P. "Hide and seek: an introduction to steganography". In: *IEEE Security & Privacy* 1.3 (2003), pp. 32–44.

[5] Lin, E. T. and Delp, E. J. "A review of data hiding in digital images". In: *IS and TS PICS Conference. Society for Imaging Science & Technology.* 1999, pp. 274–278.

[6] Johnson, N. F. and Jajodia, S. "Steganalysis of images created using current steganography software". In: *International Workshop on Information Hiding.* Springer, 1998, pp. 273–289.

[7] Kelley, J. "Terror groups hide behind Web encryption". In: *USA Today* 5 (2001). Accessed at https://usatoday30.usatoday.com/tech/news/2001-02-05-binladen.htm.

[8] Sieberg, D. "Bin Laden exploits technology to suit his needs". In: *CNN, September* (2001). Accessed at http://edition.cnn.com/2001/US/09/20/inv.terrorist.search/.

[9] Shamir, A. "How to share a secret". In: *Communications of the ACM* 22.11 (1979), pp. 612–613.

[10] Ratha, N. K., Connell, J. H., and Bolle, R. M. "Enhancing security and privacy in biometrics-based authentication systems". In: *IBM Systems Journal* 40.3 (2001), pp. 614–634.

[11] Jain, A. K., Nandakumar, K., and Nagar, A. "Biometric template security". In: *EURASIP Journal on Advances in Signal Processing 2008* (2008), p. 113.

[12] Juels, A. and Sudan, M. "A fuzzy vault scheme". In: *Designs, Codes and Cryptography* 38.2 (2006), pp. 237–257.

[13] Barman, S., Samanta, D., and Chattopadhyay, S. "Fingerprint-based cryptobiometric system for network security". In: *EURASIP Journal on Information Security* 2015.1 (2015), p. 1.

[14] Jin, A. T. B., Ling, D. N. C., and Goh, A. "Biohashing: two factor authentication featuring fingerprint data and tokenised random number". In: *Pattern Recognition* 37.11 (2004), pp. 2245–2255.

[15] Teoh, A. B., Goh, A., and Ngo, D. C. "Random multispace quantization as an analytic mechanism for biohashing of biometric and random identity inputs". In: *IEEE Transactions on Pattern Analysis and Machine Intelligence* 28.12 (2006), pp. 1892–1901.

[16] Yang, W., Hu, J., and Wang, S. "A Delaunay quadrangle-based fingerprint authentication system with template protection using topology code for local registration and security enhancement". In: *IEEE Transactions on Information Forensics and Security* 9.7 (2014), pp. 1179–1192.

[17] Dodis, Y., Reyzin, L., and Smith, A. "Fuzzy extractors: how to generate strong keys from biometrics and other noisy data". In: *International Conference on the Theory and Applications of Cryptographic Techniques.* Springer, 2004, pp. 523–540.

[18] Yang, W., Hu, J., Wang, S., and Yang, J. "Cancelable fingerprint templates with Delaunay triangle-based local structures". In: *Cyberspace Safety and Security.* Springer, 2013, pp. 81–91.

[19]   Yang, W., Hu, J., Wang, S., and Stojmenovic, M. "An alignment-free fingerprint bio-cryptosystem based on modified Voronoi neighbor structures". In: *Pattern Recognition* 47.3 (2014), pp. 1309–1320.

[20]   Wang, S., Deng, G., and Hu, J. "A partial Hadamard transform approach to the design of cancelable fingerprint templates containing binary biometric representations". In: *Pattern Recognition* 61 (2017), pp. 447–458.

[21]   Wang, S. and Hu, J. "A blind system identification approach to cancelable fingerprint templates". In: *Pattern Recognition* 54 (2016), pp. 14–22.

[22]   Wang, S. and Hu, J. "Alignment-free cancelable fingerprint template design: A densely infinite-to-one mapping (DITOM) approach". In: *Pattern Recognition* 45.12 (2012), pp. 4129–4137.

[23]   Wang, S. and Hu, J. "Design of alignment-free cancelable fingerprint templates via curtailed circular convolution". In: *Pattern Recognition* 47.3 (2014), pp. 1321–1329.

[24]   Deshmukh, M. P. and Kale, N. "Survey on biometric image sharing using cryptography and diverse image media". In: *International Journal on Recent and Innovation Trends in Computing and Communication* 2, 11 (2014), pp. 3689–3691.

[25]   Patil, S., Tajane, K., and Sirdeshpande, J. "Secret sharing schemes for secure biometric authentication". In: *International Journal of Scientific & Engineering Research* 4.6 (2013).

[26]   Jain, A. K. and Uludag, U. "Hiding biometric data". In: *IEEE Transactions on Pattern Analysis and Machine Intelligence* 25.11 (2003), pp. 1494–1498.

[27]   Ross, A. and Othman, A. "Visual cryptography for biometric privacy". In: *IEEE Transactions on Information Forensics and Security* 6.1 (2011), pp. 70–81.

[28]   Whitelam, C., Osia, N., and Bourlai, T. "Securing multimodal biometric data through watermarking and steganography". In: *Technologies for Homeland Security (HST), 2013 IEEE International Conference on. IEEE.* 2013, pp. 61–66.

[29]   Lin, C.-C. and Tsai, W.-H. "Secret image sharing with steganography and authentication". In: *Journal of Systems and Software* 73.3 (2004), pp. 405–414.

[30]   Yang, C.-N., Chen, T.-S., Yu, K. H., and Wang, C.-C. "Improvements of image sharing with steganography and authentication". In: *Journal of Systems and Software* 80.7 (2007), pp. 1070–1076.

[31]   Lin, P.-Y. and Chan, C.-S. "Invertible secret image sharing with steganography". In: *Pattern Recognition Letters* 31.13 (2010), pp. 1887–1893.

[32]   Adams, C. "Trusted third party". In: *Encyclopedia of Cryptography and Security*. Edited by Tilborg, H. C. A. van and Jajodia, S. Boston, MA: Springer US, 2011, pp. 1335–1335. ISBN: 978-1-4419-5906-5. DOI: 10.1007/978-1-4419-5906-5_98.

[33]   *Large Free Pictures.* https://www.free-pictures-photos.com (visited on 06/30/2016).

[34]   FVC Database 2002. 2002. http://bias.csr.unibo.it/ fvc2002.

Part III

Security and privacy issues inherent to biometrics

Chapter 8

Biometric template protection: state-of-the-art, issues and challenges

Christian Rathgeb[1] and Christoph Busch[1]

Nowadays, biometric recognition represents an integral component of identity management and access control systems, replacing PINs or passwords. However, the wide deployment of biometric recognition systems in the past decades has raised numerous privacy concerns regarding the storage and use of biometric data. Due to the fact that the link between individuals and their biometric characteristics, e.g. fingerprints or iris, is strong and permanent, biometric reference data (templates) need to be protected in order to safeguard individuals' privacy and biometric systems' security. In particular, unprotected biometric templates can be abused to crossmatch biometric databases, i.e. tracking individuals without consent, and to launch presentation attacks employing specific inversion techniques.

Technologies of biometric template protection offer solutions to privacy-preserving biometric authentication, which improves the public confidence and acceptance of biometric systems. While biometric template protection has been an active research topic over the last 20 years, proposed solutions are still far from gaining practical acceptance. The existing gap between theory and practice, which results in a trade-off between privacy protection and recognition accuracy, is caused by numerous factors. To facilitate a standardized design and application of biometric template protection schemes, researchers have to solve several open problems. This chapter provides an overview of state-of-the-art technologies of biometric template protection and discusses main issues and challenges.

8.1 Introduction

Biometric samples, e.g. fingerprint or iris images, underlie a natural intra-class variance. Such variance can be caused by several factors depending on the biometric characteristic, e.g. variations in translation or rotation in the case of fingerprints or pupil dilation or partial closure of eyelids in the case of iris. Example of these variations is depicted in Figure 8.1. To tolerate a certain variance, biometric systems

[1]Hochschule Darmstadt, Germany

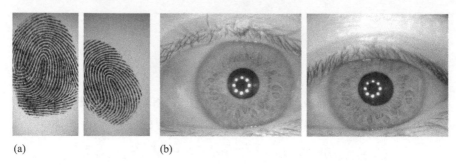

(a)   (b)

Figure 8.1   Examples for intra-class variations between pairs of fingerprints and NIR iris images: (a) fingerprints and (b) iris. Images taken from FVC'02 fingerprint database [1] and CASIA-v4-Interval iris database [2]

compare obtained (dis)similarity scores (between a probe and a reference template) against an adequate decision threshold yielding acceptance or rejection. This vital processing step of biometric recognition systems prevents from a secure application of conventional cryptographic techniques, such as the usage of hash functions for secure template handling, as common in secure password authentication. The desirable 'avalanche effect' property of cryptographic algorithms, i.e. a small change in either the key or the plaintext should cause a drastic change in the ciphertext, obviously obviates a comparison of protected templates in the encrypted domain resulting in two major drawbacks [3]: (1) on the one hand, an encrypted template needs to be decrypted before each authentication attempt, i.e. an attacker can glean the biometric template by simply launching an authentication attempt and (2) on the other, an encrypted template will be secure only as long as the corresponding decryption key is unknown to the attacker, i.e. the problem is shifted yielding two-factor authentication.

In contrast to traditional conventional cryptography, *biometric template protection* schemes enable a permanent protection of biometric reference data. Hence, such technologies will enable biometric systems to become compliant to the recently issued EU regulation 2016/679 (General Data Protection Regulation) [4]. Moreover, recently, the European Banking Authority (EBA) proposed a draft [5] for regulatory technical standards on strong customer authentication and common and secure communication under PSD2 (European Directive 2015/2366 on payment services in the internal market). According to Article 5.1 of said draft regulation, requirements for devices and software are formulated that demand security features for biometric sensors but also demand template protection ensuring resistance against the risk of sensitive information being disclosed to unauthorized parties. Based on the standardized architecture [6,7], extracted biometric features serve as input for the *pseudonymous identifier (PI) encoder* of a biometric template protection scheme. The PI encoder generates a PI and corresponding auxiliary data (AD) which constitute the protected template, which should allow for a privacy-preserving authentication. PI and AD are generated during enrolment, at the completion of which the unprotected feature vector

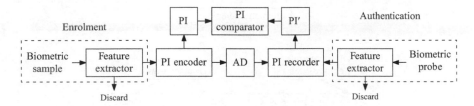

Figure 8.2   Generic framework of biometric template protection schemes

is deleted. At authentication, the *PI recorder* takes a feature vector and a queried AD as input and calculates a PI. Finally, a *PI comparator* is used to compare the generated PI to the stored one. Depending on comparators, the comparison result is either a binary decision (yes/no) or a similarity score which is then compared against a threshold, in order to obtain a binary decision. Figure 8.2 shows an illustration (adapted from [7]) of the standardized biometric template protection framework.

While the ISO/IEC 24745:2011 standard on *'Biometric information protection'* [7] does not limit or propose specific algorithms for the PI and AD functions, it constitutes two major requirements of biometric template protection:

1. *Irreversibility*: knowledge of a protected template cannot be exploited to reconstruct a biometric sample which is equal or close (within a small margin of error) to an original captured sample of the same source;
2. *Unlinkability*: different versions of protected biometric templates can be generated based on the same biometric data (renewability), while protected templates should not allow crossmatching.

Apart from satisfying these main properties, an ideal biometric template protection scheme shall not cause a decrease in biometric performance, i.e. recognition accuracy, with respect to the corresponding unprotected system [3]. However, up until now, the vast majority of proposed approaches have been found to either suffer from a significant drop in biometric performance or to reveal insufficient privacy protection. That is, biometric template protection remains an active and emerging field of research. This chapter will provide the reader with a summary of major achievements proposed in the field of biometric template protection as well as open issues and challenges.

The remainder of this chapter is organized as follows: Section 8.2 provides a review of most relevant template protection schemes representing instances of biometric cryptosystems or cancellable biometrics. Open issues and challenges are discussed in Section 8.3. Finally, conclusions are drawn in Section 8.4.

8.2   Biometric template protection

As shown in Figure 8.3, biometric template protection schemes are commonly categorized as *biometric cryptosystems*, also referred to as helper data schemes, and *cancellable biometrics*, also referred to as feature transformation approaches.

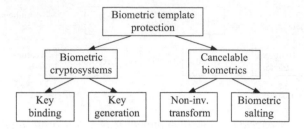

Figure 8.3   Generic categorization of biometric template protection schemes

Biometric cryptosystems are designed to securely bind a digital key to a biometric characteristic or generate a digital key from a biometric signal [8]. Biometric cryptosystems can be further divided into the groups of *key-binding* and *key-generation* schemes [9]. Within a key-binding scheme, helper data is obtained by binding a chosen key to a biometric template. As a result of the binding process, a fusion of the secret key and the biometric template is stored, which ideally does neither reveal any information about the key nor about the original biometric template. Applying an appropriate key retrieval algorithm, the key is reconstructed from the stored helper data during authentication. An update of the key usually requires re-enrolment in order to generate new helper data. In a key-generation scheme, the helper data is derived from the biometric template so that the key is generated from the helper data and a given biometric template. While the storage of helper data is not obligatory, the majority of proposed key-generation schemes do store helper data, which allow for an update of keys. Key-generation schemes in which helper data are applied are also referred to as 'fuzzy extractors' [10]. A fuzzy extractor should reliably extract a uniformly random string, i.e. key, from a biometric template or sample while helper data is used to reconstruct that string from another biometric template or sample. Hence, biometric cryptosystems either retrieve/generate a key or return a failure message.

Cancellable biometrics consists of intentional, repeatable distortions of biometric signals based on transforms that provide a comparison of biometric templates in the transformed domain [11,12]. Non-invertible transforms are either applied in image domain or to the biometric template. In order to provide renewable templates, parameters of the applied transforms are modified. Non-invertible transforms are designed in a way that potential impostors are not able to reconstruct the original biometric image or template even if these are compromised. However, applying non-invertible transforms mostly implies a loss in recognition accuracy, which might be caused by information loss or infeasible feature alignment at comparison [9]. Biometric salting usually denotes transforms which are selected to be invertible. Hence, the parameters of an applied transform have to be kept secret, since impostors may be able to recover the original biometric template in case transform parameters are compromised. Generic approaches to biometric salting maintain the biometric performance of the underlying unprotected biometric systems. In case, subject-specific transforms are applied, the parameters of the transform have to be presented at each authentication. It is important to note that the majority of cancellable biometric systems employ the biometric comparator of the original unprotected system, i.e. (dis)similarity scores are obtained, in contrast to biometric cryptosystems.

In the following subsections, key approaches to biometric cryptosystems and cancellable biometrics are briefly summarized. Further, some works on multi-biometric template protection are reviewed. Due to the fact that reporting of biometric performance, privacy protection and security is found to be inconsistent across the variety of approaches (especially for early proposals), reported performance indicators are omitted. The issue of performance reporting will be discussed in the subsequent section (Section 8.3).

8.2.1 Biometric cryptosystems

The very first sophisticated approach to biometric key-binding designed for fingerprint images was proposed by Soutar *et al.* [13,14], which is commonly referred to as *biometric encryption*. The algorithm was summarized in a patent [15], which includes explanations of how to apply the algorithm to other biometric characteristics. Despite the fact that performance measurements have been omitted the system has been shown to be impractical in terms of privacy protection. In particular, it has been demonstrated that the scheme is vulnerable to trivial linkage attacks [16].

A first conceptual proposal of an iris-based key-generation scheme has been presented in [17,18]. In the *private template scheme*, the iris-biometric feature vector itself serves as a key. During enrolment, several iris codes are combined using a majority decoder, and error-correction check bits are generated for the resulting bit string, which represent the helper data. At authentication, faulty bits of a given iris codes should be corrected using the stored helper data. Experimental results are omitted, and it is commonly expected that the proposed system reveals poor performance due to the fact that the authors restrict to the assumption that only 10% of bits change among different genuine iris codes. Moreover, the direct use of biometric features as key contradicts with the requirement of renewability and un-linkability.

The *fuzzy commitment scheme* by Juels and Wattenberg [19] represents a cryptographic primitive, which combines techniques from the area of error-correcting codes and cryptography. At key-binding, a pre-chosen key is prepared with error-correcting codes and bound to a binary biometric feature vector of the same length by XORing both resulting in a difference vector. In addition, a hash of the key is stored together with the difference vector forming the commitment. During key retrieval, another binary probe feature vector is XORed with the stored difference vector, and error-correction decoding is applied. In case the presented feature vector is sufficiently 'close' to the stored one, then the correct key is returned, which can be tested using the stored hash. In [20], iris codes are bound to keys prepared with Hadamard and Reed–Solomon error-correction codes. In order to provide a more efficient error-correction decoding in an iris-based fuzzy commitment scheme, two-dimensional iterative min-sum decoding was introduced in [21]. These initial approaches revealed promising results in terms of recognition accuracy. Representing a more sophisticated key-binding scheme, the fuzzy commitment scheme has been applied to binary feature vectors obtained from numerous biometric characteristics, e.g. fingerprints [22] or online signatures [23]. In addition, diverse improvements to the original concept have been proposed, e.g. [24,25]. Nevertheless, different attacks and security analyses [26–28] leave the privacy protection provided by fuzzy commitment schemes rather questionable.

The *fuzzy vault scheme* [29,30] by Juels and Sudan enables a protection and error-tolerant verification with feature sets. In the key-binding process, a chosen key serves as polynomial, based on which the set-based (integer-valued) biometric feature vector is encoded. The resulting genuine points are dispersed among a set of chaff points not lying on the graph of the polynomial. At the time of authentication, a given feature vector is intersected with the vault resulting in a set of unlocking pairs. If the unlocking pairs are mostly genuine, it is possible to distinguish the genuine pairs from the chaff pairs so that the polynomial, and hence the key, can be reconstructed by tolerating errors within certain limits of an error-correction code. This approach offers order invariance, meaning that the sets used to lock and unlock the vault are unordered. It is due to this property that led researchers to consider the fuzzy vault scheme as a promising tool for protecting fingerprint minutiae sets [31]. This preliminary analysis was followed by a series of working implementations for fingerprints [32,33]. In addition, fuzzy vault schemes based on other biometric characteristics have been proposed, e.g. iris [34] or palmprint [35]. However, the fuzzy vault scheme as proposed by Juels and Sudan is vulnerable to a certain kind of linkage attacks, the correlation attack [36,37], that very clearly conflicts with the un-linkability requirement and (even worse) with the irreversibility requirement of effective biometric template protection. Moreover, the use of public unprotected auxiliary alignment data can ease an attacker in performing linkage attacks. As a countermeasure, an implementation for absolutely pre-aligned minutiae that avoids any correlation between related records of the fuzzy vault scheme has been proposed [38]. Another advantage from the construction in [38] is that the improved fuzzy vault scheme by Dodis *et al.* [10] can be used for template protection which results in significantly more compact size of the reference record. It is important to note that other linkage attacks can be applied to the improved fuzzy vault scheme but they can, however, be effectively avoided [39].

Different key-generation schemes, which might be referred to as *quantization schemes*, quantize biometric features in order to obtain a stable 'biometric key'. While the quantization process should ensure irreversibility, un-linkability can be achieved by encoding of quantized features. This means quantized features are substituted by predefined bitstrings. The according mapping is stored as AD while the encoding scheme might be explicitly designed for distinct feature elements depending on their stability and discriminativity. This yields an improved adjustment of the stored AD to the nature of the employed biometric characteristic. While quantization represents an integral processing step in many template protection schemes in order to decrease intra-class variance, such approaches have frequently been applied to behavioural biometric characteristics, e.g. online signatures [40]. Since coarse quantization reduces biometric information, its application might cause a significant drop in biometric performance.

Table 8.1   List of major concepts proposed for biometric cryptosystems and biometric modalities these schemes have mainly been applied to

Method	Modality	Description	Remarks
Biometric encryption [14]	Fingerprint	Key binding using correlation function and random seed	Only theoretical concept, processes entire images, requires multiple enrolment samples
Private template [17]	Iris	Key generation using binary templates and error correction	Only theoretical concept, requires multiple enrolment samples
Fuzzy commitment [19]	Iris, fingerprint	Key binding using binary templates and error correction	Requires feature adaptation for non-binary input
Fuzzy vault [29]	Fingerprint, iris	Key binding using un-ordered set of integers and error correction	Requires feature adaptation for noninteger-set input

Besides the above-mentioned schemes, further approaches to biometric cryptosystems have been proposed, e.g. *password hardening* [41] or *shielding functions* [42]. Table 8.1 summarizes key concepts to biometric cryptosystems.

8.2.2   Cancellable biometrics

Ratha *et al.* [11] were the first to introduce the concept of cancellable biometrics applying *non-invertible geometric transforms* in image domain to obtain cancellable facial images and fingerprints, including block permutations and surface folding. Un-linkability should be achieved by changing the parameters of these transforms for different applications or subjects. In further work [43], diverse techniques such as cartesian, polar and surface-folding transformations are employed in the feature domain, i.e. minutiae positions, in order to generate cancellable fingerprint templates. Prior to applying these transformations, the images are registered by first estimating the position and orientation of the singular points (core and delta) and expressing the minutiae positions and angles with respect to these points. Since (obscured) cancellable templates cannot be aligned at the time of comparison, registration represents a vital step in such methods, which have been applied to various biometric characteristics, e.g. iris [44].

Random projections, proposed by Pillai *et al.* [45], represent another prominent class of non-invertible transform. In these methods, biometric features are projected onto a random space, i.e. unlinkability is achieved by employing different random spaces for different subjects or applications. In [45], this concept is specifically adapted to iris images using so-called sectored random projections. The well-known

BioHashing scheme [46], which has been applied to numerous biometric characteristics, might be seen as an extension of random projections [12]. In this approach,
iterated inner products are calculated between tokenized pseudorandom numbers and
the biometric feature vector, which is generated from the integrated wavelet and
Fourier-Mellin transform.

Savvides *et al.* [47] proposed *cancellable biometric filters* based on a random convolution method for generating cancellable facial images. In this technique, images
are protected employing random convolution kernels. A set of training images is convolved with a random convolution kernel, i.e. un-linkability is granted by changing
the seed used to generate the random convolution kernel. The convolved training
images are then used to generate a minimum average correlation energy (MACE)
filter, which is stored as AD. At authentication, a random convolution kernel is presented together with the corresponding random seed. The convolution kernel is then
convolved with a given facial image and cross-correlated with the stored MACE filter,
and the resulting correlation outputs are used to authenticate the subject. In [48], a
similar approach has been applied to fingerprint images.

In the concept of *biotokens*, which has been proposed by Boult *et al.* [49,50], elements of facial and fingerprint feature vectors are separated into a stable and unstable
parts. Such a separation method highly depends on the used biometric characteristic
and the variance of extracted features, i.e. a reasonable amount of training data might
be required to reliably extract stable components of biometric feature vectors. The
stable part, which should provide a perfect matching, can be encrypted using conventional cryptographic techniques. In addition, the unstable part is obscured employing
some transform, which enables a comparison in the transformed domain.

Cancellable biometrics based on *bloom filters* was introduced by Rathgeb *et al.*
[51,52] and has been applied to different biometric characteristics. The scheme is
designed to map biometric features to an irreversible representation, i.e. bloom filters,
incorporating an application-specific secret which is employed in order to achieve
un-linkability. However, the scheme has been shown to be vulnerable to crossmatching
attacks [53]. An improvement to the original concept, in order to grant un-linkability
of protected templates, has been proposed in [54].

Random permutations have first been employed in [55] to scramble bits of a
binary iris-biometric feature vectors. Since a reference template should be permanently protected, the application-specific random permutation has to be applied at
various shifting positions in order to achieve an appropriate alignment of iris codes.
Another random permutation scheme designed to protect binary iris-based templates,
which is referred to as bin combo, was introduced in [56]. Based on some random
seed, rows of two-dimensional binary templates are circularly shifted. Subsequently,
pairs of shifted rows are combined by XORing them. A similar approach can be
applied in the image domain in order to protect normalized iris textures.

In the concept of *biometric salting,* the biometric samples or templates are
obscured based on random patterns. In [55], binary iris-based templates are XORed
with a randomly generated application-specific bit stream. In order to assure that
resulting iris codes sufficiently differ from original ones, it is suggested retaining at
least half of the bits in the randomly generated bit stream as one. In [56], this scheme

Table 8.2   List of major concepts proposed for cancellable biometrics and biometric modalities these schemes have mainly been applied to

Method	Domain	Modality	Description	Remarks
Geometric transforms [11]	Signal	Face, fingerprint, iris	Block-remapping, image warping	Requires subsequent extraction of features
BioHashing [46]	Feature	Face, palmprint	Feature projection using random matrix and thresholding	Requires pre-alignment and special comparator
Cancellable filter [47]	Signal	Face	Convolution of image with random kernel	Requires special comparator
Biotoken [49]	Feature	Fingerprint, face	Encryption of stable features and random transformation	Requires pre-alignment and special comparator
Bloom filter [51]	Feature	Iris, face, fingerprint	Combination of random permutation and many-to-one transform	Requires special comparator
Salting [55]	Feature, signal	Iris	Random permutation, invertible binding with random seed	Requires two-factor authentication or trusted third party

is evaluated, which is referred to as bin salt. Again, such techniques might as well be applied in the image domain. Like random simple permutations, salting approaches are applicable to samples or feature vectors of diverse biometric characteristics.

In addition to the above-mentioned schemes, other approaches to cancellable biometrics have been presented, e.g. *BioConvolving* [57] or *knowledge signatures* [58]. Moreover, hybrid systems combining concepts of biometric cryptosystems and cancellable biometrics have been proposed, e.g. in [59].

8.2.3   Multi-biometric template protection

A combination of multiple biometric characteristics has been found to significantly improve the accuracy and reliability especially in challenging scenarios [60], while recognition systems based on a single biometric indicator often have to cope with unacceptable error rates [61]. However, improvement in biometric performance as a result of biometric fusion should be weighed against the associated overhead involved, such as additional sensing cost [62]. Moreover, the protection of multi-biometric templates is especially crucial as they contain information regarding multiple characteristics of the same subject [63]. In contrast to conventional biometric systems, where fusion may take place at score or decision level [61], with respect to template protection schemes, feature-level fusion has been identified as most suitable, since a separate storage of two or more protected biometric templates would enable parallelized attacks. One single-protected template, which has been obtained by means of feature-level fusion, is expected to improve privacy protection, since the fused template comprises more biometric information [63]. The development of multi-biometric template protection schemes is often accompanied by further issues such as common data representation or feature alignment [64].

One of the first approaches to a multi-biometric cryptosystem based on the fuzzy commitment scheme was proposed by [65], in which binary fingerprint and face features are combined. In [66], two different feature-extraction algorithms are applied to 3D face data yielding a single instance scenario. The authors provide results for feature-level, score-level and decision-level fusion. In order to obtain a comparison score, the number of errors corrected by the error-correction code is estimated, i.e. scores are only available in case of successful decoding. In [67], rearrangement of bits in iris codes is proposed to provide a uniform distribution of error probabilities. The rearrangement allows a more efficient execution of error-correction codes combining the most reliable bits generated by different feature-extraction algorithms. A multi-biometric cryptosystem employing a fuzzy vault based on fingerprint and iris is presented in [68]. It is shown that a combination of biometric modalities leads to increased privacy protection as well as recognition accuracy and, hence, higher security. Further, in a more general framework [63], a multi-biometric fuzzy commitment schemes and fuzzy vault schemes based on fingerprint, face and iris is proposed. In order to obtain a common feature representation for each type of template protection scheme, the authors propose different embedding algorithms, e.g. for mapping a binary string to a point set. In [69], a multi-finger fuzzy vault is proposed. A multi-biometric template protection system employing decision-level fusion of multiple protected fingerprint templates is presented in [70].

Focusing on cancellable multi-biometrics, a combination of two different feature-extraction methods to achieve cancellable face biometrics is presented in [71]. PCA (principle component analysis) and ICA (independent component analysis) coefficients are extracted, and both feature vectors are randomly scrambled and added in order to create a transformed template. In [72], pairs of fingerprint images are fused in the image domain to obtain a cancellable multi-finger system. A feature-level fusion of iris codes based on bloom filter is presented in [73].

8.3   Issues and challenges

Progress in the field of biometric template protection has resulted in several challenges, which have been the focus of recent research efforts. The biometric community is actively pursuing open problems, such as biometric data representation or feature alignment. The following subsections summarize major issues and challenges with respect to (multi-)biometric template protection.

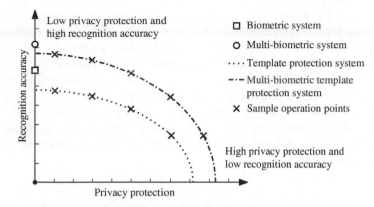

*Figure 8.4   Schematic illustration of a potential interrelation between conventional
biometric recognition systems and (multi-)biometric template
protection systems showing a frequently observed gap biometric
performance and/or privacy protection*

8.3.1   Performance decrease in template protection schemes

Focusing on proposed techniques of biometric template protection schemes, up until
now, there exists a clear performance gap between protected schemes and conventional unprotected biometric recognition schemes. On the one hand, the authentication
speed provided in traditional biometric systems remains unrivalled, especially for biometric cryptosystems which might require complex key-retrieval or key-generation
procedures. Such schemes are designed for verification purposes and may not enable
real-time identification on large-scale databases.

On the other, the vast majority of template protection scheme are operated at a
clear trade-off between recognition accuracy and privacy protection. That is, strong
privacy protection is often accompanied by a significant decrease in biometric performance while a maintenance of recognition accuracy is often achieved at the cost of
privacy protection. This interrelation is schematically illustrated in Figure 8.4, where
a measurement of privacy protection might be specific for a proposed scheme. Further, it is important to note, that favourable acquisition conditions are considered a
fundamental premise for biometric template protection, due to the sensitivity of many
schemes to high intra-class variability.

Within technologies of biometric cryptosystems and cancellable biometrics, original biometric features are transformed, i.e. obscured, which represents the main
reason of the above-mentioned trade-off. This transformation of biometric features
may cause severe information loss, e.g. due to coarse quantization or obscureness of
feature elements' neighbourhoods. In addition, the common use of error-correction
codes in biometric cryptosystems does not allow for a precise configuration of
systems' decision thresholds, while false rejection rates are lower bounded by
error-correction capacities.

With respect to desired system requirements, i.e. breaking the trade-off between accuracy and privacy protection, multi-biometric template protection systems offer significant advantages. Multi-biometrics provide low error rates compared to mono-biometric systems even under less constrained circumstances. As illustrated in Figure 8.4, the incorporation of multiple biometric characteristics in a template protection scheme might lead to reasonable recognition accuracy at practical privacy protection. For instance, in [74], it has been shown that, at reasonable level of protection, a multi-iris cryptosystem reveals biometric performance similar to that of an unprotected single-iris recognition system. Similar observations have been made for other biometric modalities [69].

8.3.2   Data representation and feature alignment

Depending on the nature of biometric characteristics, distinct feature representations have been established in the past decades, e.g. triples of minutiae coordinates and their angles or two-dimensional binary vectors for iris. Biometric template protection schemes operate on a certain form of input, e.g. point sets or binary features. Hence, incompatibility issues arise when the representation type of intended biometric features does not match the acceptable input type of a template protection scheme [75]. Ideally, biometric features can be transformed to the required feature representation in an appropriate manner. Moreover, feature-type transformations shall not cause a decrease in biometric performance compared to the original representation [9]. Furthermore, biometric feature-type transformation comes into play when features of different biometric characteristics need to be fused at the feature level.

Another important issue regarding template protection schemes is feature alignment, which significantly affects biometric performance. In a template protection system, the original biometric features are discarded, cf. Figure 8.2, such that a proper alignment at the time of authentication is non-trivial. Further, optimal alignment parameters generally differ for various biometric characteristics in a multi-biometric system. Thus, there is a great need to design alignment-free feature representations for different biometric characteristics. For instance, in [76], it has been shown that the application of feature transforms on minutiae vicinity-based feature vectors generally reveal improved biometric performance compared to protection schemes which require a registration (pre-alignment) of biometric samples.

8.3.3   Standardization and deployments

Standardization in the field of information technology is pursued by a Joint Technical Committee (JTC1) formed by the International Organization for Standardization (ISO) and the International Electrotechnical Commission (IEC). With regard to biometric template protection, there are inside JTC1 two subgroups that are actively developing related standards. These are the Sub-Committee 27 on security techniques (SC27) and the Sub-Committee 37 on biometrics (SC37). While SC27 by tradition is concerned with crypto-algorithms, procedures for security management also standards on privacy fall into the scope. More recently, in 2002, the committee SC37 was established with the intention to generate interchange standards, application programming interfaces and methodologies for biometric performance testing. These standards have been adopted for electronic passports, which have meanwhile more than 700 million implementations in the field. Moreover, SC37 defined a harmonized biometric vocabulary (ISO/IEC 2382-37) that removes contradictions in the biometric terminology. The most relevant biometric standards are the vocabulary [77], a harmonized definition of a general biometric system (ISO/IEC SC37 SD11) that describes the distributed subsystems, which are contained in deployed applications [78], a common programming interface BioAPI (ISO/IEC 19784-1) that supports ease of integration of sensors and SDKs [79], the definition of data interchange formats (ISO/IEC 19794) [80] and also the definition of methodologies for biometric performance testing and reporting (ISO/IEC 19795-1) [81]. Apart from general biometrics-related standards, there are two substantial standards focusing on biometric template protection, which are as follows:

1. ISO/IEC 24745 [7] on biometric information protection was the first standardization activity in the field of biometric template protection. The standard provides guidance and specifies requirements on secure and privacy-compliant management and processing of biometric information. A formal framework of biometric template protection schemes is defined, see Figure 8.2. Further, the standard contains a comprehensive list of threats faced by a biometric (template protection) system and of known countermeasures against these threats.
2. ISO/IEC CD 30136 [82] on performance testing of template protection schemes is (at the moment of this writing) a draft standard, which is intended to specify new metrics for evaluating template protection-based biometric verification and identification systems. Theoretical and empirical definitions are provided for each metric.

Industrial deployment of biometric template protection methods is only just beginning to take place, where major obstacles to industrial deployment is the lack of interoperability among enrolment-processing modules, database storage and verification modules [83]. This implies a shortage of end-to-end solutions for template protection. Different companies offer biometric recognition solutions employing biometric template protection in accordance with [7]. However, this does not imply the ability to evaluate and report a set of performance metrics, as is currently the case for traditional biometric recognition systems. Hence, standardized ways of measuring performance metrics regarding privacy protection are of utmost importance to leverage industrial deployments of biometric template protection technologies. A list of early deployments of biometric template protection technologies can be found in [9]. In addition, bridging the performance gap between conventional biometric recognition systems and template protection schemes is vital to ensure a user-friendly application of biometric technologies.

8.4 Conclusion

Deployments of biometric technologies have raised concerns with respect to the abuse of biometric information. Due to this fact, there is a greater need for the application of template protection systems in particular, in scenarios where the biometric data is stored in centralized repositories, in order to prevent from possible adversarial attacks on biometric reference data. Despite key requirements regarding data protection, ideal template protection schemes should maintain the recognition accuracy provided by the corresponding unprotected recognition system. However, due to numerous reasons, e.g. non-trivial feature alignment or concealment of local feature neighbourhoods, this is hardly the case for the majority of proposed template protection schemes. This chapter provides a summary of key approaches to biometric template protection and discusses open issues and challenges, such as the trade-off between protection level and recognition performance, feature alignment, as well as standardization of performance metrics or the design of multi-biometric template protection schemes.

References

[1]   FVC-onGoing: on-line evaluation of fingerprint recognition algorithms. Fingerprint verification Competition – 2002, 2016.

[2]   Chinese Academy of Sciences' Institute of Automation. CASIA Iris Image Database V4.0 – Interval, 2016.

[3]   K. Nandakumar and A. K. Jain. Biometric template protection: Bridging the performance gap between theory and practice. *IEEE Signal Processing Magazine – Special Issue on Biometric Security and Privacy*, pages 1–12, 2015.

[4]   European Parliament. Regulation (EU) 2016/679 of the European Parliament and of the Council on the protection of natural persons with regard to the processing of personal data and on the free movement of such data, and repealing Directive 95/46/EC (General Data Protection Regulation). April 2016.

[5]   European Banking Authority. Draft regulatory technical standards on strong customer authentication and common and secure communication under psd2 (directive 2015/2366), August 2016.

[6]   J. Breebart, C. Busch, J. Grave, and E. Kindt. A reference architecture for biometric template protection based on pseudo identities. In *BIOSIG 2008: Biometrics and Electronic Signatures*, Lecture Notes in Informatics, pages 25–37. GI-Edition, 2008.

[7]   ISO/IEC JTC1 SC27 Security Techniques. *ISO/IEC 24745:2011. Information Technology – Security Techniques – Biometric Information Protection*. Geneva, Switzerland: International Organization for Standardization, 2011.

[8]   U. Uludag, S. Pankanti, S. Prabhakar, and A. K. Jain. Biometric cryptosystems: Issues and challenges. *Proceedings of the IEEE*, 92(6):948–960, 2004.

[9]   C. Rathgeb and A. Uhl. A survey on biometric cryptosystems and cancelable biometrics. *EURASIP Journal on Information Security*, 2011(3):1–25, 2011.

[10] Y. Dodis, R. Ostrovsky, L. Reyzin, and A. Smith. Fuzzy extractors: How to generate strong keys from biometrics and other noisy data. *SIAM Journal on Computing*, 38(1):97–139, 2008.

[11] N. Ratha, J. Connell, and R. Bolle. Enhancing security and privacy in biometrics-based authentication systems. *IBM Systems Journal*, 40(3): 614–634, 2001.

[12] V. M. Patel, N. K. Ratha, and R. Chellappa. Cancelable biometrics: A review. *IEEE Signal Processing Magazine*, 32(5):54–65, 2015.

[13] C. Soutar, D. Roberge, A. Stoianov, R. Gilroy, and B. V. Kumar. Biometric encryption – Enrollment and verification procedures. In D. Casasent and T.-H. Chao, editors, *Optical Pattern Recognition IX*, volume 3386 of *Proc. of SPIE*, pages 24–35. SPIE, 1998.

[14] C. Soutar, D. Roberge, A. Stoianov, R. Gilroy, and B. V. Kumar. Biometric Encryption using image processing. In R. V. Renesse, editor, *Optical Security and Counterfeit Deterrence Techniques II*, volume 3314 of *Proc. of SPIE*, pages 178–188. SPIE, 1998.

[15] C. Soutar, D. Roberge, A. Stoianov, R. Gilroy, and B. V. Kumar. Method for secure key management using a biometrics, 2001. U.S. Patent 6219794.

[16] W. Scheirer and T. Boult. Cracking fuzzy vaults and biometric encryption. In *Proc. Biometrics Symposium*, pages 1–6. IEEE, 2007.

[17] G. Davida, Y. Frankel, and B. Matt. On enabling secure applications through off-line biometric identification. In *Proc. IEEE Symp. on Security and Privacy*, pages 148–157. IEEE, 1998.

[18] G. Davida, Y. Frankel, and B. Matt. On the relation of error correction and cryptography to an off line biometric based identification scheme. In *Proc. Workshop on Coding and Cryptography*, pages 129–138. IEEE, 1999.

[19] A. Juels and M. Wattenberg. A fuzzy commitment scheme. In *Proc. 6th ACM Conf. on Computer and Communications Security*, pages 28–36. ACM, 1999.

[20] F. Hao, R. Anderson, and J. Daugman. Combining cryptography with biometrics effectively. *IEEE Transactions on Computers*, 55(9):1081–1088, 2006.

[21] J. Bringer, H. Chabanne, G. Cohen, B. Kindarji, and G. Zemor. Theoretical and practical boundaries of binary secure sketches. *IEEE Transactions on Information Forensics and Security*, 3:673–683, 2008.

[22] K. Nandakumar. A fingerprint cryptosystem based on minutiae phase spectrum. In *Proc. IEEE Workshop on Information Forensics and Security*, pages 1–6. IEEE, 2010.

[23] E. A. Rua, E. Maiorana, J. L. A. Castro, and P. Campisi. Biometric template protection using universal background models: An application to online signature. *IEEE Transactions on Information Forensics and Security*, 7(1):269–282, 2012.

[24] L. Zhang, Z. Sun, T. Tan, and S. Hu. Robust biometric key extraction based on iris cryptosystem. In M. Tistarelli and M. Nixon, editors, *Proc. 3rd Int'l Conf. on Biometrics*, volume 5558 of *LNCS*, pages 1060–1070. Springer, 2009.

[25] E. R. C. Kelkboom, J. Breebaart, T. A. M. Kevenaar, I. Buhan, and R. N. J. Veldhuis. Preventing the decodability attack based cross-matching in a fuzzy commitment scheme. *IEEE Transactions on Information Forensics and Security*, 6(1):107–121, 2011.

[26] C. Rathgeb and A. Uhl. Statistical attack against fuzzy commitment scheme. *IET Biometrics*, 1(2):94–104, 2012.

[27] B. Tams. Decodability attack against the fuzzy commitment scheme with public feature transforms. *CoRR*, abs/1406.1154, 2014.

[28] T. Ignatenko and F. M. J. Willems. Information leakage in fuzzy commitment schemes. *IEEE Transactions on Information Forensics and Security*, 5(2): 337–348, 2010.

[29] A. Juels and M. Sudan. A fuzzy vault scheme. In *Proc. IEEE Int'l Symp. on Information Theory*, page 408. IEEE, 2002.

[30] A. Juels and M. Sudan. A fuzzy vault scheme. *Des. Codes Cryptography*, 38(2):237–257, 2006.

[31] T. C. Clancy, N. Kiyavash, and D. J. Lin. Secure smartcard-based fingerprint authentication. In *Proc. ACM SIGMM workshop on Biometrics methods and applications*, WBMA'03, pages 45–52, New York, NY, USA, 2003. ACM.

[32] K. Nandakumar, A. K. Jain, and S. Pankanti. Fingerprint-based fuzzy vault: Implementation and performance. *IEEE Transactions on Information Forensics and Security*, 2(4):744–757, 2007.

[33] A. Nagar, K. Nandakumar, and A. K. Jain. Securing fingerprint template: Fuzzy vault with minutiae descriptors. In *Proc. 19th Int'l Conf. on Pattern Recognition*, pages 1–4. IEEE, 2008.

[34] Y. J. Lee, K. R. Park, S. J. Lee, K. Bae, and J. Kim. A new method for generating an invariant iris private key based on the fuzzy vault system. *IEEE Transactions on Systems, Man, and Cybernetics, Part B: Cybernetics*, 38(5):1302–1313, 2008.

[35] L. Leng and A. B. J. Teoh. Alignment-free row-co-occurrence cancelable palmprint fuzzy vault. *Pattern Recognition*, 48(7):2290–2303, 2015.

[36] W. J. Scheirer and T. E. Boult. Cracking fuzzy vaults and biometric encryption. In *Proc. of Biometrics Symp.*, pages 1–6, 2007.

[37] A. Kholmatov and B. Yanikoglu. Realization of correlation attack against the fuzzy vault scheme. In *Proc. SPIE*, volume 6819, 2008.

[38] B. Tams, P. Mihăilescu, and A. Munk. Security considerations in minutiae-based fuzzy vaults. *IEEE Transactions on Information Forensics and Security*, 10(5):985–998, 2015.

[39] J. Merkle and B. Tams. Security of the improved fuzzy vault scheme in the presence of record multiplicity (full version). *CoRR*, http://arxiv.org/abs/1312.5225abs/1312.5225, 2013.

[40] C. Vielhauer, R. Steinmetz, and A. Mayerhöfer. Biometric hash based on statistical features of online signatures. In *Proc. 16th Int'l Conf. on Pattern Recognition*, pages 123–126. IEEE, 2002.

[41] F. Monrose, M. K. Reiter, and S. Wetzel. Password hardening based on keystroke dynamics. In *Proc. 6th ACM Conf. on Computer and Communications Security*, pages 73–82. ACM, 1999.

[42] J.-P. Linnartz and P. Tuyls. New shielding functions to enhance privacy and prevent misuse of biometric templates. In J. Kittler and M. Nixon, editors, *Proc. 4th Int'l Conf. on Audio- and Video-Based Biometric Person Authentication*, volume 2688 of *LNCS*, pages 393–402. Springer, 2003.

[43] N. K. Ratha, J. H. Connell, and S. Chikkerur. Generating cancelable fingerprint templates. *IEEE Transactions on Pattern Analysis and Machine Intelligence*, 29(4):561–572, 2007.

[44] J. Hämmerle-Uhl, E. Pschernig, and A. Uhl. Cancelable iris biometrics using block re-mapping and image warping. In P. Samarati, M. Yung, F. Martinelli, and C. Ardagna, editors, *Proc. 12th Int'l Information Security Conf.*, volume 5735 of *LNCS*, pages 135–142. Springer, 2009.

[45] J. K. Pillai, V. M. Patel, R. Chellappa, and N. K. Ratha. Secure and robust iris recognition using random projections and sparse representations. *IEEE Transactions on Pattern Analysis and Machine Intelligence*, 33(9):1877–1893, 2011.

[46] A. B. J. Teoh, D. C. L. Ngo, and A. Goh. Biohashing: two factor authentication featuring fingerprint data and tokenised random number. *Pattern Recognition*, 37(11):2245–2255, 2004.

[47] M. Savvides, B. Kumar, and P. Khosla. Cancelable biometric filters for face recognition. In *Proc. 17th Int'l Conf. on Pattern Recognition*, 3:922–925, 2004.

[48] K. Takahashi and S. Hirata. Generating provably secure cancelable fingerprint templates based on correlation-invariant random filtering. In *Proc. IEEE 3rd Int'l Conf. on Biometrics: Theory, Applications, and Systems*, pages 1–6. IEEE, 2009.

[49] T. Boult. Robust distance measures for face-recognition supporting revocable biometric tokens. In *Proc. 7th Int'l Conf. on Automatic Face and Gesture Recognition*, pages 560–566. IEEE, 2006.

[50] T. Boult, W. Scheirer, and R. Woodworth. Revocable fingerprint biotokens: Accuracy and security analysis. In *Proc. IEEE Conf. on Computer Vision and Pattern Recognition*, pages 1–8. IEEE, 2007.

[51] C. Rathgeb, F. Breitinger, and C. Busch. Alignment-free cancelable iris biometric templates based on adaptive bloom filters. In *Proc. of the 6th IAPR Int'l Conf. on Biometrics (ICB '13)*, pages 1–8, 2013.

[52] C. Rathgeb, F. Breitinger, C. Busch, and H. Baier. On the application of bloom filters to iris biometrics. *IET Biometrics*, 3(1):207–218, 2014.

[53] J. Hermans, B. Mennink, and R. Peeters. When a bloom filter is a doom filter: Security assessment of a novel iris biometric. In *Proc. BIOSIG*, 2014.

[54] M. Gomez-Barrero, C. Rathgeb, J. Galbally, C. Busch, and J. Fierrez. Unlinkable and irreversible biometric template protection based on bloom filters. *Information Sciences*, 370–371: 18–32, 2016.

[55] M. Braithwaite, U. von Seelen, J. Cambier, *et al.* Applications-specific biometric template. In *Proc. Workshop on Automatic Identification Advanced Technologies*, pages 167–171. IEEE, 2002.

[56] J. Zuo, N. K. Ratha, and J. H. Connel. Cancelable iris biometric. *Proc. Int'l Conf. on Pattern Recognition (ICPR'08)*, pages 1–4, 2008.

[57] E. Maiorana, P. Campisi, J. Fierrez, J. Ortega-Garcia, and A. Neri. Cancelable templates for sequence-based biometrics with application to on-line signature recognition. *IEEE Transactions on Systems, Man, and Cybernetics – Part A: Systems and Humans*, 40(3):525–538, 2010.

[58] W. Xu, Q. He, Y. Li, and T. Li. Cancelable voiceprint templates based on knowledge signatures. In *International Symposium on Electronic Commerce and Security*, pages 412–415, 2008.

[59] J. Bringer, H. Chabanne, and B. Kindarji. The best of both worlds: Applying secure sketches to cancelable biometrics. *Science of Computer Programming*, 74(1–2):43–51, 2008.

[60] A. Ross, and S. Shah. Segmenting non-ideal irises using geodesic active contours. In *Proc. Biometric Consortium Conf.*, pages 1–6. IEEE, 2006.

[61] A. Ross and A. K. Jain. Information fusion in biometrics. *Pattern Recognition Letters*, 24(13):2115–2125, 2003.

[62] A. K. Jain, B. Klare, and A. A. Ross. Guidelines for best practices in biometrics research. In *Proc. Int'l Conf. on Biometrics (ICB'15)*, pages 1–5, 2015.

[63] A. Nagar, K. Nandakumar, and A. Jain. Multibiometric cryptosystems based on feature-level fusion. *IEEE Transactions on Information Forensics and Security*, 7(1):255–268, 2012.

[64] C. Rathgeb and C. Busch. Multibiometric template protection: Issues and challenges. In *New Trends and Developments in Biometrics*, pages 173–190. Rijeka, Croatia: InTech, 2012.

[65] Y. Sutcu, Q. Li, and N. Memon. Secure biometric templates from fingerprint-face features. In *Proc. IEEE Conf. on Computer Vision and Pattern Recognition*, pages 1–6. IEEE, 2007.

[66] E. J. C. Kelkboom, X. Zhou, J. Breebaart, R. N. J. Veldhuis, and C. Busch. Multi-algorithm fusion with template protection. In *IEEE 3rd International Conference on Biometrics: Theory, Applications, and Systems, 2009. BTAS'09*, pages 1–8, 2009.

[67] C. Rathgeb, A. Uhl, and P. Wild. Reliability-balanced feature level fusion for fuzzy commitment scheme. In *Proc. Int'l Joint Conf. on Biometrics*, pages 1–7. IEEE, 2011.

[68] K. Nandakumar and A. K. Jain. Multibiometric template security using fuzzy vault. In *Proc. IEEE 2nd Int'l Conf. on Biometrics: Theory, Applications, and Systems*, pages 1–6. IEEE, 2008.

[69] B. Tams. Unlinkable minutiae-based fuzzy vault for multiple fingerprints. *IET Biometrics*, 5(3):170–180, 2015.

[70] B. Yang, C. Busch, K. de Groot, H. Xu, and R. N. J. Veldhuis. Decision level fusion of fingerprint minutiae based pseudonymous identifiers. In *2011 International Conference on Hand-Based Biometrics (ICHB)*, pages 1–6, 2011.

[71]   M. Y. Jeong, C. Lee, J. Kim, J. Y. Choi, K. A. Toh, and J. Kim. Changeable biometrics for appearance based face recognition. In *Proc. Biometric Consortium
Conf.*, pages 1–5. IEEE, 2006.

[72]   A. Othman and A. Ross. On mixing fingerprints. *IEEE Transactions on
Information Forensics and Security*, 8(1):260–267, 2013.

[73]   C. Rathgeb and C. Busch. Cancelable multi-biometrics: Mixing iris-codes
based on adaptive bloom filters. *Elsevier Computers and Security*, 42(0):1–12,
2014.

[74]   C. Rathgeb, B. Tams, J. Wagner, and C. Busch. Unlinkable improved
multi-biometric iris fuzzy vault. *EURASIP Journal on Information Security*,
2016(1):26, 2016.

[75]   M.-H. Lim, A. B. J. Teoh, and J. Kim. Biometric feature-type transformation:
Making templates compatible for template protection. *IEEE Signal Processing
Magazine*, 32(5):77–87, 2015.

[76]   M. Ferrara, D. Maltoni, and R. Cappelli. Noninvertible minutia cylinder-code
representation. *IEEE Transactions on Information Forensics and Security*,
7(6):1727–1737, 2012.

[77]   ISO/IEC JTC1 SC37 Biometrics. *ISO/IEC 2382-37:2012 Information Technology – Vocabulary – Part 37: Biometrics*. Geneva, Switzerland: International
Organization for Standardization, 2012.

[78]   ISO/IEC JTC1 SC37 Biometrics. *ISO/IEC SC37 SD11 General Biometric
System*. Geneva, Switzerland: International Organization for Standardization,
May 2008.

[79]   ISO/IEC TC JTC1 SC37 Biometrics. *ISO/IEC 19784-1:2006. Informa-
tion Technology – Biometric Application Programming Interface – Part 1:
BioAPI Specification*. Geneva, Switzerland: International Organization for
Standardization, March 2006.

[80]   SO/IEC JTC1 SC37 Biometrics. *ISO/IEC 19794-1:2011 Information Technology – Biometric Data Interchange Formats – Part 1: Framework*. Geneva,
Switzerland: International Organization for Standardization, June 2011.

[81]   ISO/IEC TC JTC1 SC37 Biometrics. *ISO/IEC 19795-1:2006. Information
Technology – Biometric Performance Testing and Reporting – Part 1: Prin-
ciples and Framework*. Geneva, Switzerland: International Organization for
Standardization and International Electrotechnical Committee, March 2006.

[82]   ISO/IEC JTC1 SC37 Biometrics. *ISO/IEC CD 30136. Information Technology – Security Techniques – Performance Testing of Template Protection
Schemes*. Geneva, Switzerland: International Organization for Standardization, 2016.

[83]   S. Rane. Standardization of biometric template protection. *IEEE MultiMedia*,
21(4):94–99, 2014.

Chapter 9

Handwriting biometrics – feature-based optimisation

Tobias Scheidat[1]

Designing a biometric system, the identification of appropriate features is mostly done based either on expert knowledge or intuition. But there is no guaranty that the extracted features are leading to an optimal authentication performance. In this chapter, statistic methods are proposed to analyse biometric features to select those having a high impact to authentication or hash generation performance and discard those having no or bad impact. Therefore, a short overview on recent related work is given, and appropriate feature-selection strategies are suggested. An exemplary experimental evaluation of the suggested methods is carried out based on two algorithms performing verification as well as biometric hash generation using online handwriting. Test results are determined in terms of equal error rate (EER) to score-verification performance as well as collision reproduction rate to assess hash generation performance. Experimental evaluation shows that the feature subsets based on sequential backward selection provide the best results in 39 out of 80 studied cases in sum for verification and hash generation. In the best case regarding verification, the same feature analysis method leads to a decrease of the EER from 0.07259 down to 0.03286 based on only 26 features out of 131. In hash generation mode, the best results can be determined by only 26 features. Here, the collision reproduction rate decreases from 0.26392 using all 131 features to 0.03142.

9.1 Introduction

The versatile objectives of this chapter comprise an introduction of handwriting as biometric modality based on offline and/or online information, a discussion of challenges of biometric feature selection in context of recognition of individual users as well as to distinguish between different persons using two kinds of feature analysis approaches: wrappers and filters. Building on this, scenarios are suggested for a practical evaluation based on biometric authentication and hash generation which are tested in experiments applying two methods, biometric hash algorithm [1] and secure sketch approach [2] for online handwriting.

[1]Department of Informatics and Media, Brandenburg University of Applied Sciences, Germany

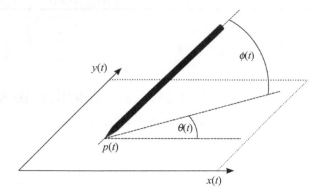

Figure 9.1   Signals provided by a handwriting sensor

Handwriting is well studied in different fields of science. It is established as authentication factor and declaration of intention since hundreds of years, e.g. for signing important documents such as laws, contracts or testaments. Thus, the examination of handwriting is an important part of forensic domain of questioned documents. Typical objectives in this field can be, for example, to find out whether a person is the originator of a signature or not (i.e. biometric signature verification), or to locate regions containing traces of writing on underlying sheets of paper caused by the pen's pressure and make those visible (i.e. forensic analysis of handwriting traces). The findings derived during the forensic examination can be used as an evidence at court of law as well as in criminal investigations (see [3]). Another considerable scientific domain dealing with handwriting is biometric user authentication. In order to obtain individual authentication information from handwriting, features can be extracted either directly from writing result (offline) or from observation of the writing process (online). Offline handwriting authentication is based on the writing trace or on digitised information using x and y coordinates. On the other side, online handwriting biometrics uses time-dependant information provided by special sensors acquiring different signals during the writing process. As shown in Figure 9.1, typical signals are pen position in terms of $x(t)$ and $y(t)$, pressure $p(t)$ of the pen tip and pen orientation angles altitude $\phi(t)$ and azimuth $\theta(t)$. Thus, the sensing results in a sample point sequence (SPS) describing the writing process.

A variety of features can be extracted from the SPS in terms of online and offline characteristics. Features which can be derived easily from such SPS are, for example:

- number of sample points – count points,
- entire writing time – last minus first time stamp,
- writing path – sum of distances between sample points from each pen down to next pen up event,
- aspect ratio of writing sample – based on writing's width and height or
- velocity and acceleration – using writing path and writing time.

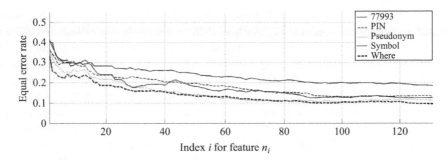

Figure 9.2 *Progression of equal error rate adding new features intuitively, all in all 131 features, values are based on biometric hash algorithm [7] in verification mode*

Besides expert knowledge, intuition plays an important role if features are developed for a biometric system. Thus, new features are often those which are most obvious regarding the available raw and/or pre-processed data. In handwriting biometrics, for example, having a sequence of time-stamped pen positions, some of the most determined features are writing duration, distance, aspect ratio and/or velocity (e.g. see [4–7]). But there is no guaranty that each of the added features is leading to the best authentication performance. In order to visualise the problem, Figure 9.2 shows exemplary progressions of the EER for five different writing contents of 53 test subjects. The rates are determined using the biometric hash algorithm for dynamic handwriting [7]. The curve progression represents the EERs calculated at each iteration of the feature set's element-wise extension. The order of features is given by the sequence of their implementation which is partly intuitive. In the exemplary progressions shown in Figure 9.2, each of the error rate characteristics displays fluctuations especially when based on the first 20 features. Thus, in most cases, the addition of one feature leads to a non-predictable reaction of the authentication performance, which may be increased, decreased or unchanged. The figure illustrates further that features' influence depends also on the written content.

One main advantage of behavioural biometrics is the possibility to change the authentication object by altering the behaviour, for example, in case, a reference information is compromised. A simple method to change the handwriting behaviour is to switch to another written content as described, for example, in [8,9] and [7]. In this work, a selection of five semantics is taken into account. For semantic *GivenPIN*, all test subjects write down an identical numerical sequence 77998, while semantic *SecretPIN* describes a freely chosen combination of five digits. Because of privacy reasons, *Pseudonym* can be seen as substitute of signature. For this semantic, the writers are asked to train a freely chosen name at least ten times before biometric data acquisition. Individual, creative and secret characteristics are represented by semantic *Symbol* where the test subjects are asked to write down a arbitrary sketch. The answer to the question *Where are you from?* represents the fifth semantic *Where*.

In 1989, Plamondon and Lorette presented an overview on handwriting-based authentication methods [4]. Authors differentiate between offline and online approaches and further classify latter in feature-based and functional methods. At the present day, there exists a high number of handwriting-based authentication approaches, and statistical features are very common in that field of biometrics. For example, in [7], Vielhauer suggests 69 offline and online features based on signals of x-, y-position, pen tip pressure and pen orientation angles. A handwriting-verification system, that is not based on statistical features obtained from sensor signals, is presented by Van *et al.* in [5]. The authors suggest an online verification using a combination of a hidden Markov model and a Viterbi path approach. For recent overviews of the biometric modality handwriting (online and offline), various review articles are to be found in literature, see, for example, [10,11] or [12].

In this chapter, we present, discuss and evaluate methods to reduce the number of elements of an initial feature set by removing those features with no or bad influence to authentication as well as hash generation performance. This is a common problem in pattern recognition, which is often referred to as feature selection/analysis. Here, the term performance describes the effectiveness of verification as well as of hash generation with regard to recognition of registered users and discrimination between different users. Vielhauer and Steinmetz investigate in [13] the influence of individual features on the intra-person stability and inter-person value space based on a feature correlation analysis. Kumar and Zhang show in [14] that feature subset selection leads to improvements of recognition of a multimodal biometric system. In [15], Farmanbar and Toygar present a biometric system based on face and palmprint and describe a feature-selection method using backtracking search. Authors report a significant improvement of both computational time and biometric recognition performance. Makrushin *et al.* study, in [16], a selection of statistical methods for feature analysis, which can be either classified as wrapper or filter regarding to John *et al.* [17]. While filters analyse features solely by its statistical relevance, wrapper takes also the targeted classification algorithm into account. The effect of feature subsets determined based on suggested analysis methods is experimental evaluated using two biometric online handwriting algorithms.

Besides feature analysis and selection, the secure handling of biometric data as well as the combination of biometrics and cryptography are recent topics in biometric research during the last years [18]. These two ideas can be realised by an approach which is able to obtain individual constant values, in the sense of cryptographic hashes, from similar biometric information provided by the same person. If this is success-ful, information to determine cryptographic keys do not have any longer memorised or securely stored on a token, but it can be generated just from a physiological or behavioural biometric. Thus, the challenge is to generate identical stable information from a biometric trait that is varying from one acquisition to the next within a certain degree without any other user to generate the same value (or at least with a small probability of doing so). Further, such information should not reveal information to give an explanation about the original biometric data.

Dodis *et al.* suggest in [19] a general principle to generate cryptographic keys from fuzzy data. Therefore, the fuzzy extractor is presented as a primary primitive for key generation. It is based on a *secure sketch* algorithm, which generates helper data (sketch), and a strong extractor to determine a secret key from presented data and sketch. The sketch as helper data is public information, and it is obtained from the reference data acquired during enrolment process. As strong extractor, an arbitrary cryptographic one-way function can be applied. Thus, the sole usage of the secure sketch allows for biometric user authentication, but a secure storage of reference data is only given after the execution of the strong extractor.

In [20], Monrose *et al.* suggest the generation of a cryptographic key using a spoken password based on 12-dimensional vector obtained from cepstral coefficients and a text and speaker-independent acoustic model. This information is used for a segmentation as basis of three kinds of features. These are applied to generate a hash value in terms of the so-called feature descriptor.

Vielhauer *et al.* present a method to determine so-called biometric hashes from online handwriting in [1,7]. Here, a set of statistical data is transferred into a hash domain by using a mapping function and helper data representing intra-class variations of the users. Since the so-called BioHash is one out of two biometric reference algorithms used in this work, in Section 9.3, it is described in more detail. The second reference algorithm is presented by Scheidat *et al.* in [2] and determines secure sketches based on biometric handwriting data.

Sutcu *et al.* present an approach to generate reference data of a biometric face-recognition system based on hash values in [21]. The method assumes that each element of the feature vector of an individual person fluctuates within a certain range. Therefore, for each user, a number of Gauss-functions are generated, and the determined values are stored as reference data. As helper data for the authentication process, the transformation parameters are stored separately (e.g. on a smartcard).

9.2 Biometric feature analysis and selection

The analysis of features suggested in this chapter is based on information extracted from a set of 131 features. According to the terminology introduced by John *et al.* in [17], feature analysis methods can be classified as wrappers and filters. Out of the variety of methods, we focus our view in this chapter on three wrappers and five filters which are described in the following subsections [16].

9.2.1 Wrappers

Using wrappers, an optimal feature subset is determined considering the corresponding biometric classification algorithm. To do so, after each selection of a feature subset, the biometric authentication is carried out, assessed and the next subset is generated based on the result (see also Figure 9.3(a)). Using wrappers, it is theoretically possible to find the optimal feature set for a given data set and classification scheme. However, given the feature set is composed of k features, there are 2^k possible subsets which have to be taken into account during a complete search. Together with running the authentication process in each iteration step, the determination of

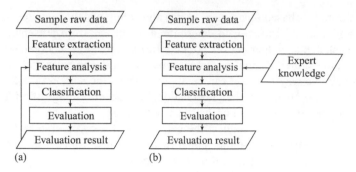

Figure 9.3   Schematic visualisation of function of wrappers (a) and filters (b)

the optimal feature set can be an np-complete problem, thus time consuming and CPU-intensive. In this work, the following three wrappers are studied:

Sequential forward selection (*sfs*). The sequential forward selection starts with an empty feature subset. Within each iteration step, the best performing single feature out of the remaining set is added to the subset. In context of this study, the best feature is the one leading to the lowest EER, if it is added to the subset.

Sequential backward selection (*sbs*). The sequential backward selection starts with the complete feature set and removes the poor features step by step. In each iteration, the currently least performance feature is the one leading to the lowest EER, if it is removed.

Simple wrapper (*simple*). Using this strategy, first step is to determine each feature's individual EER. Second, the optimal feature subset is composed by selecting the *n* features providing the lowest EER.

It should be kept in mind that the suggested three methods are not intended to determine the optimal feature set. However, *sfs*, *sbs* and *simple* provide sufficient trade-offs in acceptable runtime compared to a complete search for given classification data and algorithms.

9.2.2   Filters

In contradiction to wrappers, filters analyse the relevance of features by using statistical methods without consideration of the algorithm. Instead, filters use expert knowledge to estimate the features' relevance as shown in Figure 9.3(b). Thus, the determination of a feature subset is faster than using wrappers because it is done within one single step. In this work, we focus on five methods out of three main approaches for feature analysis: ANOVA (*anova*, *anova-2class*), *correlation* and entropy (*entropy*, *joint-entropy*) [22]. This is only a small selection out of possible analysis methods which is motivated by encouraging results from previous work.

$$
\begin{array}{cccccccc}
y & x & & & x_i & & & \\
\Downarrow & \Downarrow & & & \Downarrow & & & \\
y^1 : x^1 & = (x_1^1, & x_2^1, & \cdots, & x_i^1, & \cdots, & x_m^1) \\
y^2 : x^2 & = (x_1^2, & x_2^2, & \cdots, & x_i^2, & \cdots, & x_m^2) \\
\vdots & \vdots & \vdots & \vdots & \vdots & & \vdots \\
y^j : x^j & = (x_1^j, & x_2^j, & \cdots, & x_i^j, & \cdots, & x_m^j) \\
\vdots & \vdots & \vdots & \vdots & \vdots & & \vdots \\
y^n : x^n & = (x_1^n, & x_2^n, & \cdots, & x_i^n, & \cdots, & x_m^n)
\end{array}
\tag{9.1}
$$

To describe the filter, we define the variables shown in (9.1). Given is a set of feature vectors x and the corresponding set of user IDs y. In context of biometrics, the index j is the number of feature vectors as well as the corresponding user IDs. Thus, each feature vector x^j of the connected user y^j can be defined as $x^j = (x_1^j, x_2^j, \ldots, x_m^j)$. Corresponding to the number of extracted features, m is the dimension of each feature vector. Attention should be paid to the following assumptions: to each user, one individual positive value (user ID) is assigned, the user IDs do not have to be ordered in any kind and one ID can be assigned to more than one feature vector (i.e. multiple feature vectors for single IDs). Each feature on a random index i can be visualised for all users as vector of features $x_i = (x_i^1, x_i^2, \ldots, x_i^n)$ while the vector of identities is defined as $y = (y^1, y^2, \ldots, y^n)$. Variable n denotes the number of feature vectors within the test set. Further, we define the set of all feature vectors which can be assigned to an individual user ID k as x^{jk}. The application of a filter determines for each feature i a ranking value $R(i)$ based on $\langle x_i, y \rangle$.

Analysis of variance (*anova-2class*). The relation between inter-class variability and intra-class variability can be calculated by a set of statistical methods named ANOVA (ANalysis Of VAriance). It can be assumed that a feature provides the better discriminatory power, the smaller its variance σ within its own class and the higher the variance to other classes, at the same time. Based on two different users, the latter fact can be described as difference of averages μ_1 and μ_2 of the ith feature of both users. To generalise from a two-user to a two-class problem, it can be assumed that one class is represented by a single user which is currently considered, and the other class consists of all other persons within the test set. Thus, the ANOVA test for these two classes and the feature i can be defined as shown in (9.2) (*anova-2class*). While the number of used test samples provided by first and second class is denoted by N_1 and N_2, σ_1 and σ_2 are the corresponding variances of the values of feature i.

$$
R(i) = \frac{N_1 * N_2 * (\mu_1 - \mu_2)^2}{N_1 * \sigma_1^2 + N_2 * \sigma_2^2}
\tag{9.2}
$$

Analysis of variance (*anova*). On the other side, ANOVA can also be applied as multivariate test (*anova*). Therefore, the variation of one feature of a person with ID k is represented by the sum of deviations of this feature over all samples x^{jk} of the same user k in relation to the corresponding average value μ_k. The variation of one feature between different persons is calculated as sum of deviations of the user-specific averages μ_k related to the global average μ. The ranking value $R(i)$ of feature i can be determined as shown in (9.3), where K is the number of users, and N_k represents the number of samples provided by user k.

$$R(i) = \frac{\frac{1}{K-1} \sum_{k=1}^{K} N_k(\mu_k - \mu)^2}{\frac{1}{N-K} \sum_{k=1}^{K} \sum_{j=1}^{N_k} (x^{jk} - \mu_k)^2} \tag{9.3}$$

Correlation (*correlation*). The correlation between features and users is used as quality measure. Therefore, the correlation-based ranking value is calculated using Pearson correlation coefficient as shown in (9.4). The idea is that the discriminatory power for different classes is higher if the correlation between features and classes is small. In (9.4), are μ_x and μ_y the average values of feature i and of identities, correspondingly. In the case of maximum correlation, if ranking value $R(i)$ is equal to 1, feature i is absolute relevant in context of authentication. On the other side, a feature is absolute irrelevant, if there is no correlation between feature and user ($R(i) = 0$). Please keep in mind, the correlation considers only linear relations.

$$R(i) = \left| \frac{\sum_{x^j} \sum_{y^j} (x^j - \mu_x)(y^j - \mu_y)}{\sqrt{\sum_{x^j} (x^j - \mu_x)^2 * \sum_{y^j} (y^j - \mu_y)^2}} \right| \tag{9.4}$$

Entropy (*entropy*). On one side, with the appliance of entropy for biometric purpose it is assumed that a relevant feature provides a high inter-class entropy. On the other side, the intra-class entropy has to be very small. The calculation of *entropy* is based on the sum of entropies of the average feature values of the users $\mu = (\mu_1, \mu_2, \dots, \mu_k)$ (inter-class entropy) and on the sum of entropies of the feature values x_{jk} of individual users (intra-class entropy). The ranking $R(i)$ is determined as ratio of both sums according to (9.5). In this equation, N_μ, N_k and K are defined as number of: users' averages, samples of person k and users.

$$R(i) = \frac{\sum_{j=1}^{N_\mu} H_j(\mu) * \log_2 \frac{1}{H_j(\mu)}}{\sum_{k=1}^{K} \sum_{j=1}^{N_k} H_j(x^{jk}) * \log_2 \frac{1}{H_j(x^{jk})}} \tag{9.5}$$

Entropy (*joint-entropy*). Empirical estimation of mutual information between features and user IDs describes the quality of the individual feature. This information can be calculated as shown in (9.6). Since it is hard to guess real values of distribution $P(X = x_i)$ of feature x_i, prior class probability $P(Y = y)$ and probability of joint observation $P(x = x_i, Y = y)$, the incidences determined from the evaluation data can be applied, if features are discrete.

$$R(i) = \sum_{x_i} \sum_{y} P(X = x_i, Y = y) * \log \frac{P(X = x_i, Y = y)}{P(X = x_i) * P(Y = y)} \tag{9.6}$$

9.3 Reference algorithms

Two biometric algorithms for online handwriting are used to study the influence of feature-based optimisation: the biometric hash algorithm presented by Vielhauer *et al.* in [1] and the handwriting-based secure sketch method suggested by Scheidat *et al.* in [2], based on works of Dodis *et al.* [19]. Both algorithms can be applied to generate biometric hashes (*hash generation mode*) as well as to use these hashes for authentication purpose (*verification mode*).

9.3.1 Biometric hash algorithm

In order to generate biometric hashes, the biometric hash algorithm [1] has to calculate user-dependent helper data during the enrolment process. Based on this information, a biometric hash can be generated using a single handwriting probe which can be stored as reference or authentication data. The corresponding processes are described in this section.

As usual, in most biometric systems, during the enrolment process, reference data of a user j is generated. Therefore, user has to write n samples which are digitised each into an SPS containing spatial, temporal, pressure and angle-based information for discrete points in time (see Section 9.1). From each SPS, the feature-extraction module calculates k statistical features which are stored in the corresponding vectors $w_{j1}, \ldots, w_{jn}$. The first $(n-1)$ feature vectors $w_{j1}, \ldots, w_{jn-1}$ are used to determine the helper data in terms of the interval matrix IM_j taking into account user-specific intra-class variances. The degree of the influence of the variations can be regulated by the parameters tolerance factor tf and tolerance vector tv.

The reference biometric hash vector b_j is determined based on the interval matrix IM_j and the remaining feature vector w_{jn}. Combined with writer's ID, vector b_j can be used either as reference for biometric authentication mode or as comparative value to evaluate the hash generation. Each further biometric hash b_j' of person j can be generated by the mapping function using the individual interval matrix IM_j and the feature vector w_j' of a currently presented writing probe. In order to compare b_j and b_j', a distance function can be applied.

To describe the biometric hash algorithm formal, the following notations are used: number of persons M where j indicates the jth person, number of features N where i indicates the ith feature and f_{ij} is the value of ith feature of jth person. Since the calculation steps of all features are identical, in the following equations only one feature on index i is considered. This notation is also used in Section 9.3.2 to describe the handwriting-based secure sketch algorithm. The following depiction is based on the formal descriptions of the biometric hash algorithm given by Vielhauer *et al.* in [1,7].

In order to define the corresponding interval of feature i of user j based on its variability as shown in (9.7), the initial values $I_{\text{InitHigh } ij}$ and $I_{\text{InitLow } ij}$ have to be determined as maximum and minimum over the number of training samples S_j of user j.

$$\Delta I_{ij} = I_{\text{High } ij} - I_{\text{Low } ij} \tag{9.7}$$

$$I_{\text{InitHigh } ij} = \max_{r=1,S_j} (f_{ijr}) \tag{9.8}$$

$$I_{\text{InitLow } ij} = \min_{r=1,S_j} (f_{ijr}) \tag{9.9}$$

The initial value $\Delta I_{\text{init } ij}$ of the interval width is defined as difference of $I_{\text{InitHigh } ij}$ and $I_{\text{InitLow } ij}$:

$$\Delta I_{\text{Init } ij} = I_{\text{InitHigh } ij} - I_{\text{InitLow } ij} \tag{9.10}$$

Next, the upper interval border $I_{\text{High } ij}$ is calculated using (9.11) while the lower border $I_{\text{Low } ij}$ results from (9.12).

$$I_{\text{High } ij} = \lceil I_{\text{InitHigh } ij} + tv_{ij} * \Delta I_{\text{Init } ij} * tf \rceil \tag{9.11}$$

$$I_{\text{Low}} = \begin{cases} \lfloor I_{\text{InitLow } ij} - tv_{ij} * \Delta I_{\text{Init}} * tf \rfloor & \text{if } \lfloor I_{\text{InitLow } ij} - tv_{ij} * \Delta I_{\text{Init } ij} * tf \rfloor > 0 \\ 0 & \text{else} \end{cases} \tag{9.12}$$

In (9.11) and (9.12), tv_{ij} and tf are parameters to modify the interval width. Tolerance vector tv affects each feature by an individual parameter value tv_{ij} while tolerance factor tf is one scalar value which is applied for all users and each feature. Based on training samples, both helper data are determined a priori to obtain information about intra-class variation (see also [7]).

The following step determines the starting point of the interval as offset Ω_{ij}, using lower interval border $I_{\text{Low } ij}$ and interval width ΔI_{ij}.

$$\Omega_{ij} = I_{\text{Low } ij} \bmod \Delta I_{ij} \tag{9.13}$$

Thus, the interval matrix IM_j of a person j consists of one vector containing all feature-dependent interval widths ΔI_{ij} and one vector holding all corresponding offsets Ω_{ij}.

$$IM_j = (\Delta I_j, \Omega_j) \tag{9.14}$$

The interval matrix IM_j is stored as helper data of person j and allows the calculation of the user-specific biometric hash vector b_j. Therefore, each feature value f_{ij} is transferred to the hash domain using the mapping function and the corresponding helper data in IM_j:

$$b_{ij} = \left\lfloor \frac{(f_{ij} - \Omega_{ij})}{\Delta I_{ij}} \right\rfloor \tag{9.15}$$

Together with the interval matrix, a reference hash vector has to be stored in order to compare it to a currently generated hash vector in authentication as well as in hash generation mode.

9.3.2 Secure sketch algorithm

The theoretical concept of a fuzzy extractor based on a secure sketch is introduced by Dodis *et al.* in [19]. The basic idea of the approach is the generation of reference data (secure sketch) from input samples, that allows the reconstruction of the input if the currently presented information sufficient similar.

In the case of biometric authentication, the approach can be adopted as follows: during the generation step, a secure sketch s_j is calculated as public knowledge from the feature vector w_j, determined from reference samples of user j. Based on the feature vector w_j' of a currently presented sample of person j and the corresponding secure sketch s_j, the feature vector w_j can be reconstructed if w_j' is similar to w_j in a certain degree. The suggested secure sketch approach for online handwriting [2,16] is inspired by the face-verification system introduced by Sutcu *et al.* in [23]. The method is summarised in the following.

In order to reduce the variance of the feature values, a feature-wise quantisation is carried out using a global (i.e. same for all enrolled users) quantisation step δ_i as shown in (9.17). The number of training samples donated by user j is given in S_j while var_{ij} is in (9.16) to determine the individual quantisation step.

$$\text{var}_{ij} = \max_{r=1,S_j} (f_{ijr}) - \min_{r=1,S_j} (f_{ijr}) \tag{9.16}$$

$$\delta_i = \max\left(\min_{j=1,M} (\text{var}_{ij}), 1 \right) \tag{9.17}$$

As initial start value of each feature i of user j the average avg_{ij} is calculated based on smallest and highest value of current feature over all training sets.

$$\text{avg}_{ij} = 0.5 \cdot \left(\max_{r=1,S_j} (f_{ijr}) + \min_{r=1,S_j} (f_{ijr}) \right) \tag{9.18}$$

In the next processing steps, var_{ij} and avg_{ij} are quantised using δ_i whereby *ef* is used as expansion factor of the individual quantisation interval σ_{ij}. The expansion factor *ef* affects the output of the algorithm in verification as well as in hash generation mode.

$$\sigma_{ij} = \left\lceil \frac{ef \cdot \text{var}_{ij}}{\delta_i} \right\rceil \tag{9.19}$$

The entire reference vector w_j of a user j is determined by element-wise calculation of w_{ij} as shown in (9.20).

$$w_{ij} = \left\lfloor \frac{\text{avg}_{ij}}{\delta_i} \right\rfloor \tag{9.20}$$

The sketch s_{ij}, used as helper data during the reconstruction process, is calculated in (9.21) as difference of quantised average w_{ij} and corresponding codeword c_{ij} whose determination is shown in (9.22).

$$s_{ij} = w_{ij} - c_{ij} \tag{9.21}$$

$$c_{ij} = \text{round}\left(\frac{w_{ij}}{2\sigma_{ij} + 1} \right) \cdot (2\sigma_{ij} + 1) \tag{9.22}$$

Since the sketch s_{ij} is available as helper data, the codeword c_{ij} can be generated from a test feature vector t_j extracted from a sample written by user j if t_j is sufficient similar to the vector containing the average feature values avg_j.

$$w_{ij}' = \left\lfloor \frac{t_{ij}}{\delta_i} \right\rfloor \tag{9.23}$$

In the next step, the sketch is subtracted and the user-dependent quantisation is carried out:

$$c'_{ij} = \text{round}\left(\frac{w'_{ij} - s_{ij}}{2\sigma_{ij} + 1}\right) \cdot (2\sigma_{ij} + 1) \tag{9.24}$$

$$w_{ij} = c'_{ij} + s_{ij} \tag{9.25}$$

Based on the assumption that vectors w_j and w'_j are similar to a certain degree, vectors c_j and c'_j determined by described transformation have to be identical. Therefore, reference w_{ij} can be calculated as sum of the reconstructed codeword and the sketch as shown (9.25).

9.3.3   Handwriting-based features

Table 9.1 summarises all 131 features which are applied as initial feature vector for evaluation of verification as well as hash generation mode of both algorithms. Features are based on time-stamped SPS of x and y coordinates, pressure and pen orientation angles. List's order corresponds to chronological order of implementation of the features.

9.4   Experimental evaluation

This section presents the methodology of experimental evaluation of feature analysis methods introduced in Section 9.2. Therefore, the underlying database, the experimental setup and the measures are described to assess verification as well as hash generation performance of both algorithms (see Section 9.3). Further, the evaluation results are presented and discussed.

9.4.1   Methodology

To assess the performance of both algorithms' verification mode based on the feature subsets determined by the methods suggested in Section 9.2, common biometric error rates are applied: false match rate, false non-match rate (FNMR) and EER. The hash generation mode is evaluated using the special hash-related measurements suggested by Scheidat *et al.* in [24] and Makrushin *et al.* in [25]: reproduction rate (RR), collision rate (CR) and collision reproduction rate (CRR). While RR is calculated as relative sum of genuine hash reproduction trials of identical users, CR is determined as relative sum of non-genuine hash reproduction trials of different users. CRR is used as trade-off to compare biometric hash generation systems and is calculated as follows:

$$\text{CRR} = \frac{\text{CR} + (1 - \text{RR})}{2} \tag{9.26}$$

In order to generate the database, 53 persons were asked to donate ten samples for five semantics each in three sessions. The distance between two subsequent sessions

Table 9.1 *Short description of biometric features calculated for biometric hash and secure sketch*

Feature	Description
1	Total writing time in ms
2	Total number of event pixels
3	Image width DIV height
4, 5	Average velocity in x, y direction in pixels/ms
6	Number of consecutive pen-down segments
7, ..., 10	Minimum, maximum absolute x-, y-velocity
11, 12	Centroid of horizontal, vertical pen position in bounding box
13	Distance of centroid from origin
14	Maximum absolute pressure
15	Centroid of horizontal pen position normalised to bounding box width
16	Centroid of vertical pen position normalised to bounding box height
17	Distance of centroid from origin normalised to bounding box diameter
18	Horizontal azimuth of centroid from origin normalised to $\pi/2$
19, ..., 22	Maximum, minimum absolute altitude, azimuth of pen
23	Ratio of average and maximum pressure
24, 25	Average azimuth, altitude
26	Normalised average velocity in x direction in pixels/VxMax
27	Normalised average velocity in y direction in pixels/VyMax
28	Absolute cumulated pen-up time in ms
29	Ratio of pen-up time by total writing time
30	Total number of sample values
31	Total absolute path length in pixels
32, ..., 43	Number of pixels in rth row; cth column ($1 \leq r \leq 3, 1 \leq c \leq 4$)
44, 45	Numeric integration of normalised x, y values
46, ..., 55	Numeric integration of x, y values for ath one-fifth time period ($1 \leq a \leq 5$)
56, 57	Average pen-down, pen-up pressure
58	Baseline angle of the sample
59, ..., 61	Histogram of y for zone a in % ($1 \leq a \leq 3$)
62	Ratio of area (convex hull) and area (bounding box)
63	Ratio of area [convex hull (segments)] and area [convex hull (sample)]
64	Ratio of area [convex hull (segments)] and area (bounding box)
65	Ratio of path length (convex hull) and path length (bounding box)
66	Ratio of path length [convex hull (segments)] and path length [convex hull (sample)]
67	Ratio of path length [convex hull (segments)] and path length (bounding box)
68, 69	Histogram of x for left, right in %
70, 71	Number of maximum, minimum points in the X signal
72, 73	Number of maximum, minimum points in the Y signal
74	Ratio of number of maximums of X and Y
75	Ratio of number of minimums of X and Y
76	Number of intersections within writing trace itself
77, ..., 80	Number of intersections of the vertical line Xa with the sample ($1 \leq a \leq 4$)
81, ..., 83	Number of intersections of the horizontal line Ya with the sample ($1 \leq a \leq 3$)
84, 85	Number of intersections of the diagonal line Da with the sample ($1 \leq a \leq 2$)
86	Ratio of distance (start point; end point) and path length
87	Ratio of distance (maximum X point; minimum X point) and path length
88	Ratio of distance (maximum Y point; minimum Y point) and path length
89	Ratio of distance (start point; centroid) and distance (end point; centroid)
90, ..., 93	Mapped function of x-, y-, p-, acceleration extrema

(*continues*)

Table 9.1 (continued)

Feature	Description
94	Average distance between segments
95, ..., 97	Minimum, maximum, average number of neighbour points
98, ..., 100	Average angle between 0–30, 30–60, 60–90 degrees
101, ..., 103	Count of angles between 0–30, 30–60, 60–90 degrees
104, ..., 106	Maximum, minimum, average distance between two strokes
107	Mapped size of enclosed areas
108, ..., 119	Normalised x-, y-coordinate, each of cluster a ($0 \le a \le 5$)
120, ..., 125	Average pressure of xy cluster a ($0 \le a \le 5$)
126, ..., 129	Average altitude of pressure cluster a ($0 \le a \le 3$)
130	Mapped speed at the inflection points of the sample
131	Standard deviation of the pressure

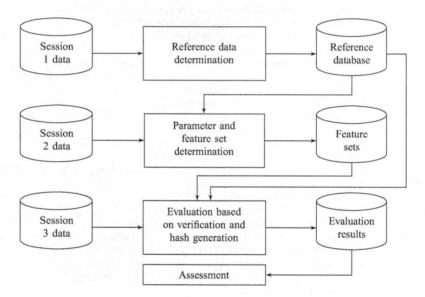

Figure 9.4   Usage of data acquired in three sessions for experimental evaluation

amounts approximately 1 month. The biometric sensor is a Toshiba Portege M200 tablet PC. As shown in Figure 9.4, data of the three sessions is used as follows:

Data acquired in first session is used for determination of reference data by biometric hash method (interval matrix IM, reference biometric hash) as well as by secure sketch method (secure sketch, reference feature vector).
Data acquired in second session is utilised to estimate parameters for biometric hash (tolerance factor *tf*) and secure sketch (expansion factor *ef*) in order to optimise the methods for either verification or hash generation. Further data is

used to carry out feature analysis based on the approaches presented in Section 9.2 for biometric hash method and secure sketch method.

Data acquired in third session is applied to calculate performance measures EER and collision reproduction rate for both biometric hash and secure sketch based on feature subsets determined a priori.

The references generated using first session's data are applied in the second session to determine the best feature subsets based on the feature selection by wrapper and filters. Further, data obtained in the third session is used as test data for verification and hash generation which is compared to reference data from first session.

In order to find the best optimised feature subset for verification and hash generation, two scenarios in terms of optimisation goals are taken into account. For scenario *bestEER* that feature subset is determined which leads to the smallest equal error in verification mode for the currently considered combination of algorithm, feature analysis method and semantic. The evaluation result of such combination consists of the values of EER as well as collision reproduction rate. The procedure for scenario *bestCRR* is the same in context of hash generation. Here, the feature subset is identified having the smallest collision reproduction rate for a certain combination, and as a result, the corresponding values of EER and CRR are stored.

Within the experimental evaluation, there are all in all 80 tests carried out based on the determined feature sets which are built from two algorithms, parameterised to generate minimal EER as well as minimal CRR, using five semantics in two scenarios while EER and CRR are measured for each combination.

Based on described database, experimental setup and evaluation methodology, in the next section, the following results are presented and discussed for both biometric algorithms, feature-selection strategies and scenarios:

Analysis of feature-selection methods. Identification of those feature-selection strategies leading to the best verification and hash generation results based on the different scenarios and semantics.

Error rate determination. EERs and CRRs determined before and after feature selection are presented and compared in context of different scenarios as well as semantics for verification and hash generation mode.

9.4.2 Evaluation results

The discussion of evaluation results focuses mainly on the comparison of the ability of suggested feature analysis methods to improve verification and/or hash generation performance. Due to the high number of tested combinations of semantics, analysis methods and reference algorithms, only a short overview on a limited selection of results in terms of biometric error rates for both, hash generation and verification, is given.

Table 9.2 shows the absolute frequencies of occurrence with that each feature analysis method (filters: columns 5–9, wrappers: columns 10–12) generates the best result for a combination of scenario and algorithm independent from semantics.

Table 9.2   Number of best results reached by the feature subsets generated by the individual feature analysis methods (values are accumulated over all five semantics)

Scenario	Algorithms	Rate	All	Anova	Anova-2class	Correlation	Entropy	Joint-entropy	Simple	Sfs	Sbs
					Filters				**Wrappers**		
bestEER	Biometric hash	EER	0	0	1	1	1	0	0	4	4
		CRR	0	0	0	0	0	0	2	3	6
	Secure sketch	EER	0	0	3	0	1	0	0	1	5
		CRR	0	0	1	0	0	0	0	2	6
bestCRR	Biometric hash	EER	4	0	0	0	0	1	0	5	0
		CRR	0	0	0	0	0	0	0	3	7
	Secure sketch	EER	4	0	0	0	0	0	0	2	4
		CRR	0	0	0	0	0	0	0	3	7

As a reference, *all* features are also taken into account and are listed in Column 4. A visualisation of the results shown in Table 9.2 is given in Figure 9.5. It can be seen from column *all* in Table 9.2 that there are four cases each for biometric hash and secure sketch where the usage of all features leads to the best results in terms of EER in scenario *bestCRR*. Using the secure sketch, this is the highest frequency compared to filters in this combination. Another important observation from Table 9.2 and Figure 9.5 is that in every studied case at least one wrapper outperforms all filters. On the other side, having a look at wrapper *simple*, it can be seen that a feature set determined by this method leads in only two cases (bestEER, biometric hash, CRR) to the best collision reproduction rate.

Results calculated based on feature sets which are determined by the five filter methods only provide 11.25% of the best results while the percentages of wrapper-based subsets and all features amount 80% and 10%, correspondingly. This observation suggests to focus on wrapper methods such as sfs and sbs in the field of handwriting-based biometric feature analysis and selection. To check this assumption, in the following sections, a selection of verification and hash generation results are discussed.

Table 9.3 shows biometric hash-based results of verification (columns *EER*) and hash generation (columns *CRR*) for both scenarios, *bestEER* and *bestCRR*, using semantic Symbol. Column *n* contains the number of features used to obtain corresponding EER and CRR. Best results for each class of analysis methods are printed bold; best result out of all methods for the individual measures is printed italic.

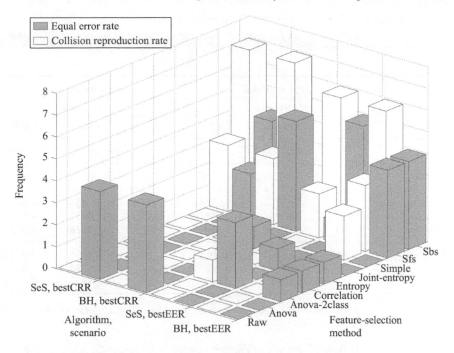

Figure 9.5 Visualisation of absolute frequencies at which each individual feature analysis method generates the best feature subset for individual algorithm and scenario (values are accumulated over all five semantics; BH – biometric hash, SeS – secure sketch)

The detailed discussion of achievable accuracy improvements is limited to semantic *Symbol*. A short overview of enhancements in context of all semantics tested is given in Tables 9.5 and 9.6.

A comparison of the verification results at scenario bestEER shows a small improvement of 2.9% based on the best filter method joint-entropy (see Table 9.3(a), $EER_{all} = 0.07366$ vs. $EER_{joint-entropy} = 0.07150$). The number of selected features amounts 111. However, using 79 features, the best wrapper sfs outperforms this value by a decrease of the EER by 41.4% ($EER_{sfs} = 0.04339$). At the same time, with a decrease of 63.1% sfs provides also the bestCRR for this scenario with 0.04338. This corresponds to a reproduction rate of 0.95660 and collision rate of 0.04336. In scenario bestCRR, the bestEER of 0.08316 is performed on 111 features identified by filter joint-entropy. Here, the improvement amounts 10.2% compared to 0.09259 using all features. With a value of 0.05994 and an improvement of 33.7%, the best CRR of this scenario is determined based on the 77 features selected by wrapper sbs. The corresponding reproduction rate amounts 0.97547 while the collision rate is equal to 0.09536.

The verification and hash generation results for semantic *Symbol* based on secure sketch algorithm are shown in Table 9.4. Here, the improvement caused by feature

Table 9.3   Biometric hash – results of feature selection for semantic Symbol and scenarios (a) bestEER using tf parameterisation to minimise EER and (b) bestCRR using tf parameterisation to minimise CRR

| Selection | (a) bestEER | | | (b) bestCRR | | |
| | Symbol | | | Symbol | | |
	n	EER	CRR	n	EER	CRR
All	131	0.07366	0.13610	131	0.09259	0.09048
Anova	98	0.07229	**0.11854**	84	0.08870	**0.06910**
Anova-2class	113	0.07151	0.13458	114	0.09187	0.08971
Correlation	118	0.07366	0.13610	84	0.09464	0.08861
Entropy	118	0.07366	0.13610	116	0.09049	0.08853
Joint-entropy	111	**0.07150**	0.11978	111	*0.08316*	0.08224
Simple	76	0.07048	0.10684	56	0.09466	0.07225
Sfs	79	*0.04339*	*0.04338*	102	**0.08784**	0.06094
Sbs	83	0.05039	0.04933	77	0.08905	*0.05994*

Bold printed values: best result of each class of analysis methods; italic printed values: best result out of all methods for individual measure.

Table 9.4   Secure sketch – results of feature selection for semantic Symbol and scenarios (a) bestEER using ef parameterisation to minimise EER and (b) bestCRR using ef parameterisation to minimise CRR

| Selection | (a) bestEER | | | (b) bestCRR | | |
| | Symbol | | | Symbol | | |
	n	EER	CRR	n	EER	CRR
All	131	0.07259	0.26392	131	**0.10021**	0.17393
Anova	98	0.06132	0.23456	68	0.14023	0.11471
Anova-2class	63	0.06212	**0.18790**	7	0.11486	**0.11464**
Correlation	115	0.06857	0.26116	61	0.12498	0.12155
Entropy	118	0.07259	0.26392	8	0.12169	0.11484
Joint-entropy	107	**0.05755**	0.25490	92	0.12974	0.13585
Simple	63	0.05430	0.20171	11	0.14395	0.10762
Sfs	76	0.04845	0.15492	81	0.07427	0.07299
Sbs	26	*0.03286*	*0.03142*	55	*0.07368*	*0.06669*

Bold printed values: best result of each class of analysis methods; italic printed values: best result out of all methods for individual measure.

analysis and selection is slightly better than using the biometric hash algorithm for scenario bestEER. Within this scenario, the smallest EER of 0.03286 is determined by the feature set provided by wrapper sbs. This EER corresponds to an improvement of 54,7% compared to the EER of 0.07259 calculated based on all 131 features.

Table 9.5   *Biometric hash – best results of EER and CRR for individual semantics and underlying feature-selection strategy, scenario and number n of features used*

Selection scenario	Measure	Given PIN	Secret PIN	Pseudonym	Symbol	Where
All features	EER	0.20082	0.12790	0.08943	0.07366	0.08559
($n = 131$)	CRR	0.49253	0.49906	0.45004	0.13610	0.50000
bestEER	EER	0.17395	0.12187	0.08515	0.04339	0.07488
	n	63	100	106	79	40
	Best strategy	Anova-2class	Sbs	Sfs	Sfs	Simple
bestCRR	CRR	0.16323	0.13944	0.08316	0.05994	0.07066
	n	75	71	60	77	80
	Best strategy	Sbs	Sbs	Sbs	Sbs	Sbs

The number of selected features amounts 26 in this constellation. Based on the same feature subset, the bestCRR within scenario bestEER is determined ($CRR_{sbs} = 0.03142$). The underlying values for RR and CR are 0.96604 and 0.02888, correspondingly. The percentage decrease of CRR compared to those calculated based on all features (0.26392) amounts 88.1%. The lowest CRR in scenario bestCRR is provided by wrapper sbs for semantic Symbol. Here, the collision reproduction rate is decreased from 0.17393 using all features to 0.06669 using 55 features with a decrease of 61.7%. The corresponding rates for reproduction as well as collision are 0.94151 and 0.07489. Applying the same feature subset for verification, the EER amounts 0.07368.

Because of limited space in this chapter, the results determined for the other semantics and parameterisations cannot be shown here completely. A short presentation of selected results of the other semantics is given in Tables 9.5 and 9.6. However, the tendency described for semantic Symbol using both reference algorithms can be observed for the other four semantics as well. As mentioned above, semantic Symbol provides the best results in context of verification and hash generation. However, in Tables 9.5 and 9.6, an overview is given on best results determined for measures EER and CRR for alternative semantics using biometric hash and secure sketch algorithms. In both tables, the first two rows summarise the rates using all 131 features. The following rows show results for scenarios bestEER and bestCRR by giving bestEER/bestCRR values, number of selected features and selection strategy used. For both algorithms, in most cases, the worst results are determined for *given PIN* and *secret PIN*. This phenomenon may cause partly by the small size of the underlying alphabet (0, . . . , 9) and the limited number of characters allowed (five digits). On the other side, semantics Pseudonym and where show that an individual self-chosen content based on a more comprehensive alphabet leads to better results in both verification and hash mode.

Table 9.6   Secure sketch – best results of EER and CRR for individual semantics and underlying feature-selection strategy, scenario and number n of features used

Selection scenario	Measure	Given PIN	Secret PIN	Pseudonym	Symbol	Where
All feature ($n = 131$)	EER	0.16211	0.10927	0.08757	0.07259	0.06817
	CRR	0.50000	0.50000	0.50000	0.26392	0.50000
bestEER	EER	0.1531	0.09452	0.08424	0.03286	0.04823
	n	41	68	33	26	70
	Best strategy	Anova-2class	Anova-2class	Anova-2class	Sbs	Entropy
bestCRR	CRR	0.16538	0.12280	0.13137	0.06669	0.10586
	n	78	86	73	55	74
	Best strategy	Sbs	Sbs	Sbs	Sbs	Sbs

Values shown in Tables 9.3–9.6 provide an overview of the results determined for biometric hash and secure sketch algorithm for verification and hash generation. The selection of the most relevant features may cause a problem in context of privacy, especially in hash generation mode, if it leads to a reduction to a very small number of features. This phenomenon can be seen, for example, in Table 9.4 semantic Symbol (lines *anova-2class*, *entropy* and *simple*). For example, feature analysis strategy anova-2class selects only seven features, and calculation of CRR results in 0.11464 (RR = 0.88491, CR = 0.11419) based on secure sketch. The determination of the corresponding feature values would be an easy task for a potential attacker, i.e. using reverse engineering attacks as described in [26,27]. On the other side, it is possible to overcome such privacy problems of too few features and/or high CRR by using an alternative semantic. For example, using the semantic Symbol, the CRR decreases to 0.05994 (RR = 0.97547, CR = 0.09536, 77 features) for biometric hash and to 0.06669 (RR = 0.94151, CR = 0.07489, 55 features) for secure sketch.

9.5   Conclusions and future work

In most cases, the designing process of features describing a biometric trait is more intuitive than based on well-grounded expert knowledge. As the result, features varying from very relevant down to worsening in the context of the biometric application. The separation of features into relevant and not relevant subsets regarding authentication as well as hash generation is an important area in biometric research. In this chapter, eight statistic methods are suggested to analyse handwriting-based features in order to select appropriate subsets for verification and hash generation modes of two reference algorithms. It has been shown that the wrappers outperform the filters as well as the complete feature set in 80% of the studied cases. Nevertheless, in 91.25%

of the tested constellations (73 out of 80), the eight suggested feature analysis methods generate feature subsets which are led to better results than the entire set of features.

The results confirm the assumed advantages of wrappers sfs and sbs towards the complete feature set, the wrapper simple and all studied filters. The results are also showing an influence of the selected feature subset on the mode of operation of the biometric system. Mostly, the best results of verification as well as hash generation are based on different feature sets, which are especially determined for the respective application, semantic and parameterisation by one of the three wrappers. One disadvantage of the wrappers is given by the fact that at any alteration within the biometric system, the time and CPU consuming feature analysis process has to be repeated. Such alterations include, for example, the addition of new users or new features, the change of algorithmic parameters or the replacement of the algorithm. At this point, filter can show their strength: the analysis is working independent from the targeted biometric system. Thus, decisions made by filters are not influenced by changes, which concerning the algorithm. Please note, the results and conclusions discussed in this chapter are evaluated only on the database and methodology described in Section 9.4.1.

The application of the findings to other handwriting databases or biometric modalities has to be studied in future work. Since the order of the features is not taken into account here, further work will examine the influence of alternative orders on authentication and hash generation performance. Another interesting point is the effect on feature analysis methods if the database is extended by additional data of already registered persons as well as of new users. Further, an individual user-based feature analysis and selection is planned using the most successful methods presented in this work. Studies on reverse engineering based on significantly reduced feature sets towards privacy analysis is another part of future work.

Acknowledgements

I would like to thank all test persons who donated the high number of handwriting samples. My special thanks goes to Jana Dittmann, Claus Vielhauer and Andrey Makrushin for their support and a lot of fruitful discussions on biometrics, security and feature analysis.

References

[1] Claus Vielhauer, Ralf Steinmetz, and Astrid Mayerhoefer, "Biometric hash based on statistical features of online signatures," *International Conference on Pattern Recognition*, vol. 1, pp. 10123, 2002.

[2] Tobias Scheidat, Claus Vielhauer, and Jana Dittmann, "Biometric hash generation and user authentication based on handwriting using secure sketches," in *Proceedings of 6th International Symposium on Image and Signal Processing and Analysis (ISPA)*, 2009.

[3]   D. Ellen, *Scientific Examination of Documents: Methods and Techniques, Third Edition,* Bosa Roca: International Forensic Science and Investigation. CRC Press, third edition, September 2005.

[4]   Rejean Plamondon and Guy Lorette, "Automatic signature verification and writer identification – the state of the art," *Pattern Recognition,* vol. 22, no. 2, pp. 107–131, 1989.

[5]   Van-Bao Ly, Sonia Garcia-Salicetti, and Bernadette Dorizzi, "Fusion of HMM's likelihood and Viterbi path for on-line signature verification," in *Biometric Authentication, ECCV 2004 International Workshop, BioAW 2004, Prague, Czech Republic, May 15, 2004, Proceedings,* 2004, pp. 318–331.

[6]   Julian Fierrez-Aguilar, Loris Nanni, Jaime Lopez-Peñalba, Javier Ortega-Garcia, and Davide Maltoni, *An On-Line Signature Verification System Based on Fusion of Local and Global Information,* pp. 523–532, Springer, Berlin, 2005.

[7]   Claus Vielhauer, *Biometric User Authentication for IT Security: From Fundamentals to Handwriting (Advances in Information Security),* Springer-Verlag New York, Inc., Secaucus, NJ, USA, 2006.

[8]   Christiane Schmidt, *On-line-Unterschriftenanalyse zur Benutzerverifikation,* Berichte aus der Elektrotechnik. Shaker Verlag GmbH, 1999.

[9]   Yosuke Kato, Takayuki Hamamoto, and Seiichiro Hangai, "A proposal of writer verification of hand written objects," in *IEEE International Conference on Multimedia and Expo, ICME 2002.* IEEE, August 2002, vol. II, pp. 585–588, IEEE.

[10]   Yazan M. Al-Omari, Siti Norul Huda Sheikh Abdullah, and Khairuddin Omar, "State-of-the-art in offline signature verification system," in *2011 International Conference on Pattern Analysis and Intelligence Robotics,* June 2011, vol. 1, pp. 59–64.

[11]   Sameera Khan and Avinash Dhole, "A review on offline signature recognition and verification techniques," *International Journal of Advanced Research in Computer and Communication Engineering,* vol. 3, no. 6, pp. 6879–6882, June 2014.

[12]   Rania A. Mohammed, Rebaz M. Nabi, Sardasht M-Raouf Mahmood, and Rebwar M. Nabi, "State-of-the-art in handwritten signature verification system," in *2015 International Conference on Computational Science and Computational Intelligence (CSCI),* December 2015, pp. 519–525.

[13]   Claus Vielhauer and Ralf Steinmetz, "Handwriting: Feature correlation analysis for biometric hashes," *EURASIP Journal on Advances in Signal Processing,* vol. 2004, pp. 542–558, January 2004.

[14]   Ajay Kumar and David Zhang, "Biometric recognition using feature selection and combination," in *Proceedings of 5th International Conference on Audio- and Video-Based Biometric Person Authentication (AVBPA),* T. Kanade, A. K. Jain, and N.K. Ratha, Eds. number 3546 in LNCS, pp. 813–822, Springer Berlin Heidelberg, 2005.

[15]   Mina Farmanbar and Önsen Toygar, "Feature selection for the fusion of face and palmprint biometrics," *Signal, Image and Video Processing,* vol. 10, no. 5, pp. 951–958, 2016.

[16] Andrey Makrushin, Tobias Scheidat, and Claus Vielhauer, "Improving reliability of biometric hash generation through the selection of dynamic handwriting features," *Transactions on Data Hiding and Multimedia Security VIII – Special Issue on Pattern Recognition for IT Security*, vol. 7228, pp. 19–41, 2012.

[17] George H. John, Ron Kohavi, and Karl Pfleger, "Irrelevant features and the subset selection problem," in *Proceedings of the Eleventh International Conference on Machine Learning*, 1994, vol. 129, pp. 121–129, Morgan Kaufmann.

[18] Anil K. Jain, Karthik Nandakumar, and Abhishek Nagar, "Biometric template security," *EURASIP Journal on Advances in Signal Processing*, vol. 2008, pp. 1–17, 2008.

[19] Yevgeniy Dodis, Leonid Reyzin, and Adam Smith, "Fuzzy extractors: How to generate strong keys from biometrics and other noisy data," in *EUROCRYPT*, 2004, pp. 523–540.

[20] Fabian Monrose, Michael K. Reiter, Qi Li, and Susanne Wetzel, "Using voice to generate cryptographic keys," in *In Proc. of Odyssey 2001, The Speaker Verification Workshop*, 2001, pp. 237–242.

[21] Yagiz Sutcu, Husrev Taha Sencar, and Nasir Memon, "A secure biometric authentication scheme based on robust hashing," in *Proceedings of the 7th workshop on Multimedia and security*. 2005, pp. 111–116, ACM New York, NY, USA.

[22] Isabelle Guyon and André Elisseeff, "An introduction to variable and feature selection," *Journal of Machine Learning Research*, vol. 3, pp. 1157–1182, 2003.

[23] Yagiz Sutcu, Qiming Li, and Nasir D. Memon, "Protecting biometric templates with sketch: Theory and practice," *IEEE Transactions on Information Forensics and Security*, vol. 2, no. 3–2, pp. 503–512, 2007.

[24] Tobias Scheidat, Claus Vielhauer, and Jana Dittmann, "Advanced studies on reproducibility of biometric hashes," in *Biometrics and Identity Management, First European Workshop, BIOID 2008*, B. Schouten, N. C. Juul, A. Drygajlo, and M. Tistarelli, Eds. 2008, number 5372 in LNCS, pp. 150–159, Springer Verlag Berlin, Heidelberg.

[25] Andrey Makrushin, Tobias Scheidat, and Claus Vielhauer, "Towards robust biohash generation for dynamic handwriting using feature selection," in *Proceedings of 17th Conference on Digital Signal Processing*, 2011.

[26] Karl Kmmel, Claus Vielhauer, Tobias Scheidat, Dirk Franke, and Jana Dittmann, "Handwriting biometric hash attack: A genetic algorithm with user interaction for raw data reconstruction," in *11th Joint IFIP TC6 and TC11 Conference on Communications and Multimedia Security*, 2010.

[27] Karl Kümmel and Claus Vielhauer, "Reverse-engineer methods on a biometric hash algorithm for dynamic handwriting," in *Proceedings of the 12th ACM Workshop on Multimedia and Security*, New York, NY, USA, 2010, MM&Sec'10, pp. 67–72, ACM.

Chapter 10

Presentation attack detection in voice biometrics

Pavel Korshunov* and Sébastien Marcel*

Recent years have shown an increase in both the accuracy of biometric systems and their practical use. The application of biometrics is becoming widespread with fingerprint sensors in smartphones, automatic face recognition in social networks and video-based applications, and speaker recognition in phone banking and other phone-based services. The popularization of the biometric systems, however, exposed their major flaw—high vulnerability to spoofing attacks [1]. A fingerprint sensor can be easily tricked with a simple glue-made mold, a face recognition system can be accessed using a printed photo, and a speaker recognition system can be spoofed with a replay of prerecorded voice. The ease with which a biometric system can be spoofed demonstrates the importance of developing efficient anti-spoofing systems that can detect both known (conceivable now) and unknown (possible in the future) spoofing attacks.

Therefore, it is important to develop mechanisms that can detect such attacks, and it is equally important for these mechanisms to be seamlessly integrated into existing biometric systems for practical and attack-resistant solutions. To be practical, however, an attack detection should have (i) high accuracy, (ii) be well generalized for different attacks, and (iii) be simple and efficient.

One reason for the increasing demand for effective presentation attack detection (PAD) systems is the ease of access to people's biometric data. So often, a potential attacker can almost effortlessly obtain necessary biometric samples from social networks, including facial images, audio and video recordings, and even extract fingerprints from high-resolution images. Therefore, various privacy-protection solutions, such as legal privacy requirements and algorithms for obfuscating personal information, e.g., visual privacy filters [2], as well as, social awareness of threats to privacy can also increase security of personal information and potentially reduce the vulnerability of biometric systems.

In this chapter, however, we focus on PAD in voice biometrics, i.e., automatic speaker verification (ASV) systems. We discuss vulnerabilities of these systems to presentation attacks (PAs), present different state-of-the-art PAD systems, give the insights into their performances, and discuss the integration of PAD and ASV systems.

*Idiap Research Institute, Switzerland

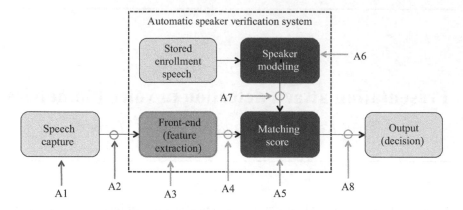

Figure 10.1   Possible attacks places in a typical ASV system

10.1   Introduction

Given the complexity of a practical speaker verification system, several different modules of the system are prone to attacks, as it is identified in ISO/IEC 30107-1 standard [3] and illustrated by Figure 10.1. Depending on the usage scenario, two of the most vulnerable places for spoofing attacks in an ASV system are marked by A1 (aka "physical access" as defined in [4] or PAs) and A2 (aka "logical access" attacks as defined in [4]) in the figure. In this chapter, we are considering A1 and A2 attacks, where the system can be attacked by presenting a spoofed signal as input. For the other points of attacks from A3 to A9, the attacker needs to have privileged access rights and know the operational details of the biometric system. Prevention of or countering such attacks is more related to system security and is thus out of the scope of this chapter.

There are three prominent methods through which A1 and A2 attacks can be carried out: (i) by recording and replaying the target speakers speech, (ii) synthesizing speech that carries target speaker characteristics, and (iii) by applying voice conversion methods to convert impostor speech into target speaker speech. Among these three, replay attack is the most viable attack, as the attacker mainly needs a recording and playback device. In the literature, it has been found that ASV systems, while immune to "zero-effort" impostor claims and mimicry attacks [5], are vulnerable to such PAs [6]. One of the reasons for such vulnerability is a built-in ability of biometric systems in general, and ASV systems in particular, to handle undesirable variabilities. Since spoofed speech can exhibit the undesirable variabilities that ASV systems are robust to, the attacks can pass undetected.

Therefore, developing mechanisms for detection of PAs is gaining interest in the speech community [7]. In that regard, the emphasis until now has been on logical access attacks, largely thanks to the "ASV Spoofing and Countermeasures Challenge" [4], which provided a large benchmark corpus containing voice conversion-based and speech-synthesis-based attacks. In the literature, development of PAD systems has largely focused on investigating short-term speech-processing-based features that can aid in discriminating genuine speech from spoofed signal. This includes cepstral-based features [8], phase information [9], and fundamental frequency-based information, to name a few.

However, having PAD methods is not enough for practical use. Such PAD systems should be seamlessly and effectively integrated with existing ASV systems. In this chapter, we integrate speaker verification and PAD systems by using score fusion considering parallel fusion (see Figure 10.5) and cascading fusion (see Figure 10.6) schemes. The score fusion-based systems integration allows to separate *bona fide* data of the valid users, who are trying to be verified by the system, from both PAs and genuine data of the nonvalid users or so-called *zero-impostors*. For ASV system, we adopt verification approaches based on intersession variability (ISV) modeling [10] and *i-vectors* [11], as the state-of-the-art systems for speaker verification.

10.1.1 Databases

Appropriate databases are necessary for testing different PAD approaches. These databases need to contain a set of practically feasible PAs and also data for speaker-verification task, so that a verification system can be tested for both issues: the accuracy of speaker verification and the resistance to the attacks.

Currently, two comprehensive publicly available databases exist that can be used for vulnerability analysis of ASV systems, the evaluation of PAD methods, and evaluation of joint ASV-PAD systems: ASVspoof[1] and AVspoof.[2] Both databases contain logical access (LA) attacks, while AVspoof also contains PAs. For the ease of comparison with ASVspoof, the set of attacks in AVspoof is split into LA and PA subsets (see Table 10.1).

10.1.1.1 ASVspoof database

The ASVspoof[3] database contains genuine and spoofed samples from 45 male and 61 female speakers. This database contains only speech synthesis and voice conversion attacks produced via logical access, i.e., they are directly injected in the system. The attacks in this database were generated with ten different speech synthesis and voice-conversion algorithms. Only five types of attacks are in the training and development set ($S1$–$S5$), while ten types are in the evaluation set ($S1$–$S10$). Since last five attacks appear in the evaluation set only, and PAD systems are not trained on them, they are considered "unknown" attacks (see Table 10.1). This split of attacks allows to evaluate the systems on known and unknown attacks. The full description of the database and the evaluation protocol are given in [4]. This database was used for the ASVspoof 2015 Challenge and is a good basis for system comparison as several systems have already been tested on it.

---

[1] http://dx.doi.org/10.7488/ds/298
[2] https://www.idiap.ch/dataset/avspoof
[3] http://dx.doi.org/10.7488/ds/298

Table 10.1   *Number of utterances in different subsets of AVspoof and ASVspoof*
             databases

Database	Type of data	Train	Dev	Eval
AVspoof	Enroll data	780	780	868
	Impostors	54,509	54,925	70,620
	Real data	4,973	4,995	5,576
	LA attacks	17,890	17,890	20,060
	PA attacks	38,580	38,580	43,320
ASVspoof	Enroll data	–	175	230
	Impostors	–	9,975	18,400
	Real data	3,750	3,497	9,404
	Known attacks	12,625	49,875	92,000
	Unknown attacks	–	–	92,000

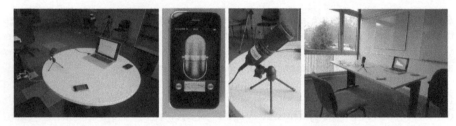

Figure 10.2   AVspoof database recording setup

10.1.1.2   AVspoof database

To our knowledge, the largest publicly available database containing speech PAs is AVspoof [6].[4]

AVspoof database contains real (genuine) speech samples from 44 participants (31 males and 13 females) recorded over the period of 2 months in four sessions, each scheduled several days apart in different setups and environmental conditions such as background noises. The recording devices, including microphone AT2020USB+, Samsung Galaxy S4 phone, and iPhone 3GS, and the environments are shown in Figure 10.2. The first session was recorded in the most controlled conditions.

From the recorded genuine data, two major types of attacks were created for AVspoof database: logical access attacks, similar to those in ASVspoof database [4], and PAs. Logical access attacks are generated using (i) a statistical parametric-based speech-synthesis algorithm [12] and (ii) a voice-conversion algorithm from Festvox.[5]

When generating PAs, the assumption is that a verification system is installed on a laptop (with an internal built-in microphone), and an attacker is trying to gain access to this system by playing back to it a prerecorded genuine data or an automatically

---

[4]https://www.idiap.ch/dataset/avspoof
[5]http://festvox.org/

generated synthetic data using some playback device. In AVspoof database, PAs consist of (i) direct replay attacks when a genuine data is played back using a laptop with internal speakers, a laptop with external high-quality speakers, Samsung Galaxy S4 phone, and iPhone 3G, (ii) synthesized speech replayed with a laptop, and (iii) converted voice attacks replayed with a laptop.

The data in AVspoof database is split into three nonoverlapping subsets: training or *Train* (real and spoofed samples from four female and ten male participants), development or *Dev* (real and spoofed samples from four female and ten male participants), and evaluation or *Eval* (real and spoofed samples from 5 female and 11 male participants). For more details on AVspoof database, please refer to [6].

10.1.2 Evaluation

Typically, in a single-database evaluation, the training subset of a given database is used for training a PAD or an ASV system. The development set is used for determining hyper-parameters of the system, and evaluation set is used to test the system. In a cross-database evaluation, the training and development sets are taken from one database, while evaluation set is taken from another database. For PAD systems, a cross-attack evaluation is also possible, when the training and development sets contain one type of attack, e.g., logical access attacks only, while evaluation set contains another type, e.g., presentation or replay attacks only.

Recent recommendations 30107-3 [13] from ISO/IEC committee specify the evaluation procedure and metrics for ASV, PAD, and joint ASV-PAD systems, which we briefly present in this chapter.

ASV and joint ASV-PAD systems are evaluated under two operational scenarios: *bona fide* scenario with no attacks and the goal to separate genuine samples from zero-effort impostors and *spoof* scenario with the goal to separate genuine samples from attacks. For bona fide scenario, we report false match rate (FMR), which is similar to false accept rate (FAR), and false nonmatch rate (FNMR), which is similar to false reject rate (FRR), while for spoof scenario, we report impostor attack presentation match rate (IAPMR), which is the proportion of attacks that incorrectly accepted as genuine samples by the joint ASV-PAD system (for details, see recommendations in ISO/IEC 30107-3 [13]).

For evaluation of PAD systems, the following metrics are recommended: attack presentation classification error rate (APCER) and bona fide presentation classification error rate (BPCER). APCER is the number of attacks misclassified as bona fide samples divided by the total number of attacks and is defined as follows:

$$\text{APCER} = \frac{1}{N} \sum_{i=1}^{N} (1 - Res_i), \tag{10.1}$$

where N represents the number of attack presentations. Res_i takes value 1 if the ith presentation is classified as an attack presentation and value 0 if classified as a bona fide presentation. Thus, APCER can be considered as the equivalent to FAR for PAD systems, as it reflects the observed ratio of falsely accepted attack samples in relation to the total number of presented attacks.

By definition, BPCER is the number of incorrectly classified bona fide (genuine) samples divided by the total number of bona fide samples:

$$\text{BPCER} = \frac{\sum_{i=1}^{N_{BF}} Res_i}{N_{BF}}, \tag{10.2}$$

where N_{BF} represents the number of bona fide presentations, and Res_i is defined similar to APCER. Thus, BPCER can be considered as the equivalent to FRR for PAD systems, as it reflects the observed ratio of falsely rejected genuine samples in relation to the total number of bona fide (genuine) samples. We compute equal error rate (EER) as the rate when APCER and BPCER are equal.

When analyzing, comparing, and especially fusing PAD and ASV systems, it is important that the scores are calibrated in a form of likelihood ratio. Raw scores can be mapped to log-likelihood ratio scores with logistic regression (LR) classifier and an associated cost of calibration C_{llr} together with a discrimination loss $C_{llr}^{\min}$ are then used as application-independent performance measures of calibrated PAD or ASV systems. Calibration cost C_{llr} can be interpreted as a scalar measure that summarizes the quality of the calibrated scores. A well-calibrated system has $0 \leq C_{llr} < 1$ and produces well-calibrated likelihood ratio. Discrimination loss $C_{llr}^{\min}$ can be viewed as the theoretically best C_{llr} value of an optimally calibrated systems. For more details on the score calibration and C_{llr} and $C_{llr}^{\min}$ metrics, please refer to [14].

Therefore, in this chapter, we report EER rates (on Eval set) when testing the considered PAD systems on each database, for the sake of consistency with the previous literature, notably [15], and BPCER and APCER of PAD systems (using the EER threshold computed on Dev set) when testing PADs in cross-database scenario. EER has been commonly used within the speech community to measure the performance of ASV and PAD systems, while BPCER and APCER are the newly standardized metrics, and we advocate for the use of the open evaluation standards in the literature. We also report calibration cost C_{llr} and the discrimination loss $C_{llr}^{\min}$ metrics for the individual PAD systems. FMR, FNMR, and IAPMR are reported for ASV and joint ASV-PAD systems on evaluation set (using EER threshold computed on the development set).

10.2 Vulnerability of voice biometrics

The research on ASV is more established with regular competitions conducted by National Institute of Standards and Technology since 1996.[6] Many techniques have been proposed with the most notable systems based on Gaussian mixture model (GMM), ISV modeling [10], joint factor analysis [16], and *i-vectors* [11].

To demonstrate vulnerability of ASV systems to PAs, we consider two systems based on ISV modeling [10] and *i-vectors* [11], which are the state-of-the-art speaker-verification systems able to effectively deal with intra-class and interclass variability. In these systems, voice activity detection is based on the modulation of the energy

---

[6]http://www.nist. gov/itl/iad/mig/sre.cfm

Table 10.2   ISV-based and i-vector ASVs on evaluation set of AVspoof database

ASV system	Zero-impostors only		PAs only
	FMR (%)	FNMR (%)	IAPMR (%)
ISV-based	4.46	9.90	92.41
i-vectors-based	8.85	8.31	94.04

around 4 Hz; the features include 20 mel-scale frequency coefficients (MFCC) and energy, with their first and second derivatives, and modeling was performed with 256 Gaussian components using 25 expectation-maximization (EM) iterations. Universal background model (UBM) was trained using training set of publicly available mobile biometry (MOBIO) database,[7] while the clients models are built using an enrollment data from the development set of AVspoof database (only genuine data).

In *i-vectors*-based system, for a given audio sample, the supervector of GMM mean components (computed for all frames of the sample) is reduced to an *i-vector* of the dimension 100, which essentially characterizes the sample. These *i-vectors* are compensated for channel variability using linear discriminative analysis and within class covariance normalization techniques (see [11] for more details).

Table 10.2 demonstrates how *i-vectors* and ISV-based ASV systems perform in two different scenarios: (i) when there are no attacks present (zero-impostors only), referred to as *bona fide* scenario (defined by ISO/IEC [13]) and (ii) when the system is being spoofed with PAs, referred to as *spoof* scenario. Histograms of score distribution in Figure 10.3(b) also illustrate the effect of attacks on *i-vectors*-based ASV system in *spoof* scenario, compared to *bona fide* scenario in Figure 10.3(a).

From Table 10.2, it can be noted that both ASV systems perform relatively well under *bona fide* scenario with ISV-based system showing lower FMR of 4.46%. However, when a spoofed data is introduced, without a PAD system in place, the IAPMR significantly increases reaching 92.41% for ISV-based and 94.04% for *i-vectors*-based systems. It means that a typical verification system is not able to correctly distinguish PAs from genuine data.

10.3   Presentation attack detection approaches

As was shown in the previous section, ASV systems are highly susceptible to PAs. This vulnerability motivated researchers to propose different systems and methods for detecting such attacks (see Figure 10.4). In this section, we present the most commonly used recent approaches and discuss feature extraction and classification components, as well as, score fusion integration technique.

---

[7]https://www.idiap.ch/dataset/mobio

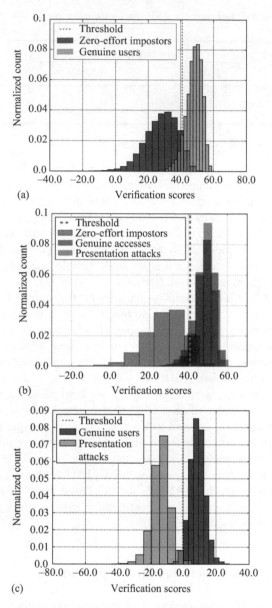

*Figure 10.3   Histogram distributions of scores from i-vector-based ASV system in
bona fide and spoof scenario and MFCC-based PAD system. (a) ASV
bona fide scenario, (b) ASV spoof scenario, and (c) PAD system*

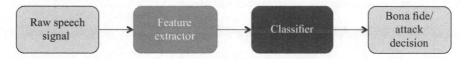

Figure 10.4 Presentation attack detection system

10.3.1 Features

A survey by Wu *et al.* [7] provides a comprehensive overview of both the existing spoofing attacks and the available attack-detection approaches. An overview of the methods for synthetic speech detection by Sahidullah *et al.* [15] benchmarks several existing feature-extraction methods and classifiers on ASVspoof database.

Existing approaches to feature extraction for speech spoofing attack detection methods include spectral- and cepstral-based features [8], phase-based features [9], the combination of amplitude and phase features of the spectrogram [17], and audio-quality-based features [18]. Features directly extracted from a spectrogram can also be used, as per the recent work that relies on local maxima of spectrogram [19].

Compared to cepstral coefficients, using phase information extracted from the signal seem to be more effective for anti-spoofing detection, as it was shown by De Leon *et al.* [9] and Wu *et al.* [20]. However, the most popular recent approaches rely on the combination of spectral-based and phase-based features [17,21–23]. Most of these features are used successfully in speaker-verification systems already, so, naturally, they are first to be proposed for anti-spoofing systems as well.

In addition to these spectral-based features, features based on pitch frequency patterns have been proposed [24,25]. There are also methods that aim to extract "pop-noise"-related information that is indicative of the breathing effect inherent in normal human speech [8].

Constant Q cepstral coefficients (CQCCs) [26] features were proposed recently, and they have shown a superior performance in detecting both known and unknown attacks in ASVspoof database. Also, a higher computational layer can be added, for instance, Alegre *et al.* [27] proposed to use histograms of local binary patterns (LBP), which can be computed directly from a set of preselected spectral, phase-based, or other features.

10.3.2 Classifiers

Besides determining "good features for detecting PAs," it is also important to correctly classify the computed feature vectors as belonging to bona fide or spoofed data. Choosing a reliable classifier is especially important given a possibly unpredictable nature of attacks in a practical system, since it is unknown what kind of attack the perpetrator may use when spoofing the verification system. The most common approach to classification is to use one of the well-known classifiers, which is usually pre-trained on the examples of both real and spoofed data. To simulate realistic environments, the classifier can be trained on a subset of the attacks, termed *known*

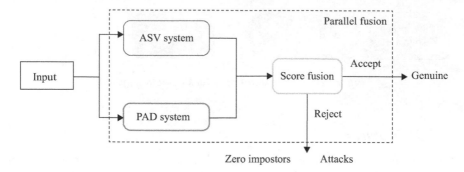

Figure 10.5   A joint ASV-PAD system based on parallel score fusion

attacks, and tested on a larger set of attacks that include both known and *unknown attacks*.

Different methods use different classifiers but the most common choices include LR, support vector machine (SVM), and classifiers. The benchmarking study on logical access attacks [15] finds GMMs to be more successful compared to two-class SVM (combined with an LBP-based feature extraction from [27]) in detecting synthetic spoofing attacks. Deep learning networks are also showing promising performance in simultaneous feature selection and classification [28].

10.3.3   Fusion

Fusion of different features or the results of different classification systems is a natural way of combining different systems, in our case, PAD and ASV systems to create a joint verification system resistant to the attacks.

In this chapter, we focus on a score-level fusion as a means to integrate different ASV and PAD systems into one joint system. Due to relative simplicity of such fusion and the evidence that it leads to a better performing combined systems, this operation has become popular among researchers. However, the danger is to rely on score fusion blindly without studying how it can affect different systems in different scenarios.

One way to fuse ASV and PAD systems at the score level is to use a parallel scheme, as it is illustrated in Figure 10.5. In this case, the scores from each of N system are combined into a new feature vector of length N that need to be classified. The classification task can be performed using different approaches, and, in this chapter, we consider three different algorithms: (i) a logistic regression classifier, denoted as "LR," which leads to a straight line separation, as illustrated by the scatter plot in Figure 10.7(a), (ii) a polynomial logistic regression, denoted as "PLR," which results in a polynomial separation line, and (iii) a simple mean function, denoted as "mean," which is taken on scores of the fused systems. For "LR" and "PLR" fusion, the classifier is pre-trained on the score-feature vectors from a training set.

Another common way to combine PAD and ASV systems is a cascading scheme, in which one system is used first, and only the samples that are accepted by this system (based on its own threshold) are then passed to the second system, which

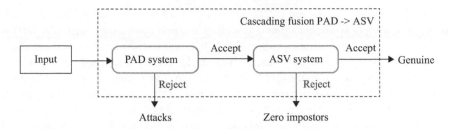

Figure 10.6   A joint PAD-ASV system based on cascading score fusion (reversed order of the systems leads to the same results)

will further filter the samples, using its own independently determined threshold. Effectively, cascading scheme can be viewed as a *logical and* of two independent systems. Strictly speaking, when considering one PAD and one ASV systems, there are two variants of cascading scheme: (i) when ASV is used first, followed by PAD, and (ii) when PAD is used first, followed by ASV (see Figure 10.6). Although these schemes are equivalent, i.e., *and* operation is commutative, and they both lead to the same filtering results (the same error rates), we consider variant (ii), since it is defined in ISO/IEC 30107-1 standard [3].

When using a score-level fusion, it is important to perform a thorough evaluation of the combined/fused system to understand how incorporating PAD system affects verification accuracy for both real and spoofed data. In the upcoming parts of this chapter, we therefore adopt and experimentally apply an evaluation methodology specifically designed for performance assessment of fusion system proposed in [29].

10.4   PADs failing to generalize

To demonstrate the performance of PAD systems in single-database and in cross-database scenario, we have selected several state of the art methods for presentation attack detection in speech, which were recently evaluated by Sahidullah *et al.* [15] on ASVspoof database with an addition of CQCC features-based method [26].

These systems rely on GMM-based classifier (two models for real and attacks, 512 Gaussians components with 10 EM iterations for each model), since it has demonstrated improved performance compared to SVM on the data from ASVspoof database. Four cepstral-based features with mel-scale, i.e., MFCC [30], rectangular frequency cepstral coefficients (RFCC), inverted mel-scale frequency cepstral coefficients (IMFCC), and linear frequency cepstral coefficients (LFCC) filters [31], were selected. These features are computed from a power spectrum (power of magnitude of 512-sized fast Fourier transform) by applying one of the above filters of a given size (we use size 20 as per [15]). Spectral flux-based features, i.e., subband spectral flux coefficients (SSFC) [32], which are Euclidean distances between power spectrums (normalized by the maximum value) of two consecutive frames, subband centroid frequency coefficient (SCFC) [33], and subband centroid magnitude coefficient (SCMC) [33], are considered as well. A discrete cosine transform (DCT-II) is applied to these above features, except for SCFC, and first 20 coefficients are taken. Before computing selected features, a given audio sample is first split into overlapping 20-ms-long speech frames with 10-ms overlap. The frames are pre-emphasized with 0.97 coefficient and preprocessed by applying Hamming window. Then, for all features, deltas and double deltas [34] are computed and only these derivatives (40 in total) are used by the classifier. Only deltas and delta-deltas are kept, because [15] reported that static features degraded performance of PAD systems.

In addition to the above features, we also consider recently proposed CQCC [26], which are computed using constant Q transform instead of fast Fourier transform (FFT). To be consistent with the other features and fair in the systems comparison, we used also only delta and delta-deltas (40 features in total) derived from 19 plus C_0 coefficients.

The selected PAD systems are evaluated on each ASVspoof and AVspoof database and in cross-database scenario. To keep results comparable with current state-of-the-art work [15,35], we computed average EER (Eval set) for single-database evaluations and APCER with BPCER for cross-database evaluations. APCER with BPCER are computed for Eval set of a given dataset using the EER threshold obtained from the Dev set from another dataset (see Table 10.4).

To avoid prior to the evaluations, the raw scores from each individual PAD system are pre-calibrated with LR based on Platts sigmoid method [36] by modeling scores of the training set and applying the model on the scores from development and evaluation sets. The calibration cost C_{llr} and the discrimination loss C_{llr}^{min} of the resulted calibrated scores are provided.

In Table 10.3, the results for known and unknown attacks (see Table 10.1) of *Eval* set of ASVspoof are presented separately to demonstrate the differences between these two types of attacks provided in ASVspoof database. The main contribution to the higher EER of unknown is given by a more challenging attack "S10" of the evaluation set.

Since AVspoof contains both LA attacks and PAs, the results for these two types of attacks are also presented separately. Hence, it allows to compare the performance on ASVspoof database (it has logical access attacks only) with an AVspoof-LA attacks.

From the results in Table 10.3, we can note that (i) LA set of AVspoof is less challenging compared to ASVspoof for almost all methods, (ii) unknown attacks for which PAD is not trained is more challenging, and (iii) PAs are also more challenging compared to LA attacks.

Table 10.4 presents the cross-database results when a given PAD system is trained and tuned using training and development sets from one database but is tested using evaluation set from another database. For instance, results in the second column of the table are obtained by using training and development sets from ASVspoof database but evaluation set from AVspoof-LA. Also, we evaluated the effect of using one type of attacks (e.g., logical access from AVspoof-LA) for training and another type (e.g., PAs of AVspoof-PA) for testing (the results are in the last column of the table).

Table 10.3 *Performance of PAD systems in terms of average EER (%), C_{llr}, and C_{llr}^{min} of calibrated scores for evaluation sets of ASVspoof [4] and AVspoof [6] databases*

PADs	ASVspoof (eval)							AVspoof (eval)					
	Known			S10	Unknown			LA			PA		
	EER	C_{llr}	C_{llr}^{min}	EER	EER	C_{llr}	C_{llr}^{min}	EER	C_{llr}	C_{llr}^{min}	EER	C_{llr}	C_{llr}^{min}
SCFC	0.11	0.732	0.006	23.92	5.17	0.951	0.625	0.00	0.730	0.000	5.34	0.761	0.160
RFCC	0.14	0.731	0.009	6.34	1.32	0.825	0.230	0.04	0.729	0.001	3.27	0.785	0.117
LFCC	0.13	0.730	0.005	5.56	1.20	0.818	0.211	0.00	0.728	0.000	4.73	0.811	0.153
MFCC	**0.47**	**0.737**	**0.023**	**14.03**	**2.93**	**0.877**	**0.435**	**0.00**	**0.727**	**0.000**	**5.43**	**0.812**	**0.165**
IMFCC	0.20	0.730	0.007	5.11	1.57	0.804	0.192	0.00	0.728	0.000	4.09	0.797	0.137
SSFC	0.27	0.733	0.016	7.15	1.60	0.819	0.251	0.70	0.734	0.027	4.70	0.800	0.160
SCMC	0.19	0.731	0.009	6.32	1.37	0.812	0.229	0.01	0.728	0.000	3.95	0.805	0.141
CQCC	0.10	0.732	0.008	1.59	0.58	0.756	0.061	0.66	0.733	0.028	3.84	0.796	0.128

Table 10.4 *Performance of PAD systems in terms of average APCER (%), BPCER (%), and C_{llr} of calibrated scores in cross-database testing on ASVspoof [4] and AVspoof [6] databases*

| PADs | ASVspoof (train/dev) | | | | | | AVspoof-LA (train/dev) | | | | | |
| | AVspoof-LA (eval) | | | AVspoof-PA (eval) | | | ASVspoof (eval) | | | AVspoof-PA (eval) | | |
	APCER	BPCER	C_{llr}	APCER	BPCER	C_{llr}	APCER	BPCER	C_{llr}	APCER	BPCER	C_{llr}
SCFC	0.10	2.76	0.751	10.20	2.76	0.809	15.12	0.00	0.887	39.62	0.35	0.970
RFCC	0.29	69.57	0.887	7.51	69.57	0.927	26.39	0.00	0.902	48.32	2.86	0.988
LFCC	1.30	0.13	0.740	21.03	0.13	0.868	17.70	0.00	0.930	37.49	0.02	0.958
MFCC	**1.20**	**2.55**	**0.764**	**17.09**	**2.55**	**0.838**	**10.60**	**0.00**	**0.819**	**19.72**	**1.22**	**0.870**
IMFCC	4.57	0.00	0.761	92.98	0.00	1.122	99.14	0.00	1.164	43.00	0.60	0.966
SSFC	4.81	64.47	0.899	18.89	64.47	0.973	71.84	0.68	1.047	63.45	23.54	1.070
SCMC	0.75	1.70	0.750	22.61	1.70	0.866	15.94	0.00	0.861	45.97	0.01	0.978
CQCC	13.99	57.05	0.968	66.29	57.05	1.191	44.65	0.61	1.009	0.86	100.00	1.009

Table 10.5   Fusing i-vector and ISV-based verification systems with the selected MFCC-based PAD (in bold in Tables 10.3 and 10.4) on evaluation set of AVspoof-PA

ASV system	Fused with PAD	Type of fusion	Zero-impostors only		PAs only
			FMR (%)	FNMR (%)	IAPMR (%)
	No fusion	–	4.46	9.90	92.41
ISV-based	**MFCC**	**Cascade**	**6.57**	**12.00**	**4.19**
	MFCC	Mean	23.05	22.73	28.98
	MFCC	LR	25.40	24.72	2.68
	MFCC	PLR	4.97	10.75	5.17
Midrule	No fusion	–	8.85	8.31	94.04
i-vectors based	**MFCC**	**Cascade**	**10.83**	**11.45**	**3.89**
	MFCC	Mean	26.33	19.44	19.47
	MFCC	LR	8.77	8.33	94.28
	MFCC	PLR	9.60	10.47	95.76

From the results in Table 10.4, we can note that all methods generalize poorly across different datasets with BPCER reaching 100%, for example, CQCC-based PAD shows poor performance for all cross-database evaluations. It is also interesting to note that even similar methods, for instance, RFCC and MFCC based, have very different accuracy in cross-database testing, even though they showed less drastic difference in single-database evaluations (see Table 10.3).

10.5   Integration of PAD and ASV

As described in Section 10.3, multiple PAD systems have been considered to detect whether a given speech sample is real or spoofed. However, the purpose of a PAD system is to work in tandem with a verification system, so that the joint system can effectively separate the genuine data from both zero-effort impostors (genuine data but incorrect identity) and spoofed attacks (spoofed data for the correct identity).

As presented in Section 10.3.3, in a score-based fusion of PAD and ASV systems, we make a decision about each speech sample using the scores from both PAD and ASV. The resulted joint system can effectively distinguish genuine data from PAs, as demonstrated in Figure 10.7(b) for ASV based on *i-vector* integrated with an example of MFCC-based PAD system. We have chosen MFFC-based system as an example for Figure 10.7, because, from the Table 10.5, it is clear that applying cascade fusion scheme to join an ASV system with *MFCC* leads to more superior performance compared to other fusion schemes and algorithms.

As results presented in Table 10.5 demonstrate, integration with PAD system can effectively reduce IAPMR from above 90% of the ASV (both ISV based and *i-vector*) down to 3.89%, which is the best performing system of *i-vector* ASV fused with *MFCC*-based PAD via cascade fusion (see Figure 10.7(c) for DET plots of different

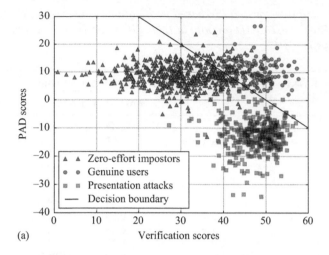

(a)

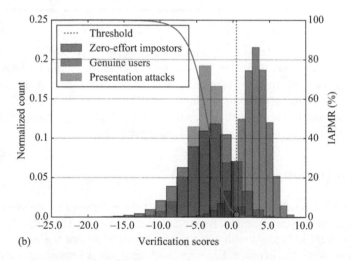

(b)

*Figure 10.7   A scatter plot, histogram distributions, and DET curves for joint
i-vector ASV and MFCC-based PAD systems. (a) Scatter plot, test set,
(b) histogram distribution, test set, and (c) DET curves, test set*

scenarios). Such drastic improvement in the attack detection comes with an increase
in FMR (from 4.46% to 6.57% when ASV is ISV and from 8.85% to 10.83% when
ASV is *i-vector*). FNMR also increases.

Please note that an important advantage of using MFCC-based PAD is that MFCC
are the most commonly used fast to compute features in speech processing, which
makes it practical to use MFCC-based PAD for fusion with an ASV.

Table 10.5 also shows that cascading fusion leads to a better overall performance
compared to parallel scheme. However, compared to a cascading scheme, where each

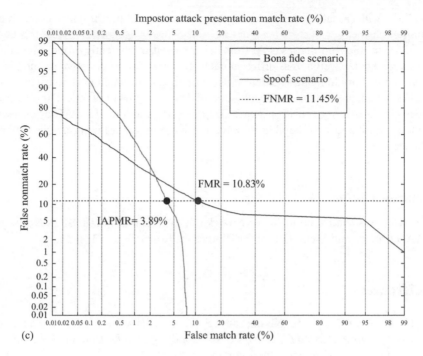

Figure 10.7 (continued)

fused system is independent and has to be tuned separately for disjoint set of parameter requirements, parallel scheme is more flexible, because it allows to tune several parameters of the fusion, as if it was one single system consisting of interdependent components. Such flexibility can be valuable in practical systems. See [29] for a detailed comparison of the different fusion schemes and their discussion.

10.6 Conclusions

In this chapter, we provide an overview of the existing PAD systems for voice biometrics and present evaluation results for selected eight systems on two most comprehensive publicly available databases, AVspoof and ASVspoof. The cross-database evaluation results of these selected methods demonstrate that state-of-the-art PAD systems generalize poorly across different databases and data. The methods generalize especially poorly, when they were trained on "logical access" attacks and tested on more realistic PAs, which means a new and more practically applicable attack-detection methods need to be developed.

We also consider score-based integration of several PAD and ASV systems following both cascading and parallel schemes. Presented evaluation results show a significantly increased resistance of joined ASV-PAD systems to PAs from AVspoof database, with cascading fusion leading to a better overall performance compared to parallel scheme.

PAD in voice biometrics is far from being solved, as currently proposed methods do not generalize well across different data. It means that no effective method is yet proposed that would make speaker-verification system resistant even to trivial replay attacks, which prevents the wide adoption of ASV systems in practical applications, especially in security sensitive areas. Deep learning methods for PAD are showing some promise and may be able to solve the issue of generalizing across different attacks.

Acknowledgments

This work was conducted in the framework of EU H2020 project TeSLA, Norwegian SWAN project, and Swiss Centre for Biometrics Research and Testing.

References

[1]   S. Marcel, M. S. Nixon, and S. Z. Li, *Handbook of Biometric Anti-Spoofing: Trusted Biometrics Under Spoofing Attacks*. London, UK: Springer Publishing Company, Incorporated, 2014.

[2]   P. Korshunov and T. Ebrahimi, "Towards optimal distortion-based visual privacy filters," in *IEEE International Conference on Image Processing*, 2014.

[3]   ISO/IEC JTC 1/SC 37 Biometrics, "DIS 30107-1, information technology—biometrics presentation attack detection," Geneva, Switzerland: American National Standards Institute, Jan. 2016.

[4]   Z. Wu, T. Kinnunen, N. Evans, *et al.*, "ASVspoof 2015: the first automatic speaker verification spoofing and countermeasures challenge," in *INTERSPEECH*, Dresden, Germany, Sep. 2015, pp. 2037–2041.

[5]   J. Mariéthoz and S. Bengio, "Can a professional imitator fool a GMM-based speaker verification system?" IDIAP, Tech. Rep. Idiap-RR-61-2005, 2005.

[6]   S. K. Ergunay, E. Khoury, A. Lazaridis, and S. Marcel, "On the vulnerability of speaker verification to realistic voice spoofing," in *IEEE International Conference on Biometrics: Theory, Applications and Systems*, Sep. 2015.

[7]   Z. Wu, N. Evans, T. Kinnunen, J. Yamagishi, F. Alegre, and H. Li, "Spoofing and countermeasures for speaker verification: A survey," *Speech Communication*, vol. 66, pp. 130–153, 2015.

[8]   S. Shiota, F. Villavicencio, J. Yamagishi, N. Ono, I. Echizen, and T. Matsui, "Voice liveness detection algorithms based on pop noise caused by human breath for automatic speaker verification," in *Sixteenth Annual Conference of the International Speech Communication Association*, 2015.

[9]   P. De Leon, M. Pucher, J. Yamagishi, I. Hernaez, and I. Saratxaga, "Evaluation of speaker verification security and detection of hmm-based synthetic

speech," *IEEE Transactions on Audio, Speech, and Language Processing*, vol. 20, no. 8, pp. 2280–2290, Oct. 2012.

[10] R. Vogt and S. Sridharan, "Explicit modelling of session variability for speaker verification," *Computer Speech and Language*, vol. 22, no. 1, pp. 17–38, Jan. 2008. [Online]. Available: http://dx.doi.org/10.1016/j.csl.2007.05.003.

[11] N. Dehak, P. J. Kenny, R. Dehak, P. Dumouchel, and P. Ouellet, "Front-end factor analysis for speaker verification," *IEEE Transactions on Audio, Speech, and Language Processing*, vol. 19, no. 4, pp. 788–798, May 2011.

[12] H. Zen, K. Tokuda, and A. W. Black, "Review: Statistical parametric speech synthesis," *Speech Communication*, vol. 51, no. 11, pp. 1039–1064, Nov. 2009.

[13] ISO/IEC JTC 1/SC 37 Biometrics, "DIS 30107-3:2016, information technology—biometrics presentation attack detection—part 3: Testing and reporting," American National Standards Institute, Oct. 2016.

[14] M. I. Mandasari, M. Günther, R. Wallace, R. Saeidi, S. Marcel, and D. A. van Leeuwen, "Score calibration in face recognition," *IET Biometrics*, vol. 3, no. 4, pp. 246–256, 2014.

[15] M. Sahidullah, T. Kinnunen, and C. Hanilçi, "A comparison of features for synthetic speech detection," in *Proc. of Interspeech*, 2015.

[16] P. Kenny, G. Boulianne, P. Ouellet, and P. Dumouchel, "Joint factor analysis versus eigenchannels in speaker recognition," *IEEE Transactions on Audio, Speech, and Language Processing*, vol. 15, no. 4, pp. 1435–1447, May 2007.

[17] T. B. Patel and H. A. Patil, "Combining evidences from mel cepstral, cochlear filter cepstral and instantaneous frequency features for detection of natural vs. spoofed speech," in *INTERSPEECH*, Dresden, Germany, Sep. 2015, pp. 2062–2066.

[18] A. Janicki, "Spoofing countermeasure based on analysis of linear prediction error," in *Sixteenth Annual Conference of the International Speech Communication Association*, 2015.

[19] J. Gałka, M. Grzywacz, and R. Samborski, "Playback attack detection for text-dependent speaker verification over telephone channels," *Speech Communication*, vol. 67, pp. 143–153, 2015.

[20] Z. Wu, X. Xiao, E. S. Chng, and H. Li, "Synthetic speech detection using temporal modulation feature," in *Acoustics, Speech and Signal Processing (ICASSP), 2013 IEEE International Conference on*, May 2013, pp. 7234–7238.

[21] M. J. Alam, P. Kenny, G. Bhattacharya, and T. Stafylakis, "Development of CRIM system for the automatic speaker verification spoofing and countermeasures challenge 2015," in *Sixteenth Annual Conference of the International Speech Communication Association*, 2015.

[22] L. Wang, Y. Yoshida, Y. Kawakami, and S. Nakagawa, "Relative phase information for detecting human speech and spoofed speech," in *Proc. of Interspeech 2015*, 2015.

[23] Y. Liu, Y. Tian, L. He, J. Liu, and M. T. Johnson, "Simultaneous utilization of spectral magnitude and phase information to extract supervectors for

speaker verification anti-spoofing," in *Proc. of Interspeech 2015*, vol. 2, 2015, p. 1.

[24] P. L. De Leon, B. Stewart, and J. Yamagishi, "Synthetic speech discrimination using pitch pattern statistics derived from image analysis," in *Proc. of Interspeech*, 2012, pp. 370–373.

[25] A. Ogihara, U. Hitoshi, and A. Shiozaki, "Discrimination method of synthetic speech using pitch frequency against synthetic speech falsification," *IEICE Transactions on Fundamentals of Electronics, Communications and Computer Sciences*, vol. 88, no. 1, pp. 280–286, 2005.

[26] M. Todisco, H. Delgado, and N. Evans, "Articulation rate filtering of CQCC features for automatic speaker verification," in *INTERSPEECH*, San Francisco, USA, Sept. 2016.

[27] F. Alegre, A. Amehraye, and N. Evans, "A one-class classification approach to generalised speaker verification spoofing countermeasures using local binary patterns," in *Proc. of BTAS*, Sep. 2013, pp. 1–8.

[28] D. Luo, H. Wu, and J. Huang, "Audio recapture detection using deep learning," in *Signal and Information Processing (ChinaSIP), 2015 IEEE China Summit and International Conference on*, Jul. 2015, pp. 478–482.

[29] I. Chingovska, A. Anjos, and S. Marcel, "Biometrics evaluation under spoofing attacks," *IEEE Transactions on Information Forensics and Security*, vol. 9, no. 12, pp. 2264–2276, Dec. 2014.

[30] S. Davis and P. Mermelstein, "Comparison of parametric representations for monosyllabic word recognition in continuously spoken sentences," *IEEE Transactions on Acoustics, Speech, and Signal Processing*, vol. 28, no. 4, pp. 357–366, Aug. 1980.

[31] S. Furui, "Cepstral analysis technique for automatic speaker verification," *IEEE Transactions on Acoustics, Speech, and Signal Processing*, vol. 29, no. 2, pp. 254–272, Apr. 1981.

[32] E. Scheirer and M. Slaney, "Construction and evaluation of a robust multifeature speech/music discriminator," in *Proc. of ICASSP*, vol. 2, Apr. 1997, pp. 1331–1334.

[33] P. N. Le, E. Ambikairajah, J. Epps, V. Sethu, and E. H. C. Choi, "Investigation of spectral centroid features for cognitive load classification," *Speech Communication*, vol. 53, no. 4, pp. 540–551, Apr. 2011.

[34] F. K. Soong and A. E. Rosenberg, "On the use of instantaneous and transitional spectral information in speaker recognition," *IEEE Transactions on Acoustics, Speech, and Signal Processing*, vol. 36, no. 6, pp. 871–879, Jun. 1988.

[35] U. Scherhag, A. Nautsch, C. Rathgeb, and C. Busch, "Unit-selection attack detection based on unfiltered frequency-domain features," in *INTERSPEECH*, San Francisco, USA, Sept. 2016, p. 2209.

[36] J. C. Platt, "Probabilistic outputs for support vector machines and comparisons to regularized likelihood methods," in *Advances in large margin classifiers*. Cambridge, MA: MIT Press, 1999, pp. 61–74.

Chapter 11

Benford's law for classification of biometric images

Aamo Iorliam[1], Anthony T. S. Ho[1,2,3], Norman Poh[1], Xi Zhao[2] and Zhe Xia[3]

It is obvious that tampering of raw biometric samples is becoming an important security and privacy concern. The Benford's law, which is also called the first digit law, has been reported in the forensic literature to be very effective in detecting forged or tampered data. In this chapter, besides an introduction to the concept and state-of-the-art reviews, the divergence values of Benford's law are used as input features for a neural network for the classification of biometric images. Experimental analysis shows that the classification of the biometric images can achieve good accuracies between the range of 90.02% and 100%.

11.1 Introduction

Biometric modalities are used for identification or verification tasks. Generally, biometric modalities are classified into physical and behavioural modalities. There exist many different types of physical biometric modalities such as fingerprint images, vein

[1] Department of Computer Science, University of Surrey, UK
[2] School of Computer Science and Information Engineering, Tianjin University of Science and Technology, China
[3] School of Computer Science and Technology, Wuhan University of Technology, China

wrist images, face images and iris images. If we consider only fingerprint images, there are also different types of fingerprints such as contact-less acquired latent fingerprints, optically acquired fingerprints and synthetic generated fingerprints. These fingerprints could be intentionally or accidentally used for a different purpose which could pose a security threat. As such, for example, there exists a possibility that fingerprints could be forged at crime scenes by potentially transferring latent fingerprint to other objects [1].

This chapter considers three classification tasks which include intra-class, interclass and mixed biometric classification. The goal of intra-class classification is to classify an intentional (attack) or unintentional mixture of biometric images of different types but the same modality, whereas inter-class classification is targeted at classifying unintentional mixture of biometric images of different types and different modality. Furthermore, mixed biometric classification is applicable where there is need to classify mixed biometric modalities (mixing different fingerprints, different faces, or different iris) in order to generate joint identities and also to preserve privacy.

Benford's law which considers the first digit of numbers (1–9) has been reported in literature to be very effective in detecting fraud in accounting and tampering in natural and biometric images. We adapt this approach to investigate data of different biometric modalities (fingerprints in different acquisition scenarios, vein wrist, face and iris images) towards different classification goals. This chapter investigates the different biometric data using Benford's law divergence metrics. To achieve the Benford's law divergence values, the luminance component of the face images is first extracted, and the first digit distributions of the JPEG coefficients are then performed on the face images. For the other datasets described in Section 11.3.2, the first digit distributions of the JPEG coefficients are performed directly on the greyscale images. Using (11.8), the Benford's law divergence values are obtained. The Benford's law divergence values are then used as features fed into a neural network (NN) classifier for classification. Therefore, an inter-class classification of biometric images using Benford's law divergence and NN is performed. By inter-class classification, we mean separability of biometric images that are not closely related (e.g. fingerprint images, vein wrist images, face images and iris images). We also perform intra-class classification of biometric images using Benford's law divergence and NN. By intra-class classification, we mean separability of biometric images that are closely related (e.g. contact-less acquired latent fingerprints, optically acquired fingerprints and synthetic generated fingerprints). Lastly, both the closely related and non-closely related biometric images are classified using Benford's law divergence and NN.

11.2   Related work

11.2.1   Benford's law

As discussed in the previous section, Benford's law has desirable characteristics used for fraud detection in accounting and for tampering detection for natural and biometric

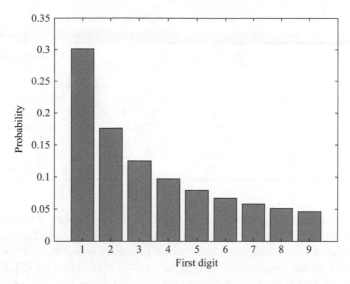

Figure 11.1   The first digit probability distribution of Benford's law

images. Furthermore, it could be investigated towards the classification of different biometric images. In this section, we want to briefly introduce the Benford's law and its related work.

Benford's law has been reported by Fu *et al.* [2], Li *et al.* [3] and Xu *et al.* [4] to be very effective in detecting tampering of images. Benford's law of "anomalous digits" was first coined by Frank Benford in 1938 [5], is also denoted as the first digit law. He considered the frequency of appearance of the most significant digit (MSD), for a broad range of natural and artificial data [6]. The Benford's law as described by Hill [7] can be expressed in the form of a logarithmic distribution, when considering the probability distribution of the first digit from 1 to 9 for a range of natural data. Naturally generated data are likely to obey this law whereas tampered or randomly guessed data do not tend to follow [8].

When considering the MSD where 0 is excluded, and the datasets satisfy the Benford's law, then the law can be expressed in the following equation:

$$P(x) = \log_{10}\left(1 + \frac{1}{x}\right) \tag{11.1}$$

where x is the first digit of the number and $P(x)$ refers to the probability distribution of x.

The first digit probability distribution of Benford's law is shown in Figure 11.1. An extension of the Benford's law by Fu *et al.* [2] is referred to as Generalised Benford's law which closely follows a logarithmic law is defined in the following equation [2]:

$$P(x) = N \log_{10}\left(1 + \frac{1}{s + x^q}\right) \tag{11.2}$$

Table 11.1   Model parameters used for generalised Benford's law [2]

Q-factor	Model parameters			Goodness-of-fit (SSE)
	N	q	s	
100	1.456	1.47	0.0372	7.104e-06
90	1.255	1.563	−0.3784	5.255e-07
80	1.324	1.653	−0.3739	3.06838e-06
70	1.412	1.732	−0.337	5.36171e-06
60	1.501	1.813	−0.3025	6.11167e-06
50	1.579	1.882	−0.2725	6.05446e-06

where N is a normalisation factor which makes $P(x)$ a probability distribution. This means that the sum of $P(x)$ over all possible occurrences of x is 1. The model parameters in this case are represented by s and q which describe the distributions for different images and different compressions quality factors (QFs) as defined in [2]. The values of s and q are data dependent and literature has shown that they are applicable for natural [2] and biometric images [9]. Through experiments, Fu *et al.* [2] provided values for N, s and q. They determined these values using the MATLAB toolbox, which returns the sum of squares due to error (SSE).

The SSE is given by the following equation:

$$\text{SSE} = \sum_{i=1}^{n} w_i(y_i - \hat{y}_i)^2 \tag{11.3}$$

where y_i is the observed data value and $\hat{y}_i$ is the predicted value from the fit. w_i is the weighting applied to each data point which is usually 1.

Considering the individual QFs and model parameters, the target is usually to achieve an SSE value that is closer to 0. A value of 0 indicates that there is no random error component and it shows a best fit for prediction. The N, s and q values with respect to the corresponding QFs gave acceptable values that were very close to 0 and as such are adopted as shown in Table 11.1 for our Generalised Benford's law experiments. There have been several developments concerning the Benford's law in image forensics [2,3,6,10–12]. For example, Fu *et al.* [2] used this law on DCT coefficients to detect unknown JPEG compression. They used 1,338 images from an uncompressed image database (UCID). They also repeated the same experiment on 198 images from the Harrison datasets. Li *et al.* [3] combined machine learning and statistical properties of the first digits based on JPEG coefficients of individual AC modes to detect tampering in JPEG images. Their tampering detection scheme is based on the fact that tampered regions undergoes a single JPEG compression, whereas the untampered regions undergoes a double JPEG compression. Gonzalez *et al.* [6] proved that when images were transformed by the discrete cosine transform (DCT), their DCT coefficients would follow this law. Moreover, they showed that the

Generalised Benford's law could detect hidden data in a natural image [6]. Digital images in the pixel domain do not follow the Benford's law because their values fall between 0 (black) and 255 (white). This means we have only 256 values to calculate the Benford's law when considering images in the pixel domain. Thus this will not conform to the Benford's law because this law works better as the dynamic values under consideration tends to be large. By transforming the pixel values using DCT, a single block matrix has 64 DCT coefficients, which when combined together in a whole image, produces dynamic values large enough to follow the Benford's law. Acebo and Sbert [10] also applied the Benford's law to image processing and showed that when synthetic images were generated using physically realistic method, they followed this law. However, when they were generated with different methods, they did not follow this law. Jolion [11] applied this law to image processing and found that it worked well on the magnitude of the gradient of an image. He also showed that it applied to the Laplacian decomposition of images as well. Qadir *et al.* [12] investigated the Discrete Wave Transform (DWT) coefficients using the Benford's law and analysing the processing history applied to JPEG2000 images, where they observed a sharp peak at the digit five of the Benford's law curve for some images. As a result, they proposed the use of this law to identify unbalanced lighting or glare effect in images with the help of DWT [12].

Recent approaches have been developed to tackle fingerprint forgeries. For example, Iorliam *et al.* [13] used the Benford's law divergence metric to separate optically acquired fingerprints, synthetically generated fingerprints and contact-less acquired fingerprints. The purpose of this study was to protect against insider attackers and against hackers that may have illegal access to such biometric data. Hildebrandt and Dittmann [14] used the differences between Benford's law and the distribution of the MSD of the intensity and topography data to detect printed fingerprints. Their experiments on 3,000 printed and 3,000 latent print samples achieved a detection performance of up to 98.85% using WEKA's Bagging classifier in a 10-fold stratified cross-validation. Iorliam *et al.* [13] performed experiments on the Benford's law divergence values of biometric data for separability purposes. The results showed that biometric datasets behave differently when analysed using the Benford's law divergence metrics. Therefore, we extend the approach in [13], to achieve the Benford's law divergence values from biometric data. We then use these divergence values as features fed into the NN for classification and source identification of biometric images. The main focus of Hildebrandt and Dittmann [14] was to differentiate between real fingerprints and printed fingerprints. However, our method focuses not only on these two types of datasets but also other biometric datasets such as face images, vein wrist datasets and iris datasets. Furthermore, we use divergence values as features for classification and source identification of biometric images. Whereas their method used the differences between Benford's law and the distribution of the MSD of the intensity and topography data to detect printed fingerprints. In addition, we use NN as a classifier, whereas their method used WEKA's Bagging classifier in a 10-fold stratified cross-validation.

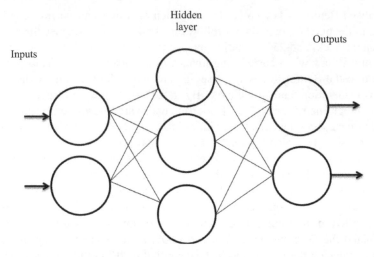

Inputs

Hidden
layer

Outputs

Figure 11.2   A multi-layer perceptron with one hidden layer

11.2.2   Neural networks

In the previous section, we explained how the Benford's law works and some related work that have been investigated using the Benford's law. In this section, we give an introduction to the NN. A NN works by accepting input data, extracting rules based on the accepted inputs and then making decisions. Multi-layer perceptrons (MLP) is a feed-forward type of NN which has one or more hidden layers between the input layer and output layer. Training is usually carried out using back-propagation learning algorithm. MLP has the advantage of modelling functions that are highly non-linear and when trained, it can correctly predict on unknown data sets. Figure 11.2 shows a pictorial representation of an MLP with two inputs, one hidden layer and two outputs.

The MLP typically uses the sigmoid hidden neurons which are described by the following formula:

$$y = \frac{1}{1 + \exp(-s)} \tag{11.4}$$

where s is an integration function defined by $s = f(x; \theta)$, x is the input variable to the hidden neurons, y is the output variable of the hidden neurons and θ is the threshold.

For classification tasks with more than two classes, softmax output layer provides a way of assigning probabilities to each class. Softmax activation of the ith output unit can be defined by the following equation:

$$y_i = \frac{\exp(s_i)}{\sum_j \exp(s_j)} \tag{11.5}$$

where s has the same meaning as above.

Equation (11.4) takes care of a single input and output, whereas (11.5) takes into consideration several inputs and outputs such that y_i is the softmax output which depends on all the input number (i), i is the input neuron number and j is the output neuron number.

After the network is trained and tested, there are performance evaluation metrics such as the cross entropy (CE) and percentage error (%E) that can be used to determine the classification errors. The CE is defined by the following equation:

$$CE = -\text{sum}(t * \log(y)) \tag{11.6}$$

where t is the target data (i.e. the ground truth of true positive values) and y is the output value from the trained network.

Whereas %E is defined by the following equation:

$$\%E = \text{sum}(t \neq y)/(\text{number of } t) \tag{11.7}$$

where t and y have the same meaning as above.

CE usually gives a faster convergence and better results when considering classification error rates as compared to other measures such as squared error (SE) [15]. Lower values of CE are better for a classification task. Zero CE means that there is no error in the classification task. The %E is measured to show the proportion of samples that are misclassified for a particular classification task. A zero %E means no misclassification and 100 %E means there exists a maximum misclassification. Based on the above reasons, we choose NN using MLP for this classification task. For performance evaluation, the CE and %E are used for our experiments in this chapter.

11.2.3 Mixed biometric data classification to preserve privacy

The concept of mixed biometric data has recently increased due to its advantages, that mixed biometric data preserve the privacy of biometric data [16]. In addition, mixed fingerprint has other advantages such as: (i) existing fingerprint algorithms have accepted mixed fingerprint for processing and (ii) attackers will find it difficult to determine whether a fingerprint is mixed or not [17].

Therefore, mixed biometric data is very helpful in cancelling the real identity of biometric images, which is often referred to as cancellable biometrics. One way to achieve mixed biometric data is by performing image level fusion. By image level fusion, we mean producing a composite image from different biometric data before operations such as classification, recognition and tracking could be performed [16].

Fusion of raw biometric data can be employed for unimodal or multimodal biometric images. By unimodal image level fusion, we mean fusing together multiple biometric samples of the same trait in order to achieve a composite biometric image. Furthermore, multimodal image level fusion means fusing together different biometric data to produce a composite biometric image [16].

Due to the fact that image level fusion has been proposed and effectively deployed, it is therefore important to classify such biometric images, i.e. detect such imagery from other data.

11.3 Experimental setup

The goal of the first experiment is to use the divergence metrics to show how the data samples used for this experiments depart from the Generalised Benford's law. The Benford's law divergence values acquired from physical biometric images (i.e. from the vein wrist images, face images, iris images, optical sensor acquired fingerprints images, artificially printed contact-less acquired latent fingerprint images and synthetic generated fingerprints) are then used as features and fed into an NN for separability purposes. There is a real need to separate biometric images as discussed in the next section.

11.3.1 Separation of different types of biometric images

Generally, biometrics is used either for verification (1-to-1 matching) where we seek to answer the question "Is this person who they say they are?" or for identification (1-to-many (n) matching) where we seek to answer the question "Who is this person?" or "Who generated this biometric?" [18]. To achieve identification or verification, biometric modalities are used. For example, fingerprints have been used for identification purposes for over a century [19]. Iris recognition systems use iris modality and it is of importance due to its high accuracy. As such, it has attracted a great deal of attention recently [20]. Vein pattern recognition, which uses vein modality, is a developing field and is a potential research area in the field of biometrics [21].

Fingerprints have different uses. For example, fingerprints captured using optical sensors are used for identification or verification, whereas synthetic generated fingerprints are used for testing fingerprint recognition algorithms. This requires a large databases and collecting real fingerprints for such a large databases is very expensive, labour intensive and problematic [22]. The contact-less acquired latent fingerprints are generated for evaluation and research purposes with the aim of avoiding privacy implications [23]. The study of different characteristics of these biometric images is of importance both for biometrics and forensic purposes. This process could assist in the source identification of captured biometric images. This can be achieved by identifying the source hardware that captured the biometric image [24]. This is necessary, for example, in cases where the "chain of custody" involving a biometric image has to be properly verified [24].

11.3.2 Data sets

As stated in Table 11.2, for the face modality, the CASIA-FACEV5 is used which consists of 2,500 coloured images of 500 subjects [25]. The face images are 16-bit colour BMP files with image resolution of 640 × 480 pixels. We use all the face images from CASIA-FACEV5 in this experiment. For the fingerprint modality, we use the FVC2000 which has four different fingerprint databases (DB1, DB2, DB3 and DB4) [26]. DB1, DB2, DB3 and DB4 have 80 grey-scale fingerprint images in (tiff format) each. DB1 fingerprint images are captured by a low-cost optical sensor 'secure desktop scanner'. While DB2 fingerprint images are captured by a low-cost optical capacitive sensor 'TouchChip'. DB3 fingerprint images are from optical sensor

Table 11.2   Summary description of data sets

Data set	Description	Number of samples
CASIA-FACEV5 [25]	Coloured face images of 500 subjects	2,500
FVC2000 [26]		
	DB1 – Fingerprint images captured by 'Secure Desktop Scanner'	80
	DB2 – Fingerprint images captured by 'TouchChip'	80
	DB3 – Fingerprint images captured by 'DF-90'	80
	DB4 – Synthetically generated fingerprint images	80
Fingerprint data set from Hildebrandt *et al.* [23]	Artificially printed contact-less acquired latent fingerprint images	48
PUT vein database [21]	Images of wrist pattern	1,200
CASIA-IrisV1 [20]	Iris images from 108 eyes	756

'DF-90'. DB4 fingerprint images are synthetically generated fingerprints. However, for the fact that these fingerprints (DB1, DB2, DB3) are captured using different sensors and DB4 is synthetically generated, we treat them as four separate fingerprint databases.

An additional fifth fingerprint data set used in this experiment which contains 48 artificially printed contact-less acquired latent fingerprint images. These are grey-scale images with 32-bit colour depth [23]. We use 1,200 images of wrist pattern from the PUT vein database [21]. The iris images used in our experiment are from CASIA-IrisV1 which contains 756 iris images which are saved in BMP format with resolution of 320 × 280 [20]. Figure 11.3 shows some sample biometric images used in these experiments. A summary description of the data sets is given in Table 11.2.

11.3.3 Divergence metric and separability of biometric databases

In order to separate biometric databases, the Benford's law divergence values are obtained from the biometric data. The divergence metric is used to show how close or far a particular data set is, either with the Standard or Generalised Benford's law. In all cases, a smaller divergence gives a better fitting. In these experiments, the luminance component of the face images is first extracted and the first digit distributions of the JPEG coefficients is then performed on the face images. The luminance component is used only for the facial images because of the coloured nature of the images. As such, the visual component of the face images is needed for JPEG quantisation due to the

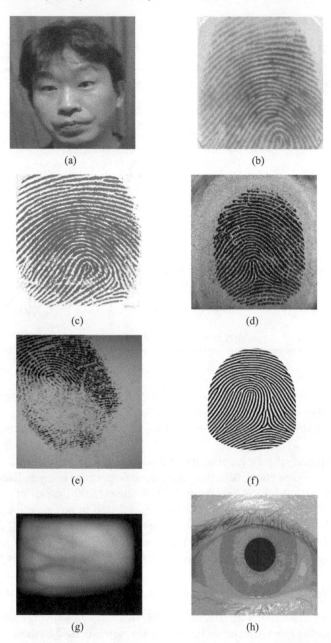

*Figure 11.3   Sample biometric images from: (a) CASIA-FACEV5, (b) FVC2000
DB1, (c) FVC2000 DB2, (d) FVC2000 DB3, (e) FVC2000 DB4, (f)
artificially printed contact-less acquired latent fingerprint, (g) PUT
vein database and (h) CASIA-IrisV1*

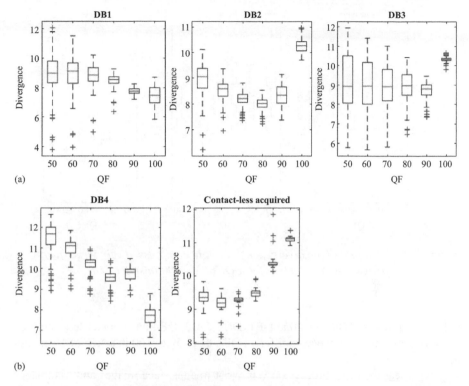

Figure 11.4 *Box plot of the divergence for single compressed: (a) DB1, DB2 and DB3 fingerprints for a QF = 50–100 in step of 10; (b) DB4 and contact-less acquired latent fingerprints for a QF = 50–100 in step of 10*

fact that the low spatial frequency regions are sensitive to little variations in intensity as compared to the high spatial frequency regions. However, for the other datasets described in Section 11.3.2, the first digit distributions of the JPEG coefficients is performed directly on the grey-scale images.

To test for conformity of a particular dataset to the Benford's law, one of the most common criteria used is the chi-square goodness-of-fit statistics test [2,3,10,12,27]. The chi-square divergence is expressed in the following equation:

$$\chi^2 = \sum_{x=1}^{9} \frac{(p'_x - p_x)^2}{p_x} \tag{11.8}$$

where p'_x is the actual first digit probability of the JPEG coefficients of the biometric images and p_x is the logarithmic law (Generalised Benford's law) as given in (11.2) for all possible digits $x = 1, \ldots, 9$. In our experiments, the fingerprint databases are single compressed at a QF of 50–100 in a step of 10. The divergence is calculated as an average on all the data sets as can be seen in Figure 11.4(a) and (b). The box plots

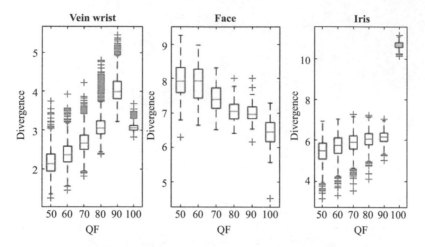

Figure 11.5   Box plot of the divergence for single compressed vein wrist, face images and iris images for a QF = 50–100 in step of 10

in Figure 11.4(a) and (b) show that DB1, DB2, DB3, DB4 and contact-less acquired latent fingerprints divergence at different QF's from 50 to 100 in a step of 10 are not the same.

We repeat the same process on vein wrist images, face images and iris images and results obtained are shown in Figure 11.5. In Figures 11.4(a) and (b) and 11.5, we observe that the different biometric databases behave differently when anal-ysed using the Benford's law divergence metrics. Generally speaking, we should expect, low divergence values for QFs = 100. However, for DB2, DB3, contact-less acquired latent fingerprints and iris images, higher divergence values were observed for QFs = 100 as seen in Figures 11.4(a) and (b) and 11.5. This further gave a clue that divergence values could be useful for this classification purposes. Again, even for the DB1, DB4, vein wrist and face images that seem to follow the expected pattern (i.e. lower divergence values for QF = 100 and higher divergence values for QF = 50), we observed that the divergence values for these data sets were not distributed in the same pattern as seen in Figures 11.4(a) and (b) and 11.5. This further gave more evidence that the divergence values could be used for this classification task.

11.4   Proposed method

As mentioned earlier, the Benford's law divergence values for different biometric images were found to be different. Hence, there is a need to improve the classification accuracy, as this may assist in source identification of the different biometric images.

Therefore, we perform an inter-class, intra-class and mixture of inter-class and intra-class classification of biometric images in this section.

11.4.1 Method description

The objective of our proposed method is to train an NN to carry out inter-class classification, intra-class classification and a mixture of inter-class and intra-class classification of biometric images. We first obtain the Benford's law divergence values from the biometric images. Divergence values are obtained using the formula in (11.8). We then apply compression using a QF of 50–100 in a step of 10 to the biometric images, resulting in six rows of Benford's law divergence values with each row representing a QF in the experiment. Thus, we have six inputs and four outputs (when considering inter-class classification) and we have six inputs and five outputs (when considering the intra-class classification). For the inter-class classification, our data set consists of a total of 4,536 instances with class 1 (DB1 fingerprints images) having 80 instances, class 2 (vein wrist images) having 1,200 instances, class 3 (face images) having 2,500 instances and class 4 (iris images) having 756 instances. So the features fed into the NN as input data is a 6 × 4,536 matrix. For our target data, we represent class 1 target as 1 0 0 0, class 2 as 0 1 0 0, class 3 as 0 0 1 0 and class 4 as 0 0 0 1. We therefore, use a 4 × 4,536 matrix as the target data. This is due to inter-class testing between the four modalities (DB1 fingerprints, vein wrist, face and iris images).

For any NN experiment, the choice of hidden layer(s) and hidden neurons is very important. Panchal *et al.* [28] noted that for most NN tasks, one hidden layer is usually adequate for a classification task especially if the data under consideration is not discontinuous.

They further suggested three rules-of-thumb to estimate the number of hidden neurons in a layer, as follows:

* Number of hidden neurons should be between the size of the input layer and size of the output layer
* Number of hidden neurons should be equal to (inputs + outputs) × 2/3
* Number of hidden neurons should be less than twice the size of the input layer

Based on the above rules-of-thumb, we use the MATLAB NN pattern recognition tool with one hidden layer and four hidden neurons for this classification task. Training of data set is carried out using 'trainscg' function because it uses less memory. The input data is randomly divided into 70% training (i.e. 3,176 training sample), 15% validation (i.e. 680 validation samples) and 15% testing the NN (i.e. 680 testing samples). The 3,176 samples are presented to the NN during the training phase and the network adjusts itself based on its errors. The 680 validation samples are used for generalisation purposes during the validation phase. Testing the NN provides a separate measure of the NN performance in the course of the training process and

Table 11.3   CE and %E for inter-class classification of biometric images

	Samples	CE	%E
Training	3,176	2.63644	0.25189
Validation	680	7.44672	0.00000
Testing	680	7.47163	0.44176

after the training process. Based on the four hidden neurons, we achieve the CE and %E for our evaluation performance as shown in Table 11.3.

11.5   Results and discussion

In this section, we present results for inter-class, intra-class and a mixture of inter-class and intra-class classification of biometric images. We further discuss these results with respect to the classification performance of biometric images.

11.5.1   Inter-class separability of biometric images

After the initial network training setup, we observe it took the network, 1 epoch and 47 iterations to complete the training process. Results for the inter-class classification of biometric images are shown in Table 11.3.

We can see from Table 11.3 that for the training set, the NN has correctly classified approximately 97.36% (when considering CE) and 99.75% (when considering %E). On the validation set, NN has correctly classified approximately 92.55% (when considering CE) and 100% (when considering %E). For the testing set, NN has correctly classified 92.53% (when considering CE) and 99.56% (when considering %E). We see that the best classification result is achieved at the validation sets based on %E (100% classification accuracy) and the least classification accuracy is achieved at the testing sets based on CE (92.53% classification). The confusion matrices for the classification of the biometric images with four classes for this experiment are shown in Figure 11.6. In this classification task, the four classes are (1) DB1 fingerprints, (2) wrist, (3) face and (4) iris. The training confusion matrix shows that 48 of the DB1 fingerprint images are correctly classified, 5 are misclassified as face images. All the vein wrist images are correctly classified. There are 1,740 face images correctly classified, whereas five face images are misclassified as DB1 fingerprints images. All the iris images are classified correctly. The overall classification accuracy for the training confusion matrix is approximately 99.7%.

To achieve the percentage of correct/incorrect classification for each biometric image class with respect to the number of total classified biometric images, a number in percentage is recorded in each cell. For example, for the Training Confusion Matrix, 1.5% is achieved as the percentage of correct classification for DB1 fingerprints (1) with respect to the total classified biometric images (i.e. $48 + 3 + 843 + 5 + 1,740 + 537 = 3,176$). The 1.5% is achieved by calculating $(48/3,176) \times 100$. Similarly, the

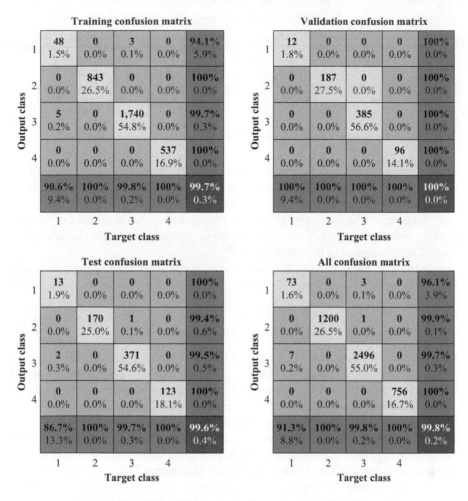

Figure 11.6 Confusion matrix for inter-class classification

0.1% at cell (1,3) is achieved by calculating (3/3,176) × 100. This same calculation is repeated to achieve the percentage of correct/incorrect classification for each of the cells.

For the Validation Confusion Matrix, all the biometric images are correctly classified at 100%. For the test confusion matrix, 13 DB1 fingerprint images are correctly classified, while two of the DB1 fingerprint images are misclassified as face images. For the vein wrist images, they are all classified correctly at 100%. However, for the face images, 371 face images are correctly classified, whereas two of the face images are misclassified as DB1 fingerprint images. All the iris images are correctly classified at 100%. The overall classification accuracy, for the test confusion matrix is approximately 99.6%. We see from the all confusion matrix that 73 DB1 fingerprint

Table 11.4   CE and %E results for intra-class classification of
biometric images

	Samples	CE	%E
Training	258	3.31436	0.00000
Validation	55	9.57847	0.00000
Testing	55	9.44136	0.00000

images are correctly classified, while seven of the DB1 fingerprint images are mis-classified as face images. All the vein wrist images are correctly classified at 100%. There are 2,496 face images accurately classified, whereas seven are misclassified as DB1 fingerprint images. Lastly, all the iris images are accurately classified. Overall, the all confusion matrix achieves approximately 99.8% accuracy for classifying the four biometric images in the experiment.

In the next section, we will investigate the intra-class separability of biometric images following the same process used for inter-class separability as mentioned in this section.

11.5.2   Intra-class separability of biometric images

In this section, we carry out an intra-class classification within the different data sets of biometric fingerprint images. These fingerprints are DB1, DB2, DB3, DB4 and contact-less acquired latent fingerprints. For the intra-class classification, our data set consist of a total of 368 instances, with class 1 (DB1 fingerprints images) having 80 instances, class 2 (DB2 fingerprints images) having 80 instances, class 3 (DB3 fingerprints images) having 80 instances, class 4 (DB4 fingerprints images) having 80 instances and class 5 (contact-less acquired fingerprints) having 48 instances. The features fed into the NN as input data is a 6×368 matrix. In this way we have 258 instances for training (70% of the data) purposes, 55 instances for validation (15% of the data) purposes and 55 instances for testing (15% of the data) purposes. For our target data, we represent class 1 target as 1 0 0 0 0, class 2 as 0 1 0 0 0, class 3 as 0 0 1 0 0, class 4 as 0 0 0 1 0 and class 5 as 0 0 0 0 1. We therefore, use a 5×368 matrix as the target data. We repeat the same experiment as in the case of inter-class classification but in this case, we use 8 hidden neurons (third rule of thumb [28]) for this classification task and we have 5 output targets. From the network training setup, we observe that it took the network, 1 epoch and 41 iterations to complete the training process. Results for the intra-class classification of biometric images are shown in Table 11.4.

With respect to the results in Table 11.4, we observe that using the %E, all the fingerprints are correctly classified at 100%. For CE, the best classification results achieved using the training set are at approximately 96.69%, followed by the testing set at approximately 90.56% and then the validation set at approximately 90.42%. Overall, for both the CE and %E, the intra-class classification shows a high evaluation performance as shown by the results achieved in Table 11.4.

Table 11.5   Experimental setup for mixed classification of biometric
            images

Classes	Instances	Target data
Class 1 (DB1 fingerprints images)	80	1 0 0 0 0 0 0 0
Class 2 (vein wrist images)	1,200	0 1 0 0 0 0 0 0
Class 3 (face images)	2,500	0 0 1 0 0 0 0 0
Class 4 (iris images)	756	0 0 0 1 0 0 0 0
Class 5 (DB2 fingerprint images)	80	0 0 0 0 1 0 0 0
Class 6 (DB3 fingerprint images)	80	0 0 0 0 0 1 0 0
Class 7 (DB4 fingerprint images)	80	0 0 0 0 0 0 1 0
Class 8 (contact-less acquired fingerprints)	48	0 0 0 0 0 0 0 1

From the confusion matrix, we observe that classes (DB1, DB2, DB3, DB4 and contact-less acquired latent fingerprints) are accurately classified at 100%. We observe from the experiments that based on the %E and confusion matrix, the Benford's law divergence values and NN are able to accurately classify all the fingerprint images. In the next section, we will further investigate a mixture of inter-class and intra-class classification of biometric images to compare with individual inter-class and intra-class classification.

11.5.3   Mixed inter-class and intra-class separability of biometric images

We mix the inter-class biometric images (DB1 fingerprints, vein wrist images, face images and iris images) and intra-class biometric images (DB2, DB3, DB4 and contact-less acquired latent fingerprints) for classification evaluation, based on the goal to separate into the resulting eight categories. For this experiment, Table 11.5 describes the experimental setup. From the table, the number of instances are the corresponding number of biometric images used for this experiment.

The features fed into the NN as input data is a 6 × 4,824 matrix. In this way, we have 3,376 instances for training (70% of the data), 724 instances for validation (15% of the data) and 724 instances for testing (15% of the data). We therefore, use a 8 × 4,824 matrix as the target data for the eight resulting classes. We again use the (inputs+outputs) × 2/3 rule-of-thumb from [28] and achieve 9 hidden neurons (i.e. $(6 + 8) \times 2/3$). So we have a total of 4,824 instances with 6 inputs, 9 hidden neurons and 8 outputs for this experiment. From the NN setup, we observe it took the network, 1 epoch and 37 iterations to complete the training process. The results for the mixed-class classification of biometric images are shown in Table 11.6.

With respect to the results in Table 11.6, we observe that the CE accurately classified the mixed biometric images with accuracies of 96.47%, 90.02% and 90.04% for the training, validation and testing, respectively. The %E achieved accuracies of 99.26%, 99.17% and 99.17% for the training, validation and testing, respectively. Based on the classification accuracies, for CE the highest performance of 96.47% is

Table 11.6   *CE and %E results for mixed-class classification of biometric images*

	Samples	CE	%E
Training	3,376	3.52958	0.740521
Validation	724	9.98217	0.828729
Testing	724	9.95969	0.828729

Table 11.7   *Classification accuracy comparison*

Classification type	Order of classification
Intra-class	1 (Best)
Inter-class	2 (Medium)
Mixture of intra-class and inter-class	3 (Least)

achieved with the training instances and the least evaluation performance of 90.02% is achieved for the validation sets. We also observe that the best classification accuracy of 99.26% is achieved for the training instances. Whereas for the validation and testing instances, our proposed method has achieved the least evaluation performance of 99.17% when considering the %E. Overall, for both CE and %E, the mixed-class classification shows a high evaluation performance.

11.5.4   Comparative analysis between inter-class, intra-class and mixture of inter-class and intra-class classification of biometric images

The need for the investigation and classification of inter-class, intra-class and mixture of inter-class and intra-class biometric images is motivated from the concept of mixed biometrics [16]. Mixed biometrics could be considered as mixing different fingerprints, different faces, or different iris in order to generate joint identities and also to preserve privacy [16]. Mixed biometrics also cover mixing two distinct biometric modalities such as fingerprint and iris to form a joint biometric modality. This means there is a possibility of mixing non-closely related biometric images (inter-class), closely related biometric images (intra-class) and a mixture of inter-class and intra-class biometric images.

Table 11.7 gives an overall summary of the classification accuracy hierarchy from the best classified class to the least using the Benford's law divergence and NN. For comparison, our work is much wider in scope than the state-of-the-art method developed by Hildebrandt and Dittmann [14], and so we only compare our classification accuracies of fingerprints to [14]. Our highest classification accuracy of 100% (for intra-class separation of fingerprint images by their sensor source) exceed the current state-of-the-art method [14] highest detection performance accuracy of 98.85%.

11.6 Conclusion and future work

In this chapter, we introduced the concept of Benford's law divergence values and their application for classifying biometric data by their underlying modalities and/or sensor sources. It was shown that the Benford's law divergence values could be successfully used with NNs for classification and source identification of biometric images (DB1 fingerprints, DB2 fingerprints, DB3 fingerprints, DB4 fingerprints, contact-less acquired latent fingerprints, vein wrist images, face images and iris images). The classification accuracy for the biometric images was between the range of 90.02% and 100%. For future work, we plan to investigate other classification techniques such as decision trees and SVM for classification and source identification of biometric images.

References

[1]   Harper, W.W.: Fingerprint "Forgery". Transferred latent fingerprints. Journal of Criminal Law and Criminology (1931–1951), 28(4), pp. 573–580, (1937).

[2]   Fu, D., Shi, Y.Q., Su, W.: A generalized Benford's law for JPEG coefficients and its applications in image forensics. In Electronic Imaging 2007. International Society for Optics and Photonics, pp. 1–11, (2007).

[3]   Li, X.H., Zhao, Y.Q., Liao, M., Shih, F.Y., Shi, Y.Q.: Detection of tampered region for JPEG images by using mode-based first digit features. EURASIP Journal on Advances in Signal Processing, 2012, pp. 1–10, (2012).

[4]   Xu, B., Wang, J., Liu, G., Dai, Y.: Photorealistic computer graphics forensics based on leading digit law. Journal of Electronics (China), 28(1), pp. 95–100, (2011).

[5]   Benford, F.: The law of anomalous numbers. Proceedings of the American Philosophical Society, 78, pp. 551–572, (1938).

[6]   Pérez-Gonález, F., Heileman, G.L., Abdallah, C.T.: Benford's law in image processing. In Image Processing, 2007. ICIP 2007. IEEE International Conference on Vol. 1, pp. I-405, (2007).

[7]   Hill, T.P.: A statistical derivation of the significant-digit law. Statistical Science, 10, pp. 354–363, (1995).

[8]   Durtschi, C., Hillison, W., Pacini, C.: The effective use of Benford's law to assist in detecting fraud in accounting data. Journal of Forensic Accounting, 5(1), pp. 17–34, (2004).

[9]   Iorliam, A., Ho, A.T.S., Poh, N., Shi, Y.Q.: Do biometric images follow Benford's law? In 2014 International Workshop on Biometrics and Forensics (IWBF), pp. 1–6, (2004).

[10] Acebo, E., Sbert, M.: Benford's law for natural and synthetic images. In Proceedings of the First Eurographics Conference on Computational Aesthetics in Graphics, Visualization and Imaging. Eurographics Association, pp. 169–176, (2005).

[11] Jolion, J.M.: Images and Benford's law. Journal of Mathematical Imaging and Vision, 14(1), pp. 73–81, (2001).

[12]   Qadir, G., Zhao, X., Ho, A.T.: Estimating JPEG2000 compression for image forensics using Benford's Law. In SPIE Photonics Europe. International Society for Optics and Photonics, 7723, pp. 77230J–1:77230J–10, (2010).

[13]   Iorliam, A., Ho, A.T.S., Poh, N.: Using Benford's Law to detect JPEG biometric data tampering. Biometrics (2014).

[14]   Hildebrandt, M., Dittmann, J.: Benford's Law based detection of latent fingerprint forgeries on the example of artificial sweat printed fingerprints captured by confocal laser scanning microscopes. In IS&T/SPIE Electronic Imaging. International Society for Optics and Photonics, pp. 94090A–94090A–10, (2015).

[15]   Golik, P., Doetsch, P., Ney, H.: Cross-entropy vs. squared error training: a theoretical and experimental comparison. In INTERSPEECH, pp. 1756–1760, (2013).

[16]   Othman, A.A.: Mixing Biometric Data For Generating Joint Identities and Preserving Privacy. PhD Thesis, West Virginia University, (2013).

[17]   Othman, A., Ross, A: Mixing fingerprints. Encyclopedia of Biometrics, 2, pp. 1122–1127, (2015).

[18]   Yan, Y., Osadciw, L.A.: Bridging biometrics and forensics. In Electronic Imaging. Boston, MA: International Society for Optics and Photonics, pp. 1–8, (2008).

[19]   Jain, A.K., Hong, L., Pankanti, S., Bolle, R.: An identity-authentication system using fingerprints. Proceedings of the IEEE, 85(9), pp. 1365–1388, (1997).

[20]   Note on CASIA-IrisV1. Biometric Ideal Test, (2006), http://biometrics.idealtest.org/dbDetailForUser.do?id=1

[21]   Vein Dataset. PUT Vein Database Description, (2011), http://biometrics.put.poznan.pl/vein-dataset/

[22]   Maltoni, D., Maio, D., Jain, A.K., Prabhakar, S.: Synthetic fingerprint generation. Handbook of Fingerprint Recognition, Springer Science & Business Media, pp. 271–302, (2009).

[23]   Hildebrandt, M., Sturm, J., Dittmann, J., Vielhauer, C.: Creation of a public corpus of contact-less acquired latent fingerprints without privacy implications. In Communications and Multimedia Security. Springer Berlin Heidelberg, pp. 204–206, (2013).

[24]   Bartlow, N., Kalka, N., Cukic, B., Ross, A.: Identifying sensors from fingerprint images. In Computer Vision and Pattern Recognition Workshops. IEEE Computer Society Conference on, pp. 78–84, (2009).

[25]   CASIA-FACEV5. Biometric Ideal Test, (2010), http://www.idealtest.org/dbDetailForUser.do?id=9

[26]   FVC2000. Fingerprint Verification Competition Databases, (2000), http://bias.csr.unibo.it/fvc2000/databases.asp

[27]   Li, B., Shi, Y.Q., Huang, J.: Detecting doubly compressed JPEG images by using mode based first digit features. In Multimedia Signal Processing, 2008 IEEE 10th Workshop on, pp. 730–735, (2008).

[28]   Panchal, G., Ganatra, A., Kosta, Y.P., Panchal, D.: Behaviour analysis of multi-layer perceptrons with multiple hidden neurons and hidden layers. International Journal of Computer Theory and Engineering, 3(2), p. 332, (2011).

Part IV

User-centricity and the future

Chapter 12

Random projections for increased privacy

Sabah Jassim[1]

The extraordinary speed with which new models of communication and computing technologies have advanced over the last few years is mind boggling. New exciting opportunities are emerging all the time to facilitate high volume of global commercial activities, enable the citizens to enjoy convenient services as well as mobile leisure activities. These exciting opportunities and benefits come with increased concerns about security and privacy due to a plethora of reasons mostly caused by blurring of control over own data. Conventional access control to personal/organisational data assets use presumed reliable and secure mechanisms including biometric authentication, but little attention is paid to privacy of participants. Moreover, digitally stored files of online transactions include traceable personal data/reference. Recent increase in serious hacking incidents deepens the perception of lack of privacy. The emerging concept of personal and biometric data de-identification seem to provide the most promising approach to deal with this challenge. This chapter is concerned with constructing and using personalised random projections (RPs) for secure transformation of biometric templates into a domain from which it is infeasible to retrieve the owner identity. We shall describe the implications of the rapid changes in communication models on the characteristics of privacy, and describe the role that RP is, and can, play within biometric data de-identification for improved privacy in general and for cloud services in particular.

12.1 Introduction

The right to privacy, before the Internet, was associated with the right to withhold information about one's own activities and wealth or lack of it. Inability to maintain reasonable control of the use of personal data constitutes invasion of privacy. The advent of the Internet and Web technology has led to a rapid growth in their deployment for e-commerce and e-services, whereby public/private organisations could legitimately acquire a variety of customer's information with incredible ease. This has eventually led to raising serious doubts about the ability of these organisations to protect customer's privacy. Indeed, these concerns are perceived to be one of the

[1]Department of Applied Computing, University of Buckingham, United Kingdom

major factors in the collapse of the early e-commerce businesses, and motivated the development of security mechanisms and protocols [such as Public Key Infrastructure (PKI) and Secure Socket Layer (SSL)] to assure customers about the secrecy of their information which eventually led to a resurrection of e-commerce with an even bigger impact.

Subsequent advances in mobile and wireless technologies at the dawn of the twenty-first century opened new opportunities for online services, although new security vulnerability and privacy concerned presented new challenges but did not impede the great benefits brought by the mobility enabling technology that have reduced reliance on infrastructure. The fast increase in adoption rate of new generations of smart mobile technology seems to have provided an added incentive, if any was needed, for the emergence and spread of new models of communication and computing technologies. New technologies facilitate high volume of global commercial activities and enable the citizens to enjoy convenient services as well as mobile leisure activities. The way social networks have become the dominant means of a person's daily activities, including anytime and anywhere access to wealth of services, is a testimony to the huge changes that has occurred over the last few years. The speed with which banking and financial services have adopted online and mobile technologies could not have been anticipated and soon high street banking may become a thing of the past. Already, similar observations are made about the security and privacy challenges when the cloud is used for healthcare services (see CSCC 2012 report [1]). The rise of more cloud services facilitates even more exciting future changes to the way people, organisations and authorities interact with each other for personal leisure or official and trading activities. These exciting opportunities and benefits come with increased legitimate concerns about security and privacy caused mostly by blurring of control over own data due to the way data are processed/transmitted in the cloud.

Finally, the availability of a variety of cheap sensors and micro-electronic units with communication had a natural implication for the deployment of sensor network technology. The Internet of Things (IoT) is one such consequence. Although the perceived impact of IoT on privacy have been investigated and developed for some time, its deployment is almost limited to wearable devices. Setting aside paranoia, IoT adverse implications for privacy are at best underestimated but at worse will not be taken seriously until it is widely adopted by the masses to more of the domestic devices in the homes.

The development of new smarter technologies is far from slowing down or following a more predictable pattern and the nature of threats to privacy seem to be escalating fast at similar speed. The changes to the sources of attacks on privacy and their implications have been influenced by the way new technologies work as well as the lack of transparency by industry. Unfortunately, security and privacy concerns are mostly an aftermath issue and often subsequent remedies/patches are not easy to implement and severely limit the adoption of the technology.

Potential solutions benefit from the close relationship between confidentiality and privacy issues and wealth of security tools amassed by decades of research and development in the basic sciences of cryptography and biometrics. In general, threats to privacy have been confronted by the security and biometrics research communities in different ways, but the emerging protection mechanisms of personal and biometrics-related data de-identification by removing/concealing user-identity seem to be among the most promising approach. Random projections (RPs) are used to transform biometric data into a domain from which it is infeasible to retrieve the owner's identity. In this chapter, we argue in favour of using personalised RPs for de-identification of biometric while authenticating the user.

The rest of this chapter is organised so that in Section 12.2, we describe the evolution of the nature and consequences of threats to privacy as the technology evolves, while Section 12.3 reviews the dual role of biometrics. Section 12.4 is concerned with approaches to de-identification of personal and biometric data and the role of cryptosystems in protecting owner privacy. Section 12.5 covers the main contribution of this chapter. It is concerned with RPs as a means of producing de-identified revocable biometric data, the focus being on the construction of pools of personalised RPs and their use for de-identification of biometrics data from different biometric traits. We also propose the uses of multiple versions of RP-transformed biometric templates and reduce the chance of linking to the owner. The performance and security of the proposed RP-based de-identification scheme will be analysed at the end.

12.2 Evolution of privacy threats

Privacy is the ability of an individual or group to seclude themselves or information about themselves and thereby express themselves selectively. The boundaries and content of what is considered private differ among cultures and individuals, but share common themes. The meaning of the right to privacy and the implications of invasion of privacy attacks have undergone significant changes as new models of communication and computing technologies have emerged and adopted over the last few decades. Changes in the nature of privacy are due to the way technology operates. Accordingly, we divide this discussion in terms of the emergence of the cloud model into pre-cloud and post-cloud.

12.2.1 Pre-cloud privacy

Prior to the advents of the Internet, the right to privacy was protected by law in most countries even if it wasn't practised. It is enshrined in the fourth amendment of the US constitution and in Article 8 of the UK Human Rights Act. The most common feature of right to privacy legislation emphasises, 'the right of a person to be left alone and obliges the state not to meddle illegally with citizen right to privacy and to criminalise individuals who invade others right to a private life'. Surveillance (lawful or otherwise) by the state or individuals are the main source of violation of privacy, while revelation or passing on information illegally about personal wealth, sexual orientation, state of health, criminal records, etc.

The emergence of Internet and Web technology and advances in database systems have led to the adoption of online electronic communication model in a variety of commercial and business activities. Recognising the benefits of this model to convenience and cost effectiveness of operations, their uses were embraced, rarely with caution, by a variety of service providers. The introduction of electronic patient records was a natural tool to be adopted in some countries, especially where the healthcare system expanded way beyond the cottage industry involving a variety of organisations including primary care, insurance companies, government departments, research centres as well as secondary industries. A growing number of public/private organisations engaged in e-commerce and e-service activities were able to freely collect customer's personal information while little if any security mechanisms were in place to protect the confidentiality of the collected data or protect customer's privacy. No policy existed on what to do with the collected data once consumed. This has eventually quickened the demise of most e-commerce businesses. The nature and implications of privacy attacks evolved to increase the possibility of passing personal information to other organisations who could misuse with consequence beyond the mere revealing of the information. For example, insurance companies could deny coverage for patients for a variety of reasons.

Analysis of the early crash of e-commerce led to the development of security mechanisms and protocols (such as PKI, SSL and digital signature) to mitigate some but not all of these concerns. Eventually, e-commerce businesses started to grow again and this model dominated the operation of many sectors of commerce, finance as well as public services. Once again technological advances came to the rescue with the emergence of mobile and wireless model of communication providing new opportunities for anytime-anywhere mobile services. In the case of online health records, it was recognised that encryption on its own is not sufficient but has to be integrated within a system that involve intervention components including deterrents (Alerts and audit trails), role/need access control obstacles and system management software [2].

While privacy concerns that could have been arising from using e-commerce and m-commerce were real, it wasn't widespread. In fact, only certain sectors of society were perceived to be targeted, and the implications of gaining access to personal data were mostly confined to getting insurance cover or spreading gossip about celebrities.

The fast-growing adoption of new generations of smart mobile phones (incorporating multisensors including camera, touch screen, and fingerprint pads) with various networking capabilities has led to the realisation of anytime-anywhere trading, service acquisition, and social networking. These exciting capabilities added incentive, if any were needed, for the development of new models of communication technologies that rendered conventional barriers irrelevant and helped spread globalisation into every aspect of human activities. Exciting opportunities are discovered all the time to facilitate high volume of global commercial activities, enabling the citizen to enjoy convenient services as well as mobile leisure activities. The speed with which banking and financial services have adopted online and mobile technologies could not have been anticipated and already high street banking is disappearing rapidly. The way social networks have become the dominant means of a person's daily activities, including anytime-anywhere access to wealth of services, is a testimony to the huge changes that occurred over the last few years. Similar observations can be made about healthcare services. Naturally, security vulnerabilities and privacy concerns got more challenging, but did not impede the great benefits brought by the mobility with reduced reliance on infrastructure. The use of profiling algorithms over platforms like Twitter and Facebook to extract information about almost anyone without strong explicit consent is becoming an obvious source of privacy violation. However, the way these platforms are embraced by individuals and organisational marketing departments raises serious doubts in some circles about the very existence of a privacy issue. However, many service providers, such as healthcare services, cannot afford to be complacent when the size of litigation they face is on the rise.

12.2.2 Post-cloud privacy

The emergence of cloud model of computing facilitates even more exciting future changes to the way people, organisations and authorities interact. These exciting opportunities and benefits come with increased legitimate concerns about security and privacy due to a variety of reasons, mostly are consequences of blurring of control over own data due to the way the cloud model operates [3]. More importantly, the cloud model and architecture have evolved fast from what was perceived to be simply a natural expansion of distributed computing systems. New business opportunities prevailed to exploit advances in technology for providing customised cloud services that involve interaction with the customer through smart devices. Customisation often requires identification of the user and profiling using biometrics technology. Examples of such scenarios include the growing use of cloud gaming as well as cloud services that involve the delivery of customised multimedia. Naturally, cloud service providers (CSPs) proactively seek to gain information about user preferences and context about the users to adapt services and applications to their preferences and needs. Many of the existing cloud architectures focus on data processing and service delivery as the essential part of a smart environment, with little or no appreciation of user's concerns about the potential loss of privacy from misuse of such information.

The loss of privacy of cloud service users (CSUs) is potentially a direct consequence of surrendering control over personal information to the CSP and the way the cloud processes, stores and transmits user information (see e.g. [1,3]). It is worth noting that many cloud-based social media revenue rely on selling customers' private information and buyers often use the information to build profile of would be customers. Hence, no one knows who else has access to their personal information or for what purposes, i.e. the boundary of access is undefined and widely open.

The ability of a CSP to maintain security of customers' data is a major concern of cloud users and lack of transparency by CSPs exacerbates these concerns. The lists of reputable CSPs that have become the source of outage of users' sensitive data include Amazon, Dropbox, Microsoft, Google Drive, etc. [4]. To help customers recover in case of service failures, data proliferation is conducted in the cloud where customers' data is replicated in multiple data centres for backup [5]. Obviously, this practice is likely to increase the risks of data breaches and inconsistency. The heterogeneity of security settings for the multiple storage devices means that the overall security is determined by the weakest link in the chain. Moreover, when customers update their data, the multiple data copies must be synchronised if data consistency/integrity is to be maintained. CSUs cannot be certain of correctness of CSP's data operations. For example, if CSU requests the deletion of all its data copies, the CSP cannot guarantee such an action completely [5]. Data de-identification, including anonymisation, is an essential mechanism to protect CSU privacy and must be adopted by CPUs and enforced by standards, regulations and laws.

The practice of virtualisation, while highly beneficial to CSUs, raises challenges to user authentication, authorisation and accountability in terms of properly defining roles and policies [6]. It enables CSUs to access their data and applications running on a single logical location which is usually the integration of multiple physical or virtual devices increasing the chance of information leakage [7]. In short, virtualisation introduces the most serious security and privacy risk in cloud computing.

Although, the concept of IoT predates the emergence of cloud technology, its full potential and impact is yet to be fully realised. The perceived impact of IoT on privacy have been investigated and developed for some time, and primarily the threat to privacy is attributed to how the various connected devices collect personal information and the blurred boundary of access to the collected data. Just as in the case of smart cloud services, threats to user privacy from IoT emanates from lack of control by the data owner. In fact, control by users will become almost impossible and continue to grow in its adverse impact. It is no longer the case that you know who could violate your privacy or what happens to your personal information. Unless something is done to mitigate, privacy protection may become a lost cause. Minimal amount of personal data should be collected by your home smart devices that are connected through IoT. Increased capabilities of smart devices for remotely collecting and transmitting data do not justify their uncontrolled use [8]. Without sounding alarmist, the consequences of breach of security of your smart household appliances results in your privacy being invaded. Identity theft would become even more widespread and loss of trust in the various organisations would have unknown damage on global economic activities.

12.2.3 Towards a privacy protection strategy

The above brief discussions reveal that the widening of concerns about loss of privacy, as a consequence of the unprecedented evolution in communication technology and environment, stems more from the potential for total loss of control as well as the enormity of adverse implications for the users. In other words, the threat to privacy benefits from the same type of loopholes but the technology has made enormous opportunity to adversaries. Hence, it is necessary to be aware of these changes when investigating possible protection mechanisms. For the cloud and IoT era, the well-established information security mechanisms of cryptography and steganography are certainly necessary to protect confidentiality and integrity of data as well as control access to data. Besides encryption, public key cryptography (including Rivest–Shamir–Adleman (RSA) and elliptic curve ciphers) enables digital signature/authentication for users and devices. However, adopting or adapting such solutions for the cloud is not as straightforward as protection of a small organisation's computing/communication system. Privacy preserving solutions that have been employed in traditional computing platforms can only mitigate the loss of control by data owner for some but not all cloud services. Even if strong encryption schemes are incorporated, privacy is not guaranteed due to many loopholes that allow unknown entities uncontrolled access to personal data on the huge number of virtual machines. ID-based cryptography has recently been proposed for the cloud computing using some user ID as the public key. This may efficiently secure cloud users' communications, but its implication for privacy is far from being obvious or understood.

Moreover, there has been an expansion in the type of online information/objects that include person's identifying content. Removing any content that could reveal personal information defeats the purpose of using the cloud and any new sophisticated technology. Privacy protection strategy must, therefore, focus on developing appropriate schemes for the de-identification of personal data/transactions (i.e. removing attributed and biometric-related references). The common practice of de-identification that masks/deletes user identity parameters from data cannot provide adequate protection for cloud services that incorporate natural interactive user interfaces because such interactions involve user biometric or attribute identifiers. In the rest of this chapter, we review the various technologies to protect security biometric templates that provide some but not full privacy protection, and we shall argue for incorporating RPs to create tools for de-identification of biometric-related data.

12.3   The dual role of biometrics in privacy protection

Biometrics are digital representation of distinguishing physical appearance or behavioural characteristics of a person which can play a significant role in securing authorised access to data and services and yet their frequent use could become a source of invasion of privacy especially as a result of widespread use of social networks, cloud computing and IoT. We shall describe existing approaches to make the use of biometrics as part of the solution rather than a source of problems for privacy invasion.

Establishing and verifying the identity of a person is a key task in a wide range of applications including the control of international border crossing, prove entitlement to a service, allowing access to a sensitive site and proof of guilt/innocence in crime investigations. Increased deployment of new technologies and the Internet in e-commerce and e-services resulted in increased cases of fraud and identity theft. A reliable identity management system has become an essential component of any reliable secure information system deployed in applications like the ones mentioned above. Throughout the different stages of technology development, three ways have been developed to establish the identity of an individual, each with own advantages and limitations.

- Something you know – An easy to remember short secret such as password or PIN.
- Something you have – Possession of a token that is lost or stolen.

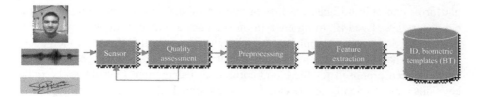

Figure 12.1   A typical enrolment stage of a biometric system

- Something you are – A physical appearance or behavioural characteristic linked with the identity of individual, i.e. a biometric. A legitimate user does not need to remember or carry anything.

Over the years, a large number of biometric traits together with a variety of feature extraction and matching schemes have been investigated. Biometrics based on physical characteristics include face, iris, fingerprint, DNA and hand geometry, while behavioural characteristics-based biometrics include speech, gait and signature. The suitability of a biometric trait for an application depends on several factors such as universality, uniqueness, invariance over time, measurability, usability and cost [9].

A typical biometric system has two stages, enrolment and recognition stage. Figure 12.1 illustrates the process of biometric enrolment stage in which a user starts by presenting their biometric data to a biometric sensor (usually in a controlled environment). If the quality of the captured biometric sample is found to be adequate, the enrolment process proceeds to a preprocessing procedure to prepare the sample for the next step. A feature extraction technique is then used to extract as set of discriminating features called biometric temple, which will be stored in a database.

At the recognition stage (verification or identification mode), a fresh biometric sample is extracted using the same steps above, which is then matched to the template in the database that is most similar to it according to an appropriate distance/similarity function.

Biometrics play a rather complex role in security. It is used for authentication and access control tasks while it could easily be perceived as a source of privacy violation. At the turn of the twenty-first century, the challenge of designing privacy preserving biometrics became an active research area resulting in a variety of techniques that are very relevant to the concept of biometric de-identification. In order to illustrate the various privacy-related techniques for biometric de-identification, it would be important to have a reasonable knowledge of the specific biometric feature vector and the corresponding recognition scheme(s). The fact that face images and speech signals can be obtained easily by others with or without consent/awareness makes all the applications that involve these two biometric traits targets of privacy violation. We shall now review common approaches to face and speech biometrics.

12.3.1 Face biometrics

The face is the most unobtrusive public identity of a person and therefore face images appear on most personal identity documents such as passports, driving licences and ID cards. Using face biometric for person identification in a wide range of security-related applications [9–11] can be attributed to the fact that the person's cooperation is not necessary. Rapid growth in identity theft and availability of high-resolution digital cameras in smart mobile phones and surveillance devices is a major driving force in the surge of research interest in face-based authentication. New opportunities as well as tough challenges are emerging for mass deployments of face-based authentications in a range of civilian and military applications. Unfortunately, these new opportunities together with the wide spread of fast exchanges of millions of personal images over a growing number of social networks are expected to have adverse impact on the security and privacy of face biometrics. To discuss the concepts/techniques needed for the mitigation of such threats, we first review some of the common models of face biometric templates that can be transformed by RPs without modification or loss of accuracy.

Face recognition, like any other biometric, begins by processing face images to extract a feature vector with good discriminating characteristics. Face images typify the well-known 'curse of dimensionality' problem. In the spatial domain, a face image is represented by a high-dimensional array (e.g. $12,000=120\times100$ pixels), the processing/analysis of which is computationally demanding. It is therefore essential to find a more compact representation of face images that encapsulates the most discriminating information in the image by eliminating redundant data without significantly reducing, if not improving, the discriminating characteristics of the resulting compact image representation. Dimension reduction is closely linked to our proposal for using RPs for de-identification biometrics data. Here, we shall describe two different solutions. The most commonly used spatial domain dimension reduction, for face images, is the principal component analysis (PCA), also known as Karhunen–Loeve transform [12]. In [13], Turk and Pentland used the PCA technique to develop the first successful and well-known Eigenface scheme for face recognition.

PCA uses a sufficiently large training set of multiple face images of the enrolled persons, and models the significant variation of the training images from their average image, by selecting unit eigenvectors corresponding to the 'most significant' eigenvalues (i.e. of largest absolute values), to be used as the basis for a projection matrix that maps the original training set, around their mean image, and align with the directions of the first few principal components which maximises the variance as much as possible. The remaining dimensions correspond to insignificant eigenvalues and tend to be highly correlated and dropped with minimal loss of information. The number of significant eigenvalues chosen by such a face biometric system determines overall tolerance to loss of information and identification accuracy. The eigenvectors corresponding to the selected significant eigenvalues are used to create eigenfaces representing the projection of all the face images in the training set and doesn't resemble the face of any one. The number of chosen significant eigenvalues represents the reduced dimension of the face model space, and each face can be recovered with

Figure 12.2   A sample of significant eigenfaces

increasing accuracy as a linear combination of a sufficient number of the most significant eigenfaces. The ten images below are the top most significant eigenfaces obtained by training face images from the Olivetti Research Laboratory (ORL) face database (Figure 12.2):

PCA does not give any consideration to class labels of the various training samples, and its performance is adversely influenced by within-class variations due to other things variation in illumination or pose. Various modifications of PCA have been developed and test with improved accuracy, e.g. see [14]. These methods require a large training dataset of face images.

Frequency transforms provide valuable tools for signal processing and analysis. Frequency information content conveys richer knowledge about features in signals/images and complements the spatial information. Fourier and wavelet transforms are two examples that have been used with significant success in image processing and analysis tasks including face recognition. These transforms significantly reduce dimension but unlike PCA are biometric data-independent where no training is needed. Discrete wavelet transforms (DWTs) have been successfully used in a variety of image processing and analysis including face recognition schemes [15].

A wavelet is simply a small waveform that, unlike the circular functions, has a finite support and rapidly diminishes outside a small interval, i.e. its energy is concentrated in time. A variety of such wavelets have been developed and used for signal processing/analysis that are referred to in the literature as mother wavelets [16]. Every mother wavelet defines an infinite nested sequence of subspaces of the $L^2(R)$ space of square integrable real-valued functions. Each subspace is generated by the shifted copies of a scaled version of the mother wavelet, and we get tools for multiresolution decomposition/filtering of most signals of interest into sub-bands of different frequency ranges at different scales. The 2D-wavelet transform works on images by successive application of the DWT on the rows of the image followed by application on its columns. At a resolution depth of k, the pyramidal scheme

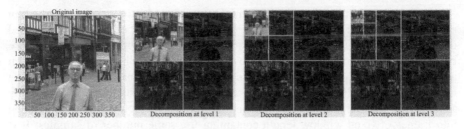

Figure 12.3   Haar wavelet decomposition of an image into depth 3

decomposes an image I into $3k + 1$ sub-bands: (LL_k, HL_k, LH_k, HH_k, ..., HL_1, LH_1, HH_1). Here, LL_k represents the k-level approximation of the image, while HL_1, LH_1, and HH_1 contain finest scale coefficients, and the coefficients get coarser as k increases, LL_k being the coarsest (see Figure 12.3).

The coefficients in each wavelet sub-band represent the face at different frequency ranges and scales (i.e. a distinct stream for recognition to be fused for improved accuracy). Our research found that the LH_3 sub-band is the most discriminating face feature vectors and the score level fusion of several sub-bands improves accuracy over single sub-bands [17].

12.3.2   Speaker recognition

Speech signal is a hybrid of physical and behavioural biometric trait that is widely used for identification and is potentially suitable for use in applications that involve interaction with users. Automatic Speaker recognition/verification schemes are popular for electronic commerce/banking, surveillance tasks in crime fighting and fraud prevention and remote access to computers and confidential information as well as services. Speech is also used for emotion recognition which is useful for improved human–computer interaction and in monitoring for mental care.

Undoubtedly, the emergence of a variety of cloud services and cloud gaming will have generated many new opportunities for deploying speech signal for various purposes including user identification. The H2020 CloudScreen EU-project (http://cordis.europa.eu/project/rcn/109182_en.html) is a developing example of application that aims to incorporate smart environment for the delivery of multimedia that also provides contextual services to users and improved human-systems-devices interaction. Natural and intuitive interaction with digital media requires better understanding of user interaction, preferences and context in media consumption. The CloudScreen project plans to improve user experience by developing a user interface that relies on natural speech and spatial sound detected by the adapted smart device(s). However, such applications are expected to further deepen the privacy concerns especially when user's biometric-related data is used for access control or for building users' profile. It is reassuring that, CloudScreen recognises the potential of adverse implications to user privacy, and are planning to adopt solutions that de-identify user's speech signals.

In order to discuss protection scheme for speaker identity-related information appropriate for cloud services, we shall briefly describe what speech biometric-related data is. The raw speech signal undergoes various signal pre-processing tasks before being identity-discriminating features are extracted. First of all, some form of speech activity detection is performed to remove non-speech portions from the signal. The number of samples in the speech clip is still very excessive and will be reduced by a truncation step using a threshold value. The filtered speech signal is subdivided into frames of fixed window length containing the same number of samples. It is well established that speaker identity-related information are encapsulated by the speech spectrum shape representing the vocal tract shape via resonances (formants) and glottal source via pitch harmonics [18]. Genoud and Chollet [19] showed that the harmonic part of a speech signal contains speaker dependent information that can be transformed to mimic other speakers. In most speaker verification systems, short-term FFT analysis, typically with 20 ms windows placed every 10 ms, is used to compute a sequence of magnitude spectra. The magnitude spectra are then converted to cepstral features after passing through a mel-frequency filter bank and time-differential (delta) cepstra are appended. Finally, some form of linear channel compensation, such as long- and short-term cepstral mean subtraction, to generate the mel-frequency cepstral coefficients feature vector which has a very high dimension (for more details see [20]).

Speaker recognition systems usually create a *speaker* model of the features extracted from the speech signal recorded at the enrolment, and an *imposter* model consisting of a set of other speaker models. The two most common speaker models are Template matching for speech-dependent recognition (whereby the dynamic time warping is used as a similarity measure) and the Gaussian Mixture Model (GMM) for speech-independent recognition. At the verification stage, two scores are computed for an input fresh feature vector: the speaker model score and the imposter model score. The imposter match score is usually computed as the max or average of the match scores from the set of non-claimant speaker models. The decision is based on a likelihood ratio (or a log likelihood) test. In some speaker recognition schemes, the imposter model is replaced with a *world* model (also called the Universal Background Model) that uses a single speaker-independent model trained on speech from a large number of speakers to represent speaker-independent speech. The idea here is to represent imposters using a general speech model, which is compared to a speaker-specific speaker model.

12.4   De-identification of biometric data

De-identifying personal non-biometric data within stored medical files and/or electronic transactions is currently done by removing/concealing/anonymising the user-identity-related subtexts [21]. Existing de-identification of biometric-related data mostly attempts to follow this approach with obvious modifications but only address privacy issues in scenarios where the identity of the person is meant to be concealed. In this section, we shall first critically review these existing techniques highlighting the unsuitability as privacy-preserving mechanisms. In Section 12.4.2, we review existing techniques that have been developed by the biometric community to protect the security of biometric templates during the process of authentication, and then show that incorporating RP within these schemes can be used for biometric data de-identification whereby authentication is achieved while the template is not only secured but the privacy of the owner is protected.

12.4.1 Existing face image de-identification

In terms of face biometrics, this approach was initially based on distorting the face image so that the person in the image could not easily be recognised. Pixe-lating/blurring/blocking the input face image was among the earliest procedures that were believed to de-identify face images. Blurring the face image using Gaussian filter was one such face de-identification. However, humans are capable of recognising faces of friends and public figures even from distorted and blurred images. In fact, it is well established that face recognition schemes tolerate Gaussian blurring of low and medium level of intensity. Moreover, compressive sensing-based super-resolution can be used to enable automatic face recognition even from a high-to-severe level of blurring [22]. Several incremental versions of face recognition schemes have been developed that concatenate feature vectors extracted from small sub-image blocks using certain criteria until an optimal accuracy is achieved beyond which the performance deteriorates, and it was found that for optimal accuracy we need no more than 30% of the face image window blocks (e.g. see [23,24]). These incremental schemes can easily defeat de-identification schemes that are based on blocking or occlusion as long as sufficient window blocks are still visible. More advanced face de-identification have been developed, that outperform the above naïve schemes, which are based on the concept of k-anonymity framework by image transformations that confuse any input face image with faces of k-different persons in a set Gallery. Initially, Gross *et al.* [25] developed a model-based face de-identification, which basically fuses/morphs the input face image with k images in a gallery. Jourabloo *et al.* [26] proposed a specialised version of the model-based scheme, called the attribute preserved face de-identification that preserves certain attributes such as gender, age and glasses. It only fuses gallery images that have similar attribute(s) of input face image.

All these techniques conceal the person's identity and/or require the use of a sufficiently large number of face images for different persons, but do not meet the requirements for applications where services are only delivered to the legitimate clients who are to be biometrically authenticated. De-identification of biometrics in this case is closely related to, and could benefit greatly from, the various sophisticated methods developed by the biometrics research community to protect biometric templates.

12.4.2 Biometric template protection schemes

The growing deployment of biometric-based authentication as a proof of identity tool for access control to physical facilities, entitlement to services and in the fight against crime and terrorism raises some privacy and security concerns. A variety of investigations revealed number vulnerabilities of biometric systems. A biometric template can be (1) replaced by an imposer's template in a system database or be stolen and replayed [11], (2) physically spoofed starting from biometric templates [11,27]. Adler [27] described a Hill Climbing attack on a biometric system that succeeded after a finite number of iterations, to generate a good approximation of the target template. Cappelli *et al.* [28] described a procedure that defeats fingerprint recognition systems by reconstructing fingerprint images from standard templates. These problems reveal that collected biometric data creates a serious source of potential misuse. In fact, the issues of template security and privacy concerns about biometric authentication are intertwined. Protecting biometric template is therefore essential for security and public acceptance of biometric systems, but it is of significant impact on privacy concerns. It is essential that we answer questions like: what can reassure users with regards to privacy intrusion and what recovery plan is there? Naturally, one may turn to cryptography and steganography for solutions.

Non-homomorphic cryptographic algorithms, whether symmetric (e.g. Advanced Encryption Standard (AES)) or public key (e.g. RSA), may not protect biometric templates for several reasons. First, matching in the encrypted domain is expected to result in significantly reduced accuracy since distances/similarities between two encrypted feature vectors should not give any indication on the distances between original feature vectors. On the other hand, decrypting the stored templates for every authentication attempt will make the templates vulnerable to eavesdropping. Finally, once the cryptographic key is compromised and the template is stolen, the system becomes insecure and the given biometric trait is rendered unusable unless the biometric is revocable. Homomorphic encryption is a good candidate for protecting biometric templates since verification/matching is conducted in the encrypting space and the template is not in the clear at any stage. Karabat *et al.* [29] proposed the THRIVE system for enrolment and authentication protocols that are based on threshold homomorphic cryptosystem where the private key is shared between a user and the verifier. Only encrypted binary biometric templates are stored in the database and verification is performed via homomorphically randomised templates, without revealing the original templates. However, hacking may undermine the security of the encrypted templates and it is not clear that authentication is robust against replay attacks. Gomez-Barrero *et al.* [30] proposed a homomorphic encryption-based protection of multi-biometric templates fused at three different levels.

Steganography and watermarking have also been suggested for protecting biometric templates [9]. However, revocability is not guaranteed. Rashid *et al.* [31] used steganography to embed local binary patterns (LBP) face features in images and the input face image cannot be revealed even if the scheme is compromised.

A great deal of research has focused on a variety of Biometric template protection schemes using other aspects of cryptography. The main strategy in these efforts is analogous to that of homomorphic encryption in that the original biometrics are to be concealed. Jain *et al.*, in [32], lists four essential properties that an ideal biometric template protection scheme must satisfy: (1) diversity: templates cannot be used for crossmatching across different databases in which users can be tracked without their permissions. (2) Revocability: templates can be revoked and new ones can be issued whenever needed; (3) security: it is computationally infeasible to reconstruct the original template starting from the transformed one; (4) performance: recognition accuracy must not degrade significantly.

The concepts of revocability of biometric templates aim to protect biometric templates so that, unlike the biometric trait, the templates could be changed and revoked any time in the same way lost or stolen credit cards are replaced. The concept of cancellable or revocable biometrics was first proposed by Ratha *et al.* [9], and such schemes are classified into two categories:

- **Feature transformation** – Matching in a secure transformed domain. Ideally, matching decisions in the secure domain must mimic that in the original domain if not more accurate.
- **Biometric cryptosystem** – Use the biometric trait to generate a key or a biometric hash to be used instead of fresh biometric sample as a proof of identity.

Success is measured in terms of level of security, computational cost and storage requirements, applicability to different biometrics and ability to handle inter-class variation i.e. maintain accuracy.

12.4.2.1 Feature transformation schemes

The basic idea is to transform the original biometric template, rather than the raw biometric trait data, to an ideally 'secure' domain. The transformation function is usually defined by parameters derived from a random key/password, generated at the enrolment stage. At the verification stage, the enrolled transformed template must be available to the verifier entity (stored in the database), and the transform parameters must be available to the user. At entry point, the transformation is applied to a fresh input biometric data and the matching process (against the stored transformed template) will take place in the secure transformed domain. The key might be a user-based or a system-based (depending on usage scenario and/or application type). Whenever necessary, the transformed template could be revoked by changing the transform parameters. Two different types of transform schemes have been developed:

Invertible transformation-based protection schemes are often referred to as salting and their security depend on secrecy of the defining parameter (called the key) of the transform. There are plenty of transformations that can be used for salting biometric data, but due to differences in the nature of data representation of different biometric traits, many salting are application and trait dependent. For example, orthonormal RPs have been used as a salting transformation of wavelet-based face biometric template [33]. Another example of salting is the random multispace quantisation technique, proposed by Jin *et al.* [34], whereby the Fisher discriminant analysis-based face templates are further projected by an orthogonal RP and binarised using a threshold that maximises the entropy of the binary template. In the above two schemes, verification/matching is done in the salted domain. In comparison to the de-identification of face images techniques that are based on distorting the image, the various salting schemes mentioned above provide stronger but not ultimate privacy protection for face biometrics.

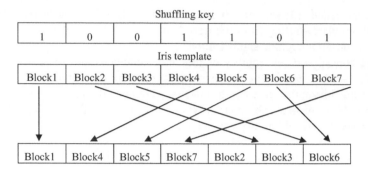

Figure 12.4   Simple secret-based shuffling for iris codes as it was proposed in [38]

A well-known example of privacy-aware feature transformation is the secret-based shuffling, which was proposed by Kanade *et al*. [35] to create revocable versions of iris template. A shuffling key of size k generated from a secret (e.g. password, PIN or a random key) is used to shuffle an iris code that is divided into k block. As illustrated in Figure 12.4, if a bit in the key is one, the corresponding iris code block is moved to the beginning in the order of their appearance; otherwise it is moved to the end. Any fresh iris code is first divided into the same number of blocks and shuffled by the same key before comparing with a stored shuffled iris template.

Such an approach can be applied to any biometric template that can be partitioned into blocks, and the transformation defines a permutation of the template blocks. As the number of blocks increases, inverting the shuffling transform by brute force becomes computationally more infeasible. But the security of the template remains dependent on the secrecy of the shuffling key.

Teoh and Chong [36] proposed a voice cancellable biometric that transforms the speech signal by a probabilistic personal RP. Like most cancellable, it maintains an internal comparison score, and has been shown to be vulnerable to Hill Climbing.

Note that the various salting schemes, described above, produce cancellable biometrics in the sense that changing the personalised transforms result in different biometric templates. However, these various schemes are vulnerable to attacks on template database, by the very fact that verification is done in the slating domain. In the cloud, this vulnerability is compounded by the fact that biometric templates may be stored in a several computing resources as a result of the virtualisation practice.

Jin *et al*. [37] developed the Voice Convergin tool to de-identify speaker voice by a special voice transformation (VT) that differs from salting in that it does not operate on the speaker voice biometric template. The baseline system was based on a GMM-mapping-based VT to convert 24 male source speakers from the LDC WSJ0 corpus to a target synthetic voice called kal-diphone (FestVox: Building Synthetic Voices http://festvox.org 2000). This VT converts speaker's voice into a synthetic speech that does not reveal the speaker's identity to unauthorised listener but a key need to be transmitted in order to enable authorised listeners to back-transform the voice to its original (i.e. speaker reidentification). While the use of synthetic voice does protect the identity of speaker but the scheme requires training with a large set of spoken samples. Moreover, VT speech tends to have artefacts and experiments revealed that in some cases the transformed speech still had features of the source speaker. In fact, Bonastre *et al.* [38] showed that imposter VT drastically increase false acceptance.

Non-invertible transformation biometric template protection schemes work in the same way as above but the feature transformation is typically a one-way function defined by a key which must be available at the time of authentication to transform the query feature vector. The difference here is that it is computationally hard/infeasible to retrieve the original feature template even with the knowledge of the defining key parameter(s). Normally, such transform functions are expected to be many-to-one functions to create confusion in that even if the key is known there are more than one original feature that is mapped to one transformed feature.

Examples of non-invertible biometric transformation schemes include the three non-invertible transforms (Cartesian, polar and functional) of fingerprint images developed by Ratha *et al.* [39]. The Cartesian transform works in similar manner to the iris code shuffling scheme, described above. The image is tessellated into an ordered grid of rectangular cells that are shuffled according to a shuffling key. The larger the number of the grid cells is, the more infeasible to reverse the shuffling. These schemes are perceived to be more secure and privacy preserving than the salting transform-based schemes, perhaps because knowledge of the transform key parameter is sufficient to break the invertible transform schemes but not the non-invertible ones.

The main challenge in designing non-invertible transform protection schemes is the difficulty in maintaining discriminability of transformed feature vectors. The impression in this case is the limited choice. However, for any such function one can obtain infinitely many such functions simply by first applying any orthonormal RP on the original feature space before applying the given transform function. Also, the topological theory of covering projections is a rich source of many-to-one functions defined on any image. However, this is outside the remit of this current chapter.

12.4.2.2 Biometric cryptosystems

Biometric cryptosystems approach to template protection differs from the transform approach by combining biometrics with cryptography usually through the generation/extraction of a cryptographic key/hash or via binding a key to the biometric. In both cases, some public information called *helper data* about the biometric template is stored with the assumption that it does not leak any significant information about the original biometric template. The biometric template is not to be stored anywhere, and matching is performed indirectly by verifying the validity of the extracted key. Existing helper schemes and secure sketches are based on a combination of quantisation and error correcting codes. Helper data are created/extracted from the original biometric data but should not leak much information about the fresh biometric sample, or used to recover/estimate the original raw biometric data.

For key generating cryptosystems, the key (i.e. the helper data) is derived only from the biometric template and the cryptographic key is directly generated from the helper data. Early biometric key generation schemes such as those by Chang *et al.* [40] and Vielhauer *et al.* [41] employed user-specific quantisation schemes. Information on quantisation boundaries is stored as helper data and used during authentication. Error correction coding techniques are typically used to handle intra-class variations. Dodis *et al.* [42] introduced the concepts of secure sketch and fuzzy extractor for key generation from biometrics. The secure sketch is (probabilistic) function that produces a public helper data from a biometric template that leaks only limited information about the template. The helper data facilitates exact reconstruction of the template if a query sample is close to the template.

For key-binding cryptosystems, the key is independent of the biometric data, and matching is meant to use the stored helper data and a fresh biometric to recover the bound key exactly. Since no two biometric samples are exact match, then the exact recovery of the bound key necessitates the use of an appropriate error correcting code that depends on the tolerance practised during verification. Key-binding cryptosystems include two different approaches: fuzzy commitment schemes (FCS) and fuzzy vaults schemes. The bound key can be used for different cryptographic tasks while authenticating the owner. Several fuzzy commitment schemes have been proposed to protect the security of different biometric templates. Hao *et al.* [43] proposed an FCS scheme for key binding with iris codes. Below we shall describe two FCS proposals for face and voice biometrics, and argue that incorporating RPs can improve the security of the biometric templates and by implication increasing their privacy protecting characteristics. These examples also provide new ideas towards an even more privacy protection which will be discussed in the last subsection.

Figure 12.5 illustrates the general format of an FCS scheme for the protection of face biometric templates. Al-Assam *et al.* [44] implemented such a scheme using two versions of PCA-based face biometric template. Users-input sparse orthonormal block diagonal RP and Reed–Solomon Error Correction Code were chosen to ensure the exact recovery at the biometric key as long as the fresh biometric sample identifies the person within an agreed tolerance error. After the enrolment the original key is discarded, but the locked version is kept in the biometric lock depository.

Billeb *et al.* [45] proposed a 2-components voice de-identification. Its first step is a binarisation procedure that works on the GMM speaker model and the Universal Background Model (UBM) world model voice features and outputs a fixed-length binary voice. The second component is an FCS scheme that incorporates the binarised template for a key-binding step. This requires the most discriminative template bits to be marked by storing additional bit-masks for high scalability at negligible cost of biometric performance. The FCS component works in a similar manner to the FCS displayed in Figure 12.5.

A fuzzy vault scheme is another key-binding scheme which is aimed at protecting the security and privacy of biometric features that are represented as an unordered set. A polynomial is selected that encodes the secret key, and evaluated on all points in the biometric features set. Then a large number of random *chaff* points which do not lie on the polynomial, which will be added to the biometric points lying on it to form the vault. Only genuine owner's biometrics samples can recover the secret and be identified. In the fuzzy vault approach, the vault decoding doesn't result in a binary

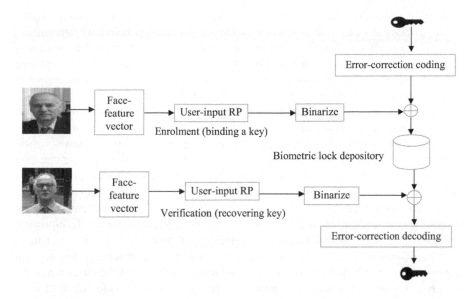

Figure 12.5   A fuzzy commitment scheme for the protection of face biometrics

decision as to whether there is a match or not, but outputs a key that is embedded in the vault. The user can retrieve the secret key from the vault by providing another biometric feature set and if substantial overlap is found between the new set and the original error-correcting codes should help to recover the bound key. There have been several fuzzy vaults proposed in the literature, mainly for fingerprint biometrics (e.g. [46–48]) but some work has been done to construct multi-biometrics fuzzy vaults. The author is not aware of any fuzzy vault for face biometric, and that may reflect the fact that most face biometric features are not represented by an unordered set. However, it is clear that the LBP face biometric features do satisfy this requirement.

Fuzzy vaults schemes have been shown to be vulnerable to various types of attacks including correlation attacks, known key attack and substitution attacks. Many attempts have been made to mitigate such attacks. The correlation attack allows linking two vaults of the same biometric. ÖRENCİK *et al.* [49] proposed the use of distance preserving hash function to mitigate correlation attacks. However, using different RPs to transform the same biometric data in different vaults can provide a perfect delinking of the different vaults. This idea need to be investigated within a general framework for biometric de-identification. It is rather essential for the privacy concerns associated with the cloud, where correlation attacks have greater chance of succeeding as a result of the virtualisation process.

12.4.2.3   Towards hybrid biometric data de-identification

Each of the above template protection schemes has its own advantages and limitations in terms of template security, computational cost, storage requirements, applicability to different kinds of biometric representations and ability to handle inter-class variations in biometric data i.e. maintaining the accuracy. The success and limitation of the various individual transformation-based and cryptosystem-based biometric de-identification schemes discussed above, raises many questions about the wisdom of searching for a single optimal solution. One may fuse the two types for improved security and privacy protection for biometric templates, i.e. using Hybrid approach to protection of user privacy and biometric de-identification.

In fact, some hybrid template protection schemes have already been proposed that make use of more than one basic approach (e.g., salting followed by key binding). The FCS face biometric template protection shown in Figure 12.5 is an example of hybrid approach. Nandakumar *et al.*, in [50], proposed a hybrid protection scheme that uses passwords to harden fingerprint fuzzy vault scheme. And the speaker de-identification of Billeb *et al.* [45] mentioned above is another example of hybrid approach.

While feature transform schemes result in revocable biometric templates, the Biometric cryptosystem allows the revocation of the helper data but not the biometric template. This is another incentive to develop hybrid protection schemes that fuses both transform-based schemes. This could be done in many different ways depending on the biometric trait and on the adopted model of biometric features template. However, for the sake of standardisation, it would be important to identify a standard procedure that could be used to efficiently generate a personalised feature transformation that is infeasible to invert even if it is invertible. This standardised procedure should also allow efficient revocation of the transform. Ideally, the transform type can act either on the raw biometric data or on the feature model.

In the earlier part of this section, we noted that several cancellable biometric schemes involve the use of random orthonormal projections (ROPs) to project biometric features onto private and personalised domains. In the final section of this chapter, we shall present arguments in support of using ROPs as the source of standard feature transformation, and attempt to demonstrate how its use can help mitigate the threat to privacy of cloud users as a result of virtualisation process as highlighted in Section 12.2.

12.5   Random projections

RPs refer to linear transformations from a linear vector space R^n into a subspace R^k whose coordinate system (i.e. generating axes) is random linear combinations of the standard axes of R^n coordinate system. When k is much smaller than n (i.e. $k \ll n$), then this concept is closely linked to the process of dimension reduction used to deal with the challenge of 'Curse of dimension' that is faced in pattern recognition (e.g. biometrics) when the objects are modelled by arrays in R^n with very large n. The PCA method used for reducing dimension in many biometric schemes is based on a structured, rather than random, projection determined by a set of carefully selected raw biometric samples, and therefore would not provide security or privacy protection for the biometric templates.

We shall now confine our discussion to the case that $k = n$. Note that given any such RP, then for any $k < n$ removing any $n - k$ of its generators produces a subspace of dimension k which defines an RP from R^n onto R^k. Mathematically, an RP is a random change of basis of the linear vector space R^n, and the corresponding linear transformations can be represented by an $n \times n$ matrix in terms of the two corresponding bases of the domain and codomain of the projection function. The projection is simply a change of the standard Cartesian base $\{e_1, e_2, \ldots, e_n\}$ of the linear space $\boldsymbol{R}^n$ into a new basis $\{v_1, v_2, \ldots, v_n\}$, where e_i is the n-dimensional vector whose ith coordinate is 1, and is the only non-zero one. In order to guarantee that distances between given dataset of points before the transformation remain the same as after the transformation, then $\{v_1, v_2, \ldots, v_n\}$ need to satisfy orthonormality property:

$$\text{i.e.} \qquad \langle v_i, v_j \rangle = \begin{cases} 1 & \text{if } i = j \\ 0 & \text{otherwise} \end{cases}. \tag{12.1}$$

Therefore, an RP defines isometry of the linear space $\boldsymbol{R}^n$. This property is very important for posttransform-based biometric security protection schemes in relation to maintaining, if not improving, the accuracy of verification in the transformed domain. Large n, such a coordinate system $\{v_1, v_2, \ldots, v_n\}$, is obtained from the standard basis axes by a combination of random rotations and reflections around the origin.

With respect to the standard basis and the new basis, the RP can be represented by an orthonormal $n \times n$ matrix (i.e. $AA^t = I$, where A^t is the transpose of A and I is the identity matrix of the same size of A) of real numbers whose columns are the unit vectors $\{v_1, v_2, \ldots, v_n\}$ representing the new base axes. The set of all such $n \times n$ matrices is a well-researched infinite group where matrix multiplication represents the sequential composition of their linear transformations. In fact, it is a finitely generated Lie Group (i.e. a differentiable manifold where matrix multiplication is also differentiable) of dimension $n(n-1)/2$.

12.5.1 RP generation schemes

There are many ways of generating RP, for use to de-identify biometric feature vectors, and in all cases one randomly selects m orthonormal vectors that have the same dimensionality of the original biometric template. The Orthogonal group is finitely generated which means that there is a finite set of special elements called generators, such that any element in the group is the product of integer powers of the elements in the generator set and their inverses. In fact, any finite random sequence of integers defines a *word* in these generators and their inverses, which defines a RP. However, given any word, finding the sequence of integers that defines it in terms of the generators is an NP hard computational problem in group theory, known as the word problem. For practical reasons, however, simple and efficient procedures are more preferable than this mathematically sound approach.

The most popular scheme to generate RPs to transform templates into secure domains works as follows:

1. Generate m pseudo random vectors in R^n of real values based on a secret key.
2. Apply the well-known Gram–Schmidt on the random vectors to produce an orthonormal matrix.

3. Transform the original template feature x to a secure domain using matrix product:

$$y = Ax \qquad (12.2)$$

where A is the $n \times n$ matrix, representing the constructed linear transformation of R^n onto itself, with respect to the standard basis and the new orthonormal basis output by the Gram–Schmidt procedure.

12.5.1.1   The Gram–Schmidt orthonormalisation (GSO) algorithm

Let $\{v_1, v_2, \ldots, v_n\}$ be a set of linearly independent vectors in the Euclidian space R^n. Generate a new basis of orthonormal vectors using the following n-steps procedure:

1. $\quad w_1 = v_1 \rightarrow u_1 = \dfrac{w_1}{\|w_1\|}$

2. $\quad w_2 = v_2 - \dfrac{v_2 \cdot w_1}{w_1 \cdot w_1} w_1 \rightarrow u_2 = \dfrac{w_2}{\|w_2\|}$

3. $\quad w_3 = v_3 - \dfrac{v_3 \cdot w_1}{w_1 \cdot w_1} w_1 - \dfrac{v_3 \cdot w_2}{w_2 \cdot w_2} w_2 \rightarrow u_3 = \dfrac{w_3}{\|w_3\|}$

$$\vdots$$

n. $\quad w_n = v_n - \sum_{i=1}^{n-1} \dfrac{v_n \cdot w_i}{w_i \cdot w_i} w_i \rightarrow u_n = \dfrac{w_n}{\|w_n\|}$

The set $\{u_1, u_2, \ldots, u_n\}$ is the new orthonormal basis of R^n.

The GSO procedure assumes that the original input set of n vectors in R^n is linearly independent which in practice may not be guaranteed although it is known that for large value of n, the probability that the random set of vector is not linearly independent is very small. Moreover, the steps of the GSO involve repeated division by the norms of the intermediate vectors $\{w_1, w_2, \ldots, w_n\}$ which must be guaranteed to be non-zero. This means that the GSO procedure is not inefficient and is ill-conditioned especially for high dimensional vector spaces.

12.5.1.2   An alternative scheme for RP generation

This procedure was proposed by Al-Assam *et al.* [33] and differs from the above RP generation in that it does not use the GSO but adopt an efficient and stable method. This scheme uses block diagonal matrices using a number of known small size orthonormal square matrices. It exploits the fact that small size orthonormal matrices can be generated from known rotation/reflection matrices. For example, for any θ, the following rotation matrices define orthonormal projections of the two-dimensional plane (resp. the three-dimensional Cartesian space):

$$R_\theta = \begin{bmatrix} \cos\theta & \sin\theta \\ -\sin\theta & \cos\theta \end{bmatrix}, \quad R'_\theta = \begin{bmatrix} \cos\theta & \sin\theta & 0 \\ -\sin\theta & \cos\theta & 0 \\ 0 & 0 & 1 \end{bmatrix}. \qquad (12.3)$$

The Al-Assam *et al.* scheme for RP generation of size $2n$ simply generates a random sequence of angles $\{\theta_1, \theta_2, \ldots, \theta_n\}$ and uses the corresponding 2×2 rotation matrices $R'_{\theta_i}s$ to construct the following block diagonal matrix:

$$
A = \begin{pmatrix}
R_{\theta_1} & 0 & \cdots & 0 \\
0 & R_{\theta_2} & \cdots & 0 \\
\vdots & \vdots & \ddots & \vdots \\
0 & 0 & \cdots & R_{\theta_n}
\end{pmatrix}
\tag{12.4}
$$

For feature vector of odd size $(2n + 1)$ simply select an i and replace R_{θ_i} with R'_{θ_i}.

This scheme facilitates an efficient way of generating personal RPs. Unlike the naïve GSO-based RP generation scheme, it is computationally stable, and completely under the control of its owner in the sense that its action on a feature template is not influenced by external procedures like the GSO determined by the easy to memorise sequence of angles $\{\theta_1, \theta_2, \ldots, \theta_n\}$. Moreover, these RP matrices being highly sparse, make the process of transforming biometric templates extremely efficient. Furthermore, should the sequence of angles be revealed to adversaries, it can easily be replaced by another sequence to revoke the undermined RP-transformed template. Note that, multiplying A on the left or the right by a permutation matrix of the same size generates a new RP that together with the original A could be used by the user in communicating with different entities. In fact, for long sequences of angles, permuting the angles could be enough to generate a different RP.

Adopting these methods of modifying an existing RP from its defining sequence of angles, allows the generation of multiple version of cancellable biometric templates that can be used for authentication associated with different organisations or purposes. More importantly, it provides a perfect mechanism to mitigate the threats to privacy in cloud services as a result of virtualisation. Simply, at the enrolment generate one version of a de-identified cancellable biometric template, using the above RP-transforming scheme, together with as many differently permuted versions as the number of virtually machines by multiplying the original RP by a permutation matrix. In this way, without knowledge of the original sequence $\{\theta_1, \theta_2, \ldots, \theta_n\}$ and the different permutation matrices, no linking can be established between these copies of templates. If the process of generating the permutations, associated with the storage of the various versions of the cancellable biometric templates, is done independently of customer enrolment process, then the linking of the different versions requires collusion of a number of entities.

The approach deployed in this scheme can be extended using other than rotation matrices but any rich pool of known small orthogonal matrices that can be combined in a random block diagonal matrices of any desired size. We shall next describe a well-known and extensively researched source of orthogonal matrices that fits our needs, and we shall show how to exploit the efficiency of construction small size orthogonal matrices, to be incorporated in a secure random process of creating personalised RP dictionaries for de-identification of biometric data suitable for privacy protection in cloud services.

12.5.2 Construction of Hadamard matrices

Hadamard matrices are constructed inductively in several efficient ways and have been used in a variety of applications including digital signal and image processing, and combinatorial designs. The class of Hadamard matrices is a useful source for the generation of RPs. All entries are $+1$ or -1 and any two-distinct vector/column are mutually perpendicular (see Agaian [51]).

An $N \times N$ matrix H_N whose entries are 1 or -1 is a **Hadamard** matrix if:

$$H_N H_N^T = N I_N = H_N^T H_N \tag{12.5}$$

I_N is the $N \times N$ identity matrix. Obviously, the square matrix $\frac{1}{\sqrt{N}} H_N$ is an orthogonal matrix so is its transpose (i.e. inverse) matrix $\frac{1}{\sqrt{N}} H_N^T$.

There are different inductive methods to construct Hadamard matrices, but all are based on using tensor product of matrices defined in terms of the Kronecker Product operation. The **Kronecker product** of two matrices $A_{m \times n}$ and $B_{p \times q}$ is the $mp \times nq$ matrix C defined as follows:

$$C = A \otimes B = \begin{bmatrix} a_{11}B & \cdots & a_{1n}B \\ \vdots & \ddots & \vdots \\ a_{m1}B & \cdots & a_{mn}B \end{bmatrix} \tag{12.6}$$

For example,

$$C = A \otimes B = \begin{bmatrix} 1 & -2 \\ 0 & 0.5 \end{bmatrix} \otimes \begin{bmatrix} 2 & 4 & -6 \\ 1 & 0 & 7 \end{bmatrix}$$

$$= \begin{bmatrix} 1B & -2B \\ 0B & 0.5B \end{bmatrix} = \begin{bmatrix} 2 & 4 & -6 & -4 & -8 & 12 \\ 1 & 0 & 7 & -2 & 0 & -14 \\ 0 & 0 & 0 & 1 & 2 & -3 \\ 0 & 0 & 0 & 0.5 & 0 & 3.5 \end{bmatrix} \tag{12.7}$$

Now we shall describe three different inductive methods for constructing Hadamard matrices of order 2^n, but all the methods use the initial step:

$$H_2 = \begin{bmatrix} 1 & 1 \\ 1 & -1 \end{bmatrix} = \begin{bmatrix} + & + \\ + & - \end{bmatrix}. \tag{12.8}$$

$+$ and $-$ stand for 1 and -1, respectively.

1. Sylvester-type Hadamard matrices (SH)

This inductive scheme generates Hadamard matrices of order $N = 2^n$, as follows:

$$H_N = H_{2^n} = H_2 \otimes H_2 \otimes \cdots \otimes H_2 = \begin{bmatrix} + & + \\ + & - \end{bmatrix} \otimes \begin{bmatrix} + & + \\ + & - \end{bmatrix} \otimes \cdots \otimes \begin{bmatrix} + & + \\ + & - \end{bmatrix} \tag{12.9}$$

For example, by using this formula, we can create H_4 from H_2 and H_8 from H_2 and H_4.

$$H_4 = H_2 \otimes H_2 = \begin{bmatrix} + & + \\ + & - \end{bmatrix} \otimes \begin{bmatrix} + & + \\ + & - \end{bmatrix} = \begin{bmatrix} + & + & + & + \\ + & - & + & - \\ + & + & - & - \\ + & - & - & + \end{bmatrix}, \quad (12.10)$$

$$H_8 == H_2 \otimes H_4 = \begin{bmatrix} + & + \\ + & - \end{bmatrix} \otimes = \begin{bmatrix} + & + & + & + \\ + & - & + & - \\ + & + & - & - \\ + & - & - & + \end{bmatrix}$$

$$= \begin{bmatrix} + & + & + & + & + & + & + & + \\ + & - & + & - & + & - & + & - \\ + & + & - & - & + & + & - & - \\ + & - & - & + & + & - & - & + \\ + & + & + & + & - & - & - & - \\ + & - & + & - & - & + & - & + \\ + & + & - & - & - & - & + & + \\ + & - & - & + & - & + & + & - \end{bmatrix}. \quad (12.11)$$

2. The Walsh–Paley matrices (WP)

This iterative procedure constructs Hadamard matrices of order $N = 2^n$ as follows:

$$WP_N = \begin{bmatrix} WP_{\frac{N}{2}} \otimes [1 & 1] \\ WP_{\frac{N}{2}} \otimes [1 & -1] \end{bmatrix} \quad WP_1 = [1], \quad N = 2^n \quad \text{for } n = 1, 2, 3, \dots$$

$$(12.12)$$

Therefore,

$$WP_4 = \begin{bmatrix} WP_2 \otimes [+ & +] \\ WP_2 \otimes [+ & -] \end{bmatrix} = \begin{bmatrix} \begin{bmatrix} + & + \\ + & - \end{bmatrix} \otimes [+ & +] \\ \begin{bmatrix} + & + \\ + & - \end{bmatrix} \otimes [+ & -] \end{bmatrix} = \begin{bmatrix} + & + & + & + \\ + & + & - & - \\ + & - & + & - \\ + & - & - & + \end{bmatrix}$$

$$(12.13)$$

3. The Walsh Hadamard matrices (W)

This iterative procedure constructs Hadamard matrices of order $N = 2^n$ as follows:

$$W_N = \begin{bmatrix} W_2 \otimes A_1, Q \otimes A_2, \dots, W_2 \otimes A_{(\frac{N}{2})-1}, Q \otimes A_{(\frac{N}{2})} \end{bmatrix} \quad (12.14)$$

where $W_2 = \begin{bmatrix} + & + \\ + & - \end{bmatrix}$,

$$Q = \begin{bmatrix} + & + \\ - & + \end{bmatrix} \quad \text{and } A_i = i\text{th column of } W_{N/2}.$$

SH_{16} WP_{16} W_{16}

Figure 12.6   Binary images of different Hadamard matrices of order 16

Therefore,

$$W_4 = [W_2 \otimes A_1, Q \otimes A_2] = \left[\begin{bmatrix} + & + \\ + & - \end{bmatrix} \otimes \begin{bmatrix} + \\ + \end{bmatrix}, \begin{bmatrix} + & + \\ - & + \end{bmatrix} \otimes \begin{bmatrix} + \\ - \end{bmatrix}\right]$$

$$= \begin{bmatrix} + & + & + & + \\ + & + & - & - \\ + & - & - & + \\ + & - & + & + \end{bmatrix} \tag{12.15}$$

It is customary to display Hadamard matrices as binary images where black and white pixels stand for $+1$ and -1, respectively. Images in Figure 12.6 are for Sylvester, Walsh=Paley and Walsh Hadamard Matrices of order 16.

12.5.3   Hadamard-based RP dictionaries for de-identification of biometric data

The availability of Hadamard orthogonal matrices of small/any sizes provide an additional source, beside the small rotation matrices, of generating sparse random orthonormal matrices without the need for the Gram–Schmidt procedure. The idea is to create a square $K \times K$ block diagonal matrix structure of different square blocks of size $2^i \times 2^i$ sizes for i, where K is the length of the original biometric feature vector. To ensure considerable sparsity, it is necessary to restrict the size of the selected blocks to be bounded, say $\leq 64 \times 64$. Also, to avoid creating an easy to invert, we must not select relatively small number of 2×2 blocks. We set each of the selected blocks as a Hadamard matrix of the same size but its type is to be randomly selected from the set {Sylvester, Walsh–Paley, Walsh}.

If K is the size of the biometric feature vector, then we partition K into sums of 2^i for $2 \leq i < r = 6$. We can include a 2×2 (or 3×3) rotation matrices within this construct to cover the cases $K = 3 + 4\,\text{m}$, or $2 + 4\,\text{m}$.

The following is a simple but effective procedure for generating a random $K \times K$ orthonormal block diagonal matrix to be used for de-identifying the given biometric data. We need to create an ordered triples of integer variables (**Bsize, Htype, T**), where

Bsize is an integer $i \leq r$, r is the upper bound on the chosen block sizes, **Htype** is an integer in the set $\{0, 1, 2\}$, where 0 stands for Sylvester, 1 stands for Walsh–Paley and 2 stands for Walsh, and **T** is a Boolean where 0 stands for the original Hadamard type while 1 stands for its transpose.

Procedure construct (Hadamard Block diagonal RP of size K); //K is even
Begin
i ← 1;
r ← 6; s ←2^r;   //s is the max allowed block size
While (K > s) do
        {b← rand(r);
        b ← 2^b;
        t ← rand (3);
        K ← K–s;
        If (K < s) then s ← s/2;
        (**Bsize, Htype, T**) ← (b, t Div 2, t mod 2);
        A(i) ← Hadamard (**Bsize, Htype, T**);
        i ← ++;
        }
e ← i-1; Output (Block-diagonal (A(1), A(2), ..., A(e)))
End

This procedure outputs the diagonal:

$$A = \begin{pmatrix} A(1) & 0 & \cdots & 0 \\ 0 & A(2) & \cdots & 0 \\ \vdots & \vdots & \ddots & \vdots \\ 0 & 0 & \cdots & A(e) \end{pmatrix}$$

(12.16)

which is an orthonormal sparse RP suitable for de-identification of biometric templates.

In the cloud, the use of different de-identified versions of de-identified templates on different virtual machines must ensure the infeasibility of linking them for tracing the owner of the template. Instead of constructing a different block diagonal Hadamard RP for each virtual machine, we simply multiply A by a permutation matrix. There are different ways of constructing such a permutation. For efficiency purposes, one can simply use different permutations of the $A(i)$ blocks. The use of relatively small size blocks ensures the number of possible permutations of the blocks is sufficiently large to ensure the creation of different version of de-identified templates for storage in different virtual machines.

12.5.4　*Analytical notes on the performance and security of block diagonal Hadamard RP*

We close this chapter by discussing the effect of using block diagonal RP transforms on the performance of the given biometric system, and analyse the security of our promoted type of RPs in terms of the probability of an adversary guessing the generated Hadamard RP.

The fact that the block diagonal RP constructions, whether based on sequences of angles or Hadamard matrices are orthonormal transformations by design then Euclidean distances, and in general the geometry of the space generated by feature vectors that correspond to the chosen biometric trait, are preserved. Therefore, the use of these transformations cannot have adverse effect on the accuracy of the original biometric scheme. In fact, as a very pleasing side effect of using these transformations for privacy preserving, the fact that the user-selected RPs are meant to be kept as private secret then the chance of an imposter succeeding in impersonating the client becomes highly unlikely without collusion. Indeed, experimental work conducted in [33], on the use of the block diagonal sequences of random rotations to transform face biometrics templates revealed significant reduction in false acceptance rates. It was shown that even when using the wavelet LL3 sub-band to model face biometric, a scheme not known for its accuracy, the use of those RPs led to significantly improved accuracy. When tested the performance of the RP transformed LL3-based recognition scheme on the Yale database, the EER has fallen from 21% to 2%, while testing on the ORL database improved EER from 17% to 0.2%.

To discuss the security strength of the proposed RP approach, we need to demonstrate the infeasibility of guessing the diagonal sub-blocks of a client RP by showing that as the size of the biometric feature vector increases, the size of the key space becomes prohibitive for brute force search. Estimating the size of the key space for K-dimensional feature vectors involves known combinatorial problems:

1. The *Integer partition* problem. For any integer K, this is the number $p(K)$ of unordered sequences of positive integers (also known as parts) which sum to K. Counting the number of certain class of partitions of K is equivalent to counting the number of ways of structural splitting of the would be suitable RP diagonal into non-decreasing sequence of adjacent parts. This problem was first studied by the eighteenth century mathematician *Leonard Euler* in terms of their generating function

$$F(x) = \sum_{n=0}^{\infty} p(n)x^n = \prod_{k=1}^{\infty} \frac{1}{(1 - x^k)} \tag{12.17}$$

For each K, $p(K)$ can be evaluated as the coefficient of x^K, when the each of the factors on RHS of the above formula is expanded to its McLaurin series. It has been shown that the integer partition values $p(i)$ grow exponentially (see [52]). For example, $p(8) = 22$ but $p(100) = 190, 569, 292$. However, for a practical reason, we have already restricted our RP construction to the case where we only use parts of the form 2^i, with $2 \le i \le k = 6$. Hence, the number of possible non-decreasing splitting of the diagonal of block diagonal Hadamard RP matrices of interest is the number $\mathrm{SP}_A(K)$ of partitions of the integer K all of whose parts are in the set $A = \{2^i: 2 \le i \le k = 6\}$. The value of $\mathrm{SP}_A(K)$ for any set A of positive integers can be evaluated by restricting the factors on the RHS of the formula above to those where $k \in A$.

2. The *number* of permutations of n symbols with repetition. For any unordered special partition of K, with our specified restriction, we can generate many different block diagonal Hadamard RPs by permuting the different parts. To determine the number of different RPs for the given partition of K, we need to take into effect the number of times the different parts in the partition are repeated. If r is the number of different parts in the given partition of K, then we have as many distinct RPs as the number of permutations with repetition of **n** elements of r different types (i.e. permutations where the ith element is repeated n_i times and $n = \sum n_i$) which is given by the Euler triangle formula:

$$P(n; \{n_i : i = 1, \ldots r\}) = \frac{n!}{n_1 n_2 \ldots n_r}. \tag{12.18}$$

Combining the computation discussed above, and recalling that for any diagonal block position with any of the selected partitions, we have a choice of three possible Hadamard matrices for that position. One gets the following formula for the size of the key-space of our RP construction:

$$|\text{BD Hadamard}(K)| = \sum_{s \in \mathrm{SP}_A(K)} 3^{|s|} P(|s|; \{n_i : i = 1, \ldots, r\}). \tag{12.19}$$

Here, $r = 5$, and n_i is the number of appearance of $2^{(i+1)}$ in the partitions. Although the set $\mathrm{SP}_A(K)$ doesn't grow as fast as $\mathrm{SP}(K)$, the value of $|\text{BD Hadamard}(K)|$ grows can be shown to grow exponentially. In fact, for $K = 16, 20, 24, 28, 32$, the value of $|\text{BD Hadamard}(K)| = 174; 666; 2{,}556; 9{,}801; 37{,}596$. Clearly as K increases, it is infeasible to deduce the user selected block diagonal Hadamard RP matrix, or the corresponding RP de-identified version of the user biometric template owner.

12.6 Conclusion

This chapter investigated the problem of privacy protection of biometric data that are exchanged for authentication or for providing personalised services over the cloud. We described the evolution of privacy concerns over the last 25 years in parallel with way the computing and communication technologies merged and evolved with the recent emergence of cloud services and IoT. Biometric-related data are regularly extracted and exchanged for various purposes, and it is clear that the way cloud services process users' data using multiple virtual machines represent a major serious threat to privacy. De-identifying biometric data, used for authentication or for improving personalised services, is an essential mean of protecting users' privacy. Transformation-based biometric cryptosystems were designed primarily to provide cancellability of biometric templates when the security of the biometrics system is undermined. Infeasible to invert, computationally efficient, personalised RPs can protect the privacy of the biometric owner when only the user stores the transformation parameters while service providers only store the transformed biometric template captured at enrolment. We developed an effective procedure for generating an extremely large pool of random orthonormal sparse matrices to be used for de-identifying biometric data. The developed procedure doesn't require the use of the unstable GSO by using block diagonal matrices of small size Hadamard and or rotational orthonormal matrices. We also described a complementary method to mitigate the privacy threat of the cloud services virtualisation practice by constructing different versions of de-identified biometric templates for storage on different virtual machines. This will ensure the infeasibility of linking them for tracing the owner of the biometric template. This method simply multiplies the orthonormal block diagonal matrix A by different permutations of the $A(i)$ blocks.

The proposed method, on its own, may not completely prevent privacy breaches. For that we need a tool that dynamically changes the personalised RP. This is related to the ongoing work being carried out at Buckingham to enhance security of modern symmetric ciphers by changing admissible cipher parameters. This will be done in the future and is expected to feed into a standard tool for de-identification of biometric data.

References

[1] Cloud Standards Customer Council, "Impact of Cloud Computing on Healthcare", 2012, http://docshare01.docshare.tips/files/25805/258050979.pdf.

[2] T. C. Rindfleisch, "Privacy, information technology, and health care", Communications of the ACM, vol. 40, no. 8, pp. 93–100, 1997.

[3] Y. Liu, Y. Sun, J. Ryoo, S. Rizvi, and A.V. Vasilaskos, "A survey of security and privacy challenges in cloud computing: solutions and future directions", Journal of Computing Science and Engineering, vol. 9, no. 3, pp. 119–133, 2015.

[4] J. R. Raphael, "The Worst Cloud Outages of 2013", http://www.infoworld.com/article/2606768/cloud-computing/107783-The-worst-cloud-outages-of-2013-so-far.html.

[5] S. Pearson and A. Benameur, "Privacy, security and trust issues arising from cloud computing," in Proceedings of IEEE 2nd International Conference on Cloud Computing Technology and Science (CloudCom), Indianapolis, IN, 2010, pp. 693–702.

[6] Virtualization Special Interest Group and PCI Security Standards Council, "PCI DSS Virtualization Guidelines", 2011, https://www.pcisecuritystandards.org/documents/Virtualization_nfoSupp_v2.pdf.

[7] X. Luo, L. Yang, L. Ma, S. Chu, and H. Dai, "Virtualization security risks and solutions of cloud computing via divide conquer strategy," in Proceedings

of 2011 3rd International Conference on Multimedia Information Networking and Security (MINES), Shanghai, China, 2011, pp. 637–641.

[8]   CDT – Comments after November 2013 Workshop on the "Internet of Things", https://cdt.org/files/pdfs/iot-comments-cdt-2014.pdf.

[9]   N. K. Ratha, J. H. Connell, and R. M. Bolle, "Enhancing security and privacy in biometrics-based authentication systems", IBM Systems Journal, vol. 40, no. 1, pp. 614–634, 2004.

[10] U. Uludag, S. Pankanti, S. Prabhakar, and A. K. Jain, "Biometric cryptosystems: issues and challenges," in Proceedings of IEEE, Special Issue on Multimedia Security for Digital Rights, 2004.

[11] K. Nandakumar, "Multibiometric Systems: Fusion Strategies and Template Security", PhD thesis, Michigan State University, 2008.

[12] M. Kirby and L. Sirovich, "Application of the Karhunen–Loeve procedure for the characterization of human faces", IEEE Transactions on Pattern Analysis and Machine Intelligence, vol. 12, no. 1, pp. 103–108, 1990.

[13] M. Turk and A. Pentland, "Eigenfaces for recognition", Journal of Cognitive Neuroscience, vol. 3, no. 1, pp. 71–86, 1991.

[14] W. Zhao, R. Chellappa, A. Rosefeld, and P. J. Phillips, "Face Recognition: A Literature Survey," Technical Report, Computer Vision Lab, University of Maryland, 2000.

[15] J.-T. Chien and C.-C. Wu, "Discriminant wavelet faces and nearest feature classifiers for face recognition", IEEE Transaction on Pattern Analysis and Machine Intelligence, vol. 24, no. 12, pp. 1644–1649, December 2002.

[16] I. Daubechies, "The wavelet transform, time-frequency localization and signal analysis", IEEE Transactions Information Theory, vol. 36, no. 5, pp. 961–1004, 1990.

[17] H. Sellahewa and S. Jassim, "Wavelet-based face verification for constrained platforms", Proceedings SPIE Biometric Technology for Human Identification II, vol. 5779, pp. 173–183, 2005.

[18] D. A. Reynolds and L. P. Heck, "Speaker verification: from research to reality," in ICASSP, 2001.

[19] D. Genoud and G. Chollet, "Speech pre-processing against intentional imposture in speaker recognition," in ICASLP, 1998.

[20] A. Singh, T. Panchal, and M. Saharan, "Review on automatic speaker recognition system", International Journals of Advanced Research in Computer Science and Software Engineering, vol. 3, no. 2, pp. 350–354, 2013.

[21] J. Jaćimović, C. Krstev, and D. Jelovac, "Automatic de-identification of protected health information," in 9th Language Technologies Conference Information Society, 2014, pp. 73–78.

[22] N. AL-Hassan, S. Jassim, and H. Sellahewa "Enhancing face recognition at a distance using super resolution," in 14th ACM Workshop on Multimedia and Security, 2012, pp. 123–132.

[23] A. Aboud, H. Sellahewa, and S. A. Jassim, "Quality based approach for adaptive face recognition," in Proc. SPIE, Mobile Multimedia/Image Processing, Security, and Applications, vol. 7351, Florid, April 2009.

[24]   S. A. Jassim, H. Al-Assam, A. J. Abboud, and H. Sellahewa, "Analysis of relative entropy, accuracy, and quality of face biometric," in Pattern Recognition for IT Security Workshop, Darmstadt, Germany, September 2010.

[25]   R. Gross, L. Sweeney, F. de la Torre, and S. Baker, "Model-based face de-identification," in IEEE Workshop on Privacy Research in Vision, 2006.

[26]   A. Jourabloo, X. Yin, and X. Liu, "Attribute preserved face de-identification," in ICB, 2015, pp. 278–285.

[27]   A. Adler, "Vulnerabilities in biometric encryption systems," in Proceedings of the 5th International Conference on Audio and Video-Based Biometric Person Authentication, 2005.

[28]   R. Cappelli, A. Lumini, D. Maio, and D. Maltoni, "Fingerprint image reconstruction from standard templates", IEEE Transactions on Pattern Analysis and Machine Intelligence, vol. 29, no. 9, pp. 1489–1503, 2007.

[29]   C. Karabat, M. S. Kiraz, H. Erdogan, and E. Savas, THRIVE: threshold homomorphic encryption based secure and privacy preserving biometric verification system, arXiv:1409.8212v1 [cs.CR] 29 September 2014.

[30]   M. Gomez-Barrero, E. Maiorana, J. Galbally, P. Campisi, and J. Fierrez, "Multibiometric template protection based on Homomorphic Encryption", Pattern Recognition, vol. 67, Issue C, pp. 149–163, July 2017.

[31]   R. D. Rashid, H. Sellahewa, and S. A. Jassim, "Biometric feature embedding using robust steganography technique," in Proceedings of SPIE 8755, Mobile Multimedia/Image Processing, Security, and Applications, Baltimore/USA, May 2013.

[32]   A. K. Jain, K. Nandakumar, and A. Nagar, "Biometric template security", EURASIP Journal on Advances in Signal Processing, Special Issue on Biometrics, pp. 1–17, 2008.

[33]   S. Jassim, H. Al-Assam, and H. Sellahewa, "Improving performance and security of biometrics using efficient and stable random projection techniques," in Proc. of the 6th Int. Symp. on Image and Signal Processing and Analysis (ISPA), Salzburg, September 2009, pp. 556–561.

[34]   A. T. B. Jin, D. N. C. Ling, and A. Goh, "Biohashing: two factor authentication featuring fingerprint data and tokenised random number", Pattern Recognition, vol. 37, pp. 2245–2255, 2004.

[35]   S. Kanade, D. Camara, and B. Dorizzi, "Three factor scheme for biometric based cryptographic key regeneration using iris", Proceedings of 6th Biometrics Symposium (BSYM 2008), pp. 59–64, 2008.

[36]   A. B. J. Teoh, L.-Y. Chong, "Secure speech template protection in speaker verification system", Speech Communication, vol. 52, no. 2, pp. 150–163, 2010.

[37]   Q. Jin, A. R. Toth, T. Schultz, A. W. Black, "Voice converging: speaker de-identification by voice transformation," in ICASSP, 2009, pp. 3909–3912.

[38]   J. Bonastre, D. Matrouf, and C. Fredouille, "Artificial impostor voice transformation effects on false acceptance rates," in INTERSPEECH, 2007, pp. 2053–2056.

[39]   N. K. Ratha, S. Chikkerur, J. H. Connell, and R. M. Bolle, "Generating cancelable fingerprint templates", IEEE Transactions on Pattern Analysis and Machine Intelligence, vol. 29, no. 4, pp. 561–572, 2007.

[40]   Y.-J. Chang, W. Zhang, and T. Chen, "Biometrics-based cryptographic key generation," in Proceedings of the IEEE International Conference on Multimedia and Expo (ICME '04), vol. 3, Taipei, Taiwan, June 2004, pp. 2203–2206.

[41]   C. Vielhauer, R. Steinmetz, and A. Mayerhöfer, "Biometric hash based on statistical features of online signatures," in Proceedings of the International Conference on Pattern Recognition, vol. 1, Quebec, QC, Canada, August 2002, pp. 123–126.

[42]   Y. Dodis, L. Reyzin, and A. Smith, "Fuzzy extractors: how to generate strong keys from biometrics and other noisy data," in Proceedings of International Conference on the Theory and Applications of Cryptographic Techniques: Advances in Cryptology (EUROCRYPT '04), vol. 3027 of Lecture Notes in Computer Science, Interlaken, Switzerland, May 2004, pp. 523–540.

[43]   F. Hao, R. Anderson, and J. Daugman, "Combining crypto with biometrics effectively", IEEE Transactions on Computers, vol. 55, no. 9, pp. 1081–1088, 2006.

[44]   H. Al-Assam, H. Sellahewa, and S. Jassim, "A lightweight approach for biometric template protection," in Proc. SPIE Mobile Multimedia/Image Processing, Security, and Applications, vol. 7351, April 2009, p. 73510P.

[45]   S. Billeb, C. Rathgeb, H. Reininger, K. Kasper, and C. Busch, "Biometric template protection for speaker recognition based on universal background models", IET Biometrics, vol. 4, no. 2, pp. 116–126, 2015.

[46]   A. Juels and M. Sudan, "A fuzzy vault scheme," in Proceedings of the IEEE International Symposium on Information Theory, Piscataway, NJ, USA, June–July 2002, p. 408.

[47]   U. Uludag and A. K. Jain, "Securing fingerprint template: fuzzy vault with helper data," in Proc. IEEE Workshop on Privacy Research in Vision, NY, 2006, pp. 163–171.

[48]   K. Nandakumar, A. K. Jain, and S. Pankanti, "Fingerprint-based fuzzy vault: implementation and performance", IEEE Transactions on Information Forensics and Security, vol. 2, no. 4, pp. 744–757, 2007.

[49]   C. Örencik, T. B. Pedersen, E. Savas, and M. Keskinoz, "Securing fuzzy vault schemes through biometric hashing", Turkish Journal of Electrical Engineering and Computer Sciences, vol. 18, no. 4, pp. 515–539, 2010.

[50]   K. Nandakumar, A. Nagar, and A. K. Jain, "Hardening fingerprint fuzzy vault using password," in Proceedings of 2nd International Conference on Biometrics, South Korea, 2007, pp. 927–937.

[51]   S. S. Agaian, (Ed.), 2011. Hadamard transforms. SPIE, Bellingham, Washington, 2011.

[52]   G. H. Hardy and S. Ramanujan, "Asymptotic formulæ in combinatory analysis", Proceedings of the London Mathematical Society, vol. 17, pp. 75–115, 1918.

Chapter 13

De-identification for privacy protection in biometrics

Slobodan Ribarić[1] and Nikola Pavešić[2]

De-identification, which is defined as the process of removing or concealing personal identifiers or replacing them with surrogate personal identifiers to prevent direct or indirect identification of a person, is recognized as an efficient tool for protection of a person's privacy that is one of the most important social and political issues of today's information society. The chapter "De-identification for privacy protection in biometrics" aims to review the progress of recent research on the de-identification of biometric personal identifiers. The chapter covers de-identification of physiological biometric identifiers (face, fingerprint, iris, ear), behavioural biometric identifiers (voice, gait, gesture), as well as soft biometrics identifiers (body silhouette, gender, tattoo) and discuss different threats to person's privacy in biometrics.

13.1 Introduction

Privacy is one of the most important social and political issues in our information society, characterized by a growing range of enabling and supporting technologies and services such as communications, multimedia, biometrics, big data, cloud computing, data mining, Internet, social networks and audio-video surveillance. Privacy, described as 'an integral part of our humanity' and 'the beginning of all freedom', has no unique definition; even more, according to Solove, it is a concept in disarray [1]. Its meaning depends on legal, political, societal, cultural and socio-technological contexts. From the legal point of view, the first definition of privacy was given by Warren and Brandeis more than 120 years ago [2]. They defined privacy as 'the right to be let alone', with respect to the acquisition and dissemination of information concerning the person, particularly through unauthorized publication, photography or other media. Westin defined privacy as the claim of an individual to determine what information about himself or herself should be known to others [3]. He identified four basic states of individual privacy: solitude, intimacy, anonymity and the creation of a psychological barrier against unwanted intrusion.

[1]University of Zagreb, Faculty of Electrical Engineering and Computing, Zagreb, Croatia
[2]University of Ljubljana, Faculty of Electrical Engineering, Ljubljana, Slovenia

Depending on the socio-technological contexts and real life situations, privacy, in general, can be divided into a number of separate, but related, concepts:

- *Information privacy*, which involves the establishment of rules governing the collection and handling of personal data, such as medical and tax records and credit information;
- *Privacy of communications*, which covers the security and privacy of mail, telephone, e-mail and other forms of communication;
- *Bodily privacy*, which concerns the protection of a person's physical self against invasive procedures such as genetic tests, drug testing and cavity searches.
- *Territorial privacy*, which concerns the setting of limits on intrusion into domestic and other environments, such as the workplace or public space. This includes searches, video surveillance and ID checks.

De-identification is one of the basic methods for protecting privacy. It is defined as the process of removing or concealing personal identifiable information (shorter: *personal identifiers*), or replacing them with surrogate personal identifiers, in order to prevent the recognition (identification) of a person directly or indirectly, e.g. via association with an identifier, user agent or device. In general, a person can be identified on the basis of biometric personal identifiers, but also by combination of different types of biometric personal identifiers and non-biometric personal identifiers, such as environmental and/or specific socio-political context, speech context, and dressing style.

As this chapter focuses for the most part on the de-identification of biometric personal identifiers for privacy protection, it is interesting to view some of the main concerns related to the use of biometrics [4]:

- A biometric personal identifier can be collected and shared without the user's knowledge and permission;
- A biometric personal identifier which has been collected for some specific purposes can later be used for other unintended or unauthorized purposes. This is referred to as 'functional creep';
- A biometric personal identifier can be copied or removed from the user and used for secondary purposes, e.g. in malicious attack scenarios;
- A biometric personal identifier can be used to reveal sensitive personal information, such as gender, race and ethnicity, but also mental and health status;
- A biometric personal identifier can be used to pinpoint, locate and track individuals. Further, associating biometric personal identifiers with non-biometric identifiers, e.g. name, address, ID and passport number, can lead to covert surveillance, profiling and social control;
- A biometric personal identifier can be exposed to external attacks due to improper storage and/or transmission.

A biometric system for person identification or verification is essentially a pattern recognition system which recognizes a person based on physiological and/or behavioural personal identifiers and to this end, a biometric template is a mathematical representation of one or more personal identifiers. It is obtained from raw biometric data during a feature extraction process and stored in a system's database for future comparison with templates of live biometric personal identifiers.

Biometric templates may be stolen from the system's database, and/or they may be modified and shared, and consequently the privacy and security of the persons involved may be compromised. There are three aspects of protection surrounding the person's privacy with regard to the stored biometric templates [5]:

- *Irreversibility* – it should be computationally hard to reconstruct the original biometric template from the protected stored biometric template;
- *Unlinkability* – it should be impossible to link different biometric templates to each other or to the individual who is the source of both; and
- *Confidentiality* – a person's biometric template should be protected against unauthorized access or disclosure.

Recently, efforts have been made to standardize biometric template protection [5,6]. There are four main biometric template protection schemes:

- Extracting and storing a secure sketch, i.e. an error-tolerant cryptographic primitive of a biometric template;
- Fuzzy commitment in which a biometric template is bound to a secret message;
- Encrypting the biometric template during enrolment; and
- Using of cancellable or revocable biometrics, where, instead of enrolling a person with his or her true biometric template, the biometric template is intentionally distorted in a repeatable manner and this new (distorted) biometric template is used for original biometric template generation. Cancellable biometrics includes face [7], fingerprint [8], iris [9], voice [10] and other biometric personal identifiers.

There are two different approaches to the relation between privacy and biometrics technology. The first approach is based on the assumption that biometrics protects privacy and information integrity by restricting access to personal information [11]. The second approach is based on the belief that biometrics technology introduces new privacy threats [12] by supporting *dragnet* – indiscriminate tracking where institutions stockpile data on individuals at an unprecedented pace [13]. For example, a national identification system gives the government great power and can be used as a tool for surveillance and for tracking people, as well as, for scanning different large databases in order to profile and socially control persons. In applying biometrics technology, we have to avoid the so-called Titanic phenomenon [12] which is reflected in the designer's dead certainty that the Titanic is unsinkable and there is no need to provide many lifeboats. The same is true for biometrics technology – we should not resist a new technology but must show caution in how we implement and use it. De-identification is one of the components in biometrics technology, which concerns privacy protection and provides sufficient 'lifeboats' in situations of abuse of this new technology.

This chapter is based on the authors' paper published in the international journal Signal Processing: Image Communication [14]. However, it has been revised and updated with the latest research in the field of de-identification, such as fast unconstrained face detection based on Normalized Pixel Difference (NPD) features, a two-stage cascade model for face detection (face detection in the wild is a precondition for successful face de-identification), deep convolutional networks used for detection, localization of faces or tattoo regions, head pose estimation and multi-object tracking, and deep learning-based facial identity, preserving features in the process of face de-identification.

13.2   De-identification and irreversible de-identification

This section discusses the difference between de-identification and anonymization. We introduce the term *multimodal de-identification* which is related to the de-identification of different biometric, soft-biometric and non-biometric identifiers which are simultaneously included in the acquired data. Inspired by the Safe Harbour approach [15], we present taxonomy of personal identifiers.

13.2.1   De-identification and anonymization

The terms *de-identification* and *anonymization* are often used interchangeably, but there is difference between them. De-identification refers to the reversible (two-directional) process of removing or obscuring any personally identifiable information from individual records in a way that minimizes the risk of unintended disclosure of the identity of individuals and information about them. It involves the provision of additional information (e.g. secrete key) to enable the extraction of the original identifiers in certain situations by an authorized body or by a trusted party.

Anonymization refers to an irreversible (uni-directional) process of de-identification that does not allow the original personal identifiers to be obtained from de-identified ones. Anonymized personal identifiers cannot be linked back to an original as they do not include the required translation variables to do so.

In this chapter, we use the term de-identification for both approaches, but in some cases we emphasize whether it is a case of reversible or irreversible process. In either case, the de-identification process is required to be of sufficient effectiveness, regardless of whether the recognition (identification) attempts are made by humans or by machines or their combination [14,16]. Effectiveness of the de-identification process is measured by the *de-identification rate*, which is typically stated as the ratio between number of identification attempts of de-identified personal identifiers that are not correctly identified and the total number of identification attempts.

Moreover, in many cases, the process of de-identification has to preserve the data utility, naturalness and intelligibility [16,17]. A crowdsourcing approach is used to evaluate the above mentioned characteristics [18].

13.2.2   Multimodal de-identification

It is worth noting that very often personal identifiers may simultaneously include different biometric and non-biometric personal identifiers, which all have to be de-identified in order to protect the privacy of individuals. This can be referred to as *multimodal de-identification* [14]. For example, in order to protect a person from covert biometric recognition based on iris acquisition at a distance, it is necessary to simultaneously de-identify eye and face regions.

13.2.3 Taxonomy of the personal identifiers

Personal identifiers are classified in three categories as follows [14]:

- *Biometric identifiers* are the distinctive, measurable, generally unique and permanent personal characteristics used to identify individuals. Biometric identifiers are usually categorized as physiological (face, iris, ear, fingerprint) versus behavioural (voice, gait, gesture, lip-motion, stile of typing);
- *Soft biometric identifiers* provide some vague physical, behavioural or adhered human characteristic that is not necessarily permanent or distinctive (height, weight, eye colour, silhouette, age, gender, race, moles, tattoos, birthmarks, scars). In most cases, soft biometric identifiers alone cannot provide a reliable personal identification, but they can be used for improving the performance of recognition [19,20], or to classify people into particular categories, which is also privacy intrusive;
- *Non-biometric identifiers* are information items, like given and family names, personal identification number, social security number, driver licence number, text, speech and/or specific socio-political and environmental context, licence plate, dressing style and hairstyle, which allow his or her direct or indirect identification.

In this chapter, our attention will be focused on the de-identification of main representatives of first two categories of personal identifiers: biometric and soft biometric.

13.3 De-identification of physiological biometric identifiers

Physiological biometric identifiers, such as deoxyribonucleic acid, the face, fingerprint, iris, palmprint, hand geometry, hand vein infrared thermograms and the ear, offer highly accurate recognition of a person. This section describes de-identification of the most frequently used physiological biometric personal identifiers: the face, fingerprint, iris and ear. Special attention is devoted to face de-identification in video surveillance applications on account of certain problems, such as the detection and de-identification of the faces in the wild.

13.3.1 Face de-identification in still images

There is no doubt that the main physiological biometric identifier which can be collected at a distance and use to covert person recognition is face [21]. Obviously, this fact strongly urges de-identification for privacy preservation. The early research, in period 2000–04, into face de-identification was focused on *face still images*, and recommended the use of ad hoc or so-called naive approaches such as 'black box', 'blurring' and 'pixelation' of the image region occupied by the face [22,23]. In *the black-box approach*, after the face detection and face localization in the image, the

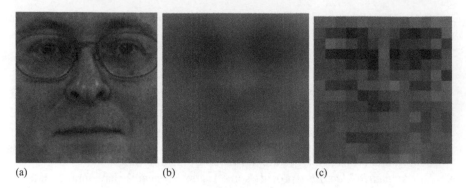

(a)   (b)   (c)

Figure 13.1   Naive methods of face de-identification: (a) original image;
(b) blurring; (c) pixelation. © 2015 IEEE. Reprinted, with permission,
from [24]

face region is simply substituted by a black (or white) rectangle, elliptical or circular cover. *Blurring* (Figure 13.1(b)) is a simple method based on smoothing the face in an image with Gaussian filters using a variety of sufficiently large variances. *Pixelation* (Figure 13.1(c)) consists of reducing the resolution (subsampling) of a face region [24]. Naive methods such as blurring and pixelation might prevent a human from recognizing subjects in the image, but they cannot thwart recognition systems.

An effective approach that subverts naive de-identification methods is called *parrot recognition* [25]. Instead of comparing de-identified images to the original images, parrot recognition is based on comparing probe (de-identified) images with gallery images, where the same distortion is applied as in the probe images. Such an approach drastically improves the recognition rate, i.e. it reduces the level of privacy protection [25].

To achieve an improved level of privacy protection, more sophisticated approaches have been proposed: In [26], an eigenvector-based de-identification method is described. The original face is substituted by a reconstructed face that is obtained by applying a smaller number of eigenfaces. The eigenvector-based method easily produces very unnatural images, but still keeps some of the facial characteristics that can be used for automatic recognition. In [27–29], the face de-identification methods referred to as *k-Same*, *k-Same-Select* and *Model-based k-Same*, respectively, are proposed. The last two methods are based on the *k-Same*, so we describe it in detail. By applying the *k-Same algorithm* to the given person-specific set of images, where each person is represented by no more than one image, a set of de-identified images is computed. Each de-identified image is represented by an average face image of the k closest face images from the person-specific set of images. The k closest face images in the person specific set are replaced by the same k de-identified face images. The *k-Same* algorithm selects the k closest images based on Euclidean distances in the image space or in the principal component analysis (PCA) coefficient space.

Figure 13.2 illustrates the *k-Same* algorithm ($k = 4$) where for a person-specific set of face images I (which consists of 12 original images I_1, ..., I_{12}), the set of

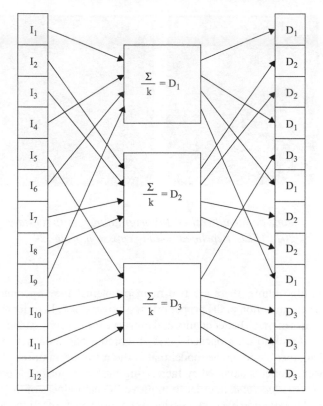

Figure 13.2   Illustration of k-Same algorithm

de-identified face images D is computed. $\sum$ represents a sum of the k closest face images from a person-specific set of images I. The set D consists of $12/k$ identical face images (i.e. D_1, D_2 and D_3), where each image is represented as an average of the $k = 4$ closest original images. For the given example, it should be noted that the original images I_1, I_4, I_6 and I_9 are represented with the same de-identified face image D_1, I_2, I_3, I_7, I_8 with D_2 and the remaining I_5, I_{10}, I_{11}, I_{12} with D_3.

Figure 13.3 gives an example of *k-Same* de-identification of an original face image for value $k = 6$.

In order to improve the data utility and the naturalness of the de-identified face images, the *k-Same-Select* is proposed [28]. The algorithm partitions the input set of face images into mutually exclusive subsets using the data-utility function and applies the *k-Same* algorithm independently to the different subsets. The data utility function is a goal-oriented function which is usually selected to preserve the gender or a facial expression in the de-identified image. Due to the use of the *k-Same* algorithm, *k-Same-Select* guarantees that the resulting face set is k-anonymized [30].

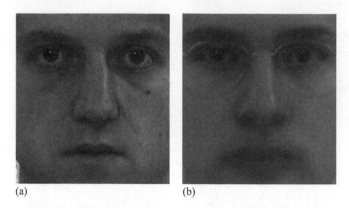

(a)   (b)

Figure 13.3   k-Same de-identification: (a) original image; (b) de-identified image.
© 2015 IEEE. Reprinted, with permission, from [24]

For both algorithms, there are two main problems: they operate on a closed set I, and the determination of the proper privacy constraint k. In order to produce de-identified images of much better quality and preserve the data utility, the *Model-based k-Same algorithms* [29] are proposed – one of which is based on Active Appearance Models and another based on the model that is the result of mixtures of identity and non-identity components obtained by factorizing the input images. Modifications to the *k-Same Select algorithm,* in order to improve the naturalness of the de-identified face images (by retaining face expression) and privacy protection, are proposed in [31,32].

In [33], the authors proposed a reversible privacy-preserving photo sharing architecture which ensures privacy and preserves the usability and convenience of online photo sharing. The architecture takes into account the content and context of a photo and utilizes a Secure JPEG framework. Visual privacy in a JPEG can be protected by using: (i) naive de-identification where the reconstruction of an original image is performed by extracting from a JPEG header, decrypting and placing back the original pixels; (ii) scrambling, which modifies the original values of the pixels and the Discrete Cosine Transform (DCT) coefficients in a reversible way. In [34], a morphing-based visual privacy protection method is described. The morphing is performed by using a set of face key points (eyes, nose, mouth), both original source and target images, the interpolation of some pixels between the key points and dividing both images using Delaunay triangulation.

An interesting approach to face de-identification is described in [35]. The authors use deep learning-based Facial Identity Preserving (FIP) features in the process of face de-identification. The main characteristic of the FIP features is that they, unlike conventional face descriptors, significantly reduce intra-identity variances, while maintaining inter-identity distinctions. The FIP features are invariant to pose and illumination variations. By replacing the original feature values in the FIP space with the averaged FIP values of k individuals and then reconstruct the de-identified facial images based on these k-averaged FIP values, k-anonymity face de-identification is achieved.

In [36], the authors propose a novel face de-identification process that preserves the important face clues which support further behaviour and emotions analysis. The approach is based on a unified model for face detection, pose estimation and face landmark localization proposed by Zhu and Ramannan [37], and applying variational adaptive filtering to preserve the most important non-verbal communication signs such as eyes, gaze, lips and lips corners. A quantitative software- and human-based analysis on the Labeled Faces in the Wild (LFW) dataset was used to validate the de-identification method. The results of validation demonstrated that this approach sufficiently prevents recognition of the de-identified face while it preserves facial expressions.

13.3.2 Face de-identification in videos

Special attention is devoted to automatic face de-identification in video surveillance systems [38–45], as well as drone-based surveillance systems [46,47], due to tremendous development and use of visual technologies such as CCTVs, visual sensor networks, camera phones and drones equipped with camera(s).

Solutions for face de-identification in videos should be robust (with False Negative Detection $= 0$ per cent), simple, and, in most applications in real-time. Unfortunately, face de-identification in videos is certainly not a complete solution. The problem lies not in the de-identification of regions of interest (ROIs), but in computer vision algorithms for the detection and localization of one or more faces in video sequences. Despite recent intensive research in computer vision, numerous problems still remain in automatic face detection. These include issues such as the detection of the face under different illumination conditions, bad lighting, different head positions, the presence of structural components (e.g. glasses, sunglasses, beards, moustaches) and occlusions. The unsolved problems are the detection of faces in crowd scenes and real-time de-identification. Privacy might be compromised in video sequences if the face detection algorithm fails in a single frame, so one of the directions of research is the development of robust and effective algorithms for privacy protection that can efficiently cope with situations when computer vision algorithms fail [17,48,49].

Recently, new methods have been proposed for face detection in the wild. In [50], the authors propose a method for unconstrained face detection based on the features called NPD. The NPD feature, which was inspired by the Weber Fraction in experimental psychology [51], is defined as $f(x_i, x_j) = (x_i - x_j)/(x_i + x_j)$, where $x_i \geq 0$ and $x_j \geq 0$ are the intensity values of two pixels and $f(0,0)$, by definition, is 0. It is computed for each pixel pair resulting in $d = p^2(p^2 - 1)/2$ features, where $(p \times p)$ are dimensions of an image patch. For example, for a (20×20) image patch, there are 79,800 NPD features. The authors use a deep quadratic tree learner to define an optimal subset of the NPD features. The final face detector contains 1,226 trees and uses 46,401 NPD features. The classifiers were trained on the Annotated Facial Landmarks in the Wild (AFLW) database (25,993 face annotated) [52]. The evaluation of the

NPD-based face detector was performed on the public-domain face databases FDDB, GENKI and CMU-MIT. The results are compared with state-of-the-art results reported on the FDDB website [53]. Evaluation on the above-mentioned face databases showed that the proposed method achieves state-of-the-art performance, and it is much faster than others.

In [54], the authors describe a two-stage cascade detector for unconstrained face detection, which can be used in the de-identification pipeline. The first stage of the cascade is the NPD face detector, while the second stage is based on the deformable part model [37]. The experimental results on AFWL and FDDB databases show that a two-stage cascade model increases precision, as well as recall. The main characteristic of the proposed model is that it keeps false negative face detections as low as possible, while it reduces false positive face detections by more than 34 times (for the AFWL database) and 15 times (for the FDDB database) in the comparison with the NPD detector, which is important for real-time de-identification and for the naturalness of the de-identified images.

Using spatial and temporal correspondence between frames, i.e. a combination of face detection and tracking, improves the effectiveness of the localization of faces in video footages. Very recently, approaches based on deep convolutional networks [55,56], hierarchical dense structures [57] and social context [58] have been used for the detection and multi-object tracking.

De-identification in drone-based surveillance systems deserves special attention due to specific problems which are, in a computer vision sense, very close to moving-camera–moving object problems and different scenarios in comparison with 'classic' CCTV surveillance. There are open problems in the detection of several ROIs (face, body silhouette, accessories, different positions and sizes) in dynamic scenes. Due to the complex problem of de-identification in drone-based surveillance systems, it is expected that the privacy-by-design approach has to be applied together with strict laws regarding the use of drones.

13.3.3 Fingerprint de-identification

Based on the Biometric Market Report [59], fingerprint-based biometric systems for overt person recognition are the leading biometric technology in terms of market share. If we take into account also the recent reports of the ongoing research [60] where the possibility of capturing fingerprint of a person at a distance up to 2 m is reported, the de-identification of the fingerprint becomes even more important issue in protecting the privacy due to the possibility of covert fingerprint-based person recognition.

Fingerprint still images may be de-identified with the usual de-identification procedures such as black box, blurring, pixelation, replacement by a synthetic fingerprint [61] or by applying privacy filters based on image morphing or block scrambling. In addition, feature perturbation and non-invertible feature transforms [61], as well as watermarking techniques, are used for hiding biometric templates [62].

Templates in a fingerprint-based person recognition system may be derived from either the fingerprint minutiae points, the pattern of the fingerprint, or simply, from the image of the fingerprint.

In order to protect a person's privacy in a fingerprint-based biometric recognition system, some possible modifications of fingerprint templates are presented below:

- A template is a binary fingerprint image obtained by thinning an original raw fingerprint image [63]. Private personal data are embedded in the template in the user enrolment phase as follows: by applying an 'embeddability criterion' [64], data are hidden into the template by adding some pixels to the boundary pixels in the original thinned fingerprint image. The result of this action is twofold: user personal information is embedded and hidden into a fingerprint template, and an original thinned fingerprint stored in the database is hidden and protected by distortion which is the result of adding pixels to the boundary pixels. In the authentication phase, the hidden personal data are extracted from the template for verifying the authenticity of the person, the added boundary pixels are removed so the original thinning fingerprint is recovered and then it is used for matching with the live thinned fingerprint.

- During the enrolment phase, fingerprint images from two different fingers of the same person [65] are captured. From one fingerprint image, the minutia positions are extracted, while orientation is taken from the other fingerprint. The reference points are extracted from both fingerprint images. Based on these extracted features, a combined minutia template is generated and stored in the database. During the authentication phase, two query fingerprints are required from the same two fingers which are used in the enrolment phase.

  By using the reconstruction approach, it is possible only to convert the combined minutiae template into a synthetic real-look fingerprint image. The authors have shown that the reconstruction of both fingerprint images based on the template which represents a new virtual identity is a tough problem. The authors report that the recognition system based on a virtual fingerprint obtained by the combination of two different fingerprints achieved a relatively low error rate with False Rejection Rate (FRR) equals 0.4 per cent and False Acceptance Rate (FAR) equals 0.1 per cent.

- Two fingerprint images of the same person are mixed in order to generate a new cancellable fingerprint image, which looks like a plausible fingerprint [66]. A mixed fingerprint creates a new entity that appears as a plausible fingerprint. A mixed fingerprint can be processed by conventional fingerprint algorithms.

Methods used for privacy enhancement based on different types of distortion of original biometric templates at the signal or feature level may also be applied to hide soft-biometric identifiers (gender, ethnicity) and/or medical information in fingerprint templates. In [67], the authors describe a relatively simple method of fingerprint de-identification for gender estimation. The proposed approach is based on image filtering in the frequency domain. The linear filtering process applies blurring by attenuating the high-frequency content. Certain frequency components are suppressed, while others are amplified. A de-identified fingerprint image is obtained by using the inverse of the Fourier transform. The experiments have shown that the recognition performance of the de-identified fingerprint images achieves FRR = 1 per cent, while FAR = 0.05 per cent.

Experiments have shown that the gender estimation accuracy in de-identified fingerprint images for 100 users is reduced from the initial 88.7 per cent (original fingerprints) to 50.5 per cent.

To the best of our knowledge, apart from [65,67], there has been no research to evaluate the degree of protection of medical or other privacy sensitive information for such distorted fingerprints and its impact on the identification performance.

13.3.4   Iris de-identification

Iris represents an important biometric identifier and it enables an efficient approach to reliable, non-invasive recognition of people due to its utmost cross-person variability, and minimal within-person variability across time [68,69]. Most iris-recognition systems require users' cooperation to collect images of adequate quality so biometric identification based on iris is classified as overt. Due to the small size of the iris (about 10 mm in diameter) and the required typical resolution between 100 and 200 pixels across the iris diameter, the images are captured at a relatively close stand-off (i.e. between 15 and 50 cm), where the stand-off is the camera-to-subject-iris distance. Most commercial iris-recognition systems operate at a stand-off between 0.1 and 0.45 m, with a verification time of 2 to 7 s [70,71]. However, the Iris at a Distance (IAD) system developed recently provides the capability to identify a person at a range of more than 1 m in less than a second [72].

Even more recent, iris-recognition technology is oriented to further reducing the need for subject cooperation, reducing the time of image acquisition and increasing the distance between the sensor and the person [71,73–78]. For example, in [77], the authors introduced the IAD prototype system, which is capable of acquiring an iris image at 30 m stand-off and perform iris recognition (Figure 13.4). In this case, we can talk about covert iris-based person recognition.

Based on the characteristics of the current iris-based recognition systems at a distance, and expected future advances in the field, it can be concluded that iris de-identification for privacy protection is a growing problem. An additional complexity to note is that most IAD systems combine face and iris image acquisition. Therefore, both biometric identifiers – iris and face – have to be simultaneously de-identified, i.e. a multimodal de-identification has to be applied.

To date, however, research into iris de-identification for privacy protection has been rather limited. A rare study related to de-identification of the eye areas, and thus the iris, is presented in [79]. The IAD systems are also capable of acquiring a face, which leads to multimodal de-identification. The proposed system [78] for the reversible de-identification of an eye region consists of two modules: an automatic eye-detection module and a privacy-enabling encoder module. The automatic eye-detection module in real time locates the human-eye region by a combination of colour-based and Haar-like/GentleBoost methods. The input to the privacy-enabling encoder module is the pixel location information of both eyes in the given input frame. Based on a JPEG XR encoder, the macroblocks consisting of (16 × 16) pixels of located eye region are scrambled. The privacy-enabling JPEG XR encoder utilized three encryption techniques (Random Level Shift, Random Permutation, Random

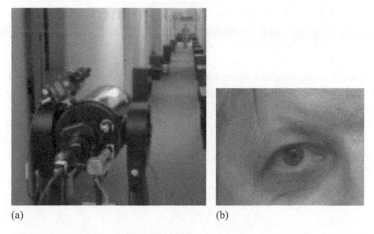

(a)   (b)

*Figure 13.4   (a) The IAD prototype system; (b) view of the eye at 30 m by iris image
acquisition camera. © 2010 IEEE. Reprinted, with permission,
from [77]*

Sign Inversion) to transform the coefficients of frequency sub-bands on a macro-block
basis. The de-identified images, due to scrambling, lose their original naturalness, but
they prevent iris recognition. Also, depending on the dimensions of the scrambling
block, the proposed scheme successfully prevents any correct face identification.

13.3.5   Ear de-identification

Despite the fact that the face and iris, in addition to fingerprints, are the most used in
biometric technologies for person recognition, they both have a number of drawbacks.
Face-based biometrics can fail due to the changes in head pose, facial expressions,
the growth of a beard, hair styles, the presence of obstacles (glasses, scarf or collar),
cosmetics, aging and/or changing the illumination conditions in unconstrained environments. An iris is stable and consistent over time, but due to the relatively small
dimension it requires a high-resolution camera and a long-distance near-infrared illu-
minator for image acquisition at a distance. Therefore, a human ear is offered as an
alternative physiological biometric identifier for non-invasive person recognition at
a distance [80–84].

A two-dimensional (2D) ear image can be easily acquired from a distance, even
without the cooperation of the subject. This fact makes ear-based recognition systems
also interesting for applications in intelligent video-surveillance systems [82–84].
Until now, ear-recognition systems were successfully tested in controlled indoor conditions [80]. There are some unsolved problems in automatic ear recognition relating
to the disruptive factors present in real-life scenes, like pose variations, scaling,
varying lighting conditions and hair occlusion, and these open up new research areas.

Despite the relatively long period of research in the field of automatic ear-based
person recognition and its maturity, as far as we know, there are only a few existing

commercial ear-based biometric systems for automatic person identification or verification, e.g. a 2D ear-based verification system for mobile applications [85]. This is
the main reason for lack of research in the field of ear de-identification for privacy
protection.

Due to the development of relatively low-cost, high-resolution, video cameras
and telescopic equipment, we can expect ear-based recognition and tracking in semi-
or non-controlled outdoors conditions. This will lead to the need for research and
development of ear de-identification methods in order to protect the privacy of individuals. Most ear-recognition systems use the combination of a profile face and
ear detection. Therefore, in the near future, ear de-identification will be a multimodal de-identification problem – the face and the ear have to be de-identified
simultaneously.

13.4   De-identification of behavioural biometric identifiers

Behavioural biometrics is less intrusive compared with most physiological-based
biometrics, but it is, generally, less accurate due to the variability of its characteristics over time. Some behavioural biometric personal identifiers (e.g. gait and
gesture) can be used for covert biometric identification at distance in video surveillance applications. The individual's psychological and physical characteristics are
reflected in behavioural personal biometric identifiers, such as handwritten signatures, keystroke dynamics, voice print, gait and gesture. Due to this reflection, the
information obtained from behavioural personal biometric identifiers (identity, emotional state, health status) is privacy sensitive and should be protected. In this section,
we describe the de-identification of three behavioural biometric identifiers: voice,
gait and gesture.

13.4.1   Voice de-identification

Biometric identifiers such as the face, iris and ear refer to *the visual identity* of a
person. However, in addition to a visual identity, a person has *an audio identity*. It is
based on human voice which is a unique pattern that identifies an individual – there
are no two individuals that sound identical [86]. Voice is a significant modality that
can be used effectively by humans and machines for the recognition of individuals.
The speech signal, besides identity information, carries privacy-sensitive information
such as gender, age, emotional state, health status, level of education and origin of the
speaker. These all set an additional requirement for voice de-identification in order
to protect personal privacy.

Voice de-identification is mainly based on the principles of Voice Transformation
(VT). Voice transformation refers to modifications of the non-linguistic characteristics
of a given utterance without affecting its textual content. The non-linguistic information of speech signals, such as voice quality and voice individuality, may be controlled
by VT [87], which is based on three types of voice modifications [88]: source, filter and their combination. Source modifications include time-scale, pitch and energy

modifications, while filter modifications refer to a modification that changes the magnitude response of the vocal tract system. Voice conversion [89–91] is a special form of VT where the characteristics of a source speaker's voice are mapped to those of a specific (target) speaker. Voice conversion may be text-dependent or text-independent. In the first case, during the learning phase, a parallel corpora (training material of source and target speaker uttering the same text) is required. This is the main limitation of using such an approach for voice de-identification in real-world applications.

Text-independent voice conversion [92–94] does not require parallel corpora in the learning phase and it is more realistic for speaker-privacy protection.

One of the earliest proposed voice-conversion methods that can be used for de-identification is described in [90]. The authors present a text-dependent voice-conversion method based on vector quantization and spectral mapping. The method produces a mapping codebook that shows correspondences between the codebook of the source and target speaker. The voice-conversion method consists of two steps: a learning step and a conversion-synthesis step. During the learning step, based on the parallel corpora, the mapping codebooks for several acoustic parameters that describe a mapping between the vector spaces of two speakers are generated. The synthesized speech from using the mapping codebooks is generated in the conversion-synthesis step. The evaluation of the proposed method (for male-to-female and male-to-male conversion) is performed subjectively. The above-described approach can be used for reversible de-identification.

In [95], the authors propose a transformation of the speaker's voice that enables the secure transmission of information via voice without revealing the identity of the speaker to unauthorized listeners. Owing to the transmitted key, which allows the authorized listeners to perform back-transformation, the voice de-identification is reversible. The authors use a strategy for de-identifying these results in the speech of various speakers to be transformed to the same synthetic (target) voice. They use the Gaussian Mixture Model (GMM)-mapping-based VT to convert a relatively small set of source speakers (24 males) to a syntactic voice. The proposed VT system has both a training and a testing, or transformation phase. During the training phase a parallel corpora of utterances is used. The authors tested different VT strategies (standard GMM-mapping-based VT, de-duration VT, double VT and transterpolated VT). The best results for de-identification are obtained with transterpolated VT (100 per cent de-identification rate for the GMM-based voice-identification system and 87.5 per cent for phonetic voice-identification system). In [96], the same authors present voice de-identification via VT, similar to [95], but de-identification with larger groups of speakers is easier and it can keep the de-identified voices distinguishable from each other, which contributes to its naturalness. They reported a 97.7 per cent de-identification rate for male and 99 per cent for female speakers.

A novel scheme for voice de-identification, where a set of precalculated VTs based on GMM mapping is used to de-identify the speech of a new speaker, is presented in [97]. The scheme enables the online de-identification of speakers whose speech has not been used in the training phase to build a VT. The scheme uses automatic voice recognition within the set that is used to build pre-calculated VTs to select the appropriate transform, which is then used to de-identify the speech of the new user.

The approach avoids the need for a parallel corpus, even for training of the initial set of transformations based on GMM mapping, and it was inspired by an approach that is used for face de-identification (e.g. *k-Same*). The preliminary experiments showed that the proposed scheme produces de-identified speech, which has satisfactory levels of naturalness and intelligibility, and a similar de-identification rate in comparison with previous VT systems [95,96].

In [98], an approach to voice de-identification based on a combination of diphone recognition and speech synthesis is proposed. De-identification is performed in two steps. First, the input speech is recognized with a diphone-based recognition system and converted into phonetic transcription. In the second step, phonetic transcription is used by a speech synthesis subsystem to produce a new speech. With this approach, the acoustic models of the recognition and synthesis subsystems are completely independent and a high level of protection of speaker identity is ensured. Two different techniques for speech synthesis are used: one is Hidden Markov Model-based and one is based on the diphone Time-Domain Pitch Synchronous Overlap and Add technique. Since every user's speech utterance is converted into the speech of the same speaker (whose data were used during the training phase of the synthesis subsystem), the described process of de-identification is irreversible. The system is applicable in different scenarios where users either want to conceal their identity or are reluctant to transmit their natural speech through the communication channel. The proposed voice de-identification system runs in real time and is language dependent and text independent. The obtained de-identified speech was evaluated for intelligibility (a measure of how comprehensible is de-identified speech). The level of privacy protection was evaluated in speaker recognition experiments by using a state-of-the-art speaker recognition system based on identity vectors and Probabilistic Linear Discriminant Analysis. The experiments showed that the speaker recognition system was unable to recognize the true speaker identities from the de-identified speech with a performance better than chance, while the de-identified speech was intelligible in most cases.

An example of privacy protection for the text-independent speaker verification system that require low-level or even no user cooperation is given in [99]. The text-independent privacy-preserving speaker verification system based on a password matching principle is proposed. The process of authentication is based on a client-server model, where the speaker verification system has the role of server, and the user executes a client program on a network-enabled computation device (e.g. computer or smart-phone). The authentication system does not observe the (raw) speech input provided by the user. Instead, speech input is represented in the form of 2,496-dimensional supervectors [(64 × 39), where 64 is the number of components of the GMM and 39 is the dimension of the Mel Frequency Cepstral Coefficients-based feature vector] on which a cryptographic hash function is applied. The speech samples, needed for matching in the verification phase, are stored in the same form in the internal storage of the system. So, the speech samples are irreversibly obfuscated from the system and this one-way transformation preserves the privacy of a user's speech utterances.

There still are several challenges in the field of online voice or speaker de-identification, such as de-identification in an environment with background noise, voice de-identification in situations where there are multiple individuals speaking simultaneously, which leads to crosstalk and overlapped speech. Additional efforts have to be made to develop more sophisticated voice de-identification systems with 'personalized' multi-target voices and the preservation of the emotional expression of a speaker.

13.4.2 Gait and gesture de-identification

Gait is defined as a manner of walking and represents a behavioural biometric characteristic [100,101]. Gait, as a body gesture, which is usually a motion without meaning, conveys information that can be used for person recognition or for diagnostics. Besides the dynamics of individual walking, gait includes information about individual appearance, such as silhouette, leg length, height, even age and gender [102,103]. By introducing visual surveillance systems in people's daily lives, and owing to the development of computer-vision techniques, it is possible to recognize non-cooperating individuals at a distance based on their walking characteristics.

Based on the state of the art for gait recognition systems [104–108], their characteristics and performances, we can conclude that gait-based technologies can be used for biometric-based person verification in controlled environments. It is technically unfeasible for large-scale surveillance systems to record all the gait parameters of individuals in public places, or to identify them by searching in a database [109]. The main reasons for limitations of the accuracy of large-scale gait-based person verification are gait changes over time affected by clothes, footwear, walking surface, walking speed, the carrying of personal belongings and emotional conditions.

Very few studies have been directly geared towards *gait de-identification*. The study in [110] presents an automated video-surveillance system designed to ensure the efficient and selective storage of data, to provide a means for enhancing privacy protection and to secure visual data against malicious attacks. The approach to the privacy enhancement of captured video sequences is based on two main steps: the first step is performed by the salient motion detector, which finds ROIs (corresponding mainly to moving individuals), and the second step applies to those regions with a procedure of information concealment based on a scrambling technique described in [40]. The DCT-based scrambling is applied to each ROI, represented by a rough binary mask, which covers a silhouette of the moving individual, so the gait information is obscured. Image regions corresponding to the involved individuals in the scene are distorted, while the scene still remains comprehensible. Owing to the reversible scrambling procedure, the authorized user can get a clear video sequence and reveal all the privacy details by using the embedding and scrambling keys. The de-identified videos, due to the scrambling procedure, do not preserve the naturalness of the original videos.

In [44,45], *gait de-identification* based on two de-identification transformations: the exponential blur of pixels of the voxel [set of pixels defined in a spatial $(x \times y)$ and temporal (t) domains] and Line Integral Convolution (LIC). These two kinds of smooth temporal blurring of the space-time boundaries of an individual aim to remove any gait information.

Gestures are defined as the movement of a body part (fingers, hands, arms, head or face) or a whole body that is made with or without the intension meaning something [111,112]. For example, the expressive and meaningful motion of fingers or hands conveys meaningful information to another human, or it can be used for interacting with a real or virtual environment (virtual reality, augmented reality).

To date, there have only been a few attempts to develop biometric verification systems based on *hand-gesture recognition* [113–116].

As far as we know, there has been no research into the problem of *hand gesture de-identification*. The problem of gesture de-identification in video surveillance is similar to the problem of gait de-identification and can be solved by approaches similar to those used for gait in future.

13.5   De-identification of soft biometric identifiers

Soft biometric identifiers, as ancillary information, can be combined by biometric identifiers to improve the overall recognition, particularly when person recognition system is designed to work in accordance with the less constrained scenarios including recognition at a distance [117].

There are three main modalities of using soft biometric identifiers for:

1. Person recognition based on verbal descriptions of soft biometric identifiers [20],
2. Person recognition in a biometric system based on the fusion of soft biometric identifiers and physiological and/or behavioural biometric identifiers in order to ensure better accuracy of the recognition process [20],
3. Retrieval of large biometric databases [118,119].

Regardless of the above-described modalities of using soft biometric identifiers, it is obvious that soft biometric identifiers, such as silhouette, gender, race, moles, tattoos, birthmarks and scars, carry privacy-intrusive information about individuals, and are subject to de-identification requirements.

13.5.1   Body silhouette de-identification

The body silhouette is an important soft biometric identifier and it can help the recognition process (on its own or in combination with other biometric identifiers). In addition to recognition, body silhouettes are used for people reidentification, i.e. tracking people across multiple cameras with non-overlapping fields of view in surveillance applications [120].

To the best of our knowledge, there are only a few papers on *body silhouette de-identification*. In [44,45], the authors showed that the masking of a silhouette is relatively easy, through the use of dilatation or Gaussian blurring. The Gaussian blurring of the silhouette is also used for the de-identification of individuals in activity videos (Figure 13.5) [121]. In [45], it has been shown that a combination of LIC and

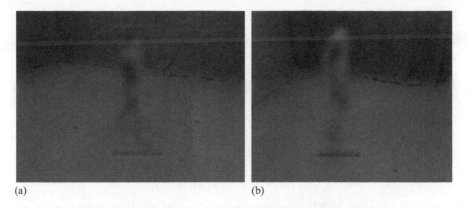

(a)   (b)

Figure 13.5   De-identification of individuals in activity videos depicting:
(a) walking; (b) jumping in place actions after 2D Gaussian filtering.
© 2014 IEEE. Reprinted, with permission, from [121]

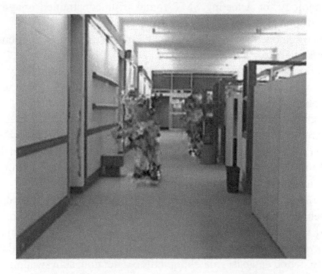

Figure 13.6   Result of the body silhouette de-identification by the scrambling
method. © 2008 IEEE. Reprinted, with permission, from [40]

the exponential blurring of pixels of a voxel gives the best results for silhouette
de-identification.

An approach to reversible body de-identification in video is based on distortion
applied to the ROI which contains the silhouette of an individual by using transform-
domain scrambling methods proposed in [40]. Figure 13.6 illustrates as an example,
the result of the body de-identification by the scrambling method described in [40].

An interesting approach to silhouette de-identification is described in [122];
it involves replacing a person with another one from a dataset gallery.

13.5.2   Gender, age, race and ethnicity de-identification

In literature, there are many papers related to the automatic recognition of gender, age, race and ethnicity, but relatively little is done on their de-identification. Information about gender, age, race and ethnicity is usually obtained from facial images [123–127], speaker utterances [128], gait and silhouette and/or silhouetted face profiles [129]. In [45], the authors have mentioned that the masking of race and gender is a difficult problem. However, they agreed that it is possible to mask skin colour (which is closely related to race) using different colour transformations at the price of destroying the naturalness of the de-identified data.

13.5.3   Scars, marks and tattoos de-identification

Scars, marks and tattoos (SMT) are imprints on skin that provide more discriminative information than age, height, gender and race to identify a person [130]. In [131], the authors have showed that facial marks, such as freckles, moles, scars and pockmarks, can improve automatic face recognition and retrieval performance.

A methodology for detecting SMT found in unconstrained imagery normally encountered in forensics scenarios is described in [132].

As far as we know, there are no published papers related to de-identification of scars and marks to date.

Tattoos are not only popular in particular groups, such as motorcyclists, sailors and members of criminal gangs; they have become very popular in the wider population. In fact, about 24 per cent of people aged from 18 to 50 in the USA have at least one tattoo, and this number is increasing [133].

Tattoos are primarily used for content-based image retrieval in law-enforcement applications [134,135], but based on the visual appearance of tattoos and their location on a body [134], they can be used for person recognition, as well as for suspect and victim identification in forensics.

There are no published papers related to SMT de-identification, except [136,137]. The experimental system for tattoo localization and de-identification for privacy protection [136] was intended to be used for still images, but it was also tested for videos. The system consists of the following modules: skin and ROI detection, feature extraction, tattoo database, matching, tattoo detection, skin swapping and quality evaluation. An image or a sequence of frames obtained by a colour camera is an input to the skin and ROI detection module. Uncovered body parts like the head, neck, hands, legs or torso are detected in two phases. In the first phase, skin-colour cluster boundaries are obtained by a pixel-based method through a series of decision rules in the Red, Green and Blue (RGB) colour space. In the second phase, geometrical constraints are used to eliminate skin-like colour regions that do not belong to the uncovered body-part areas. The Scale-invariant Feature Transform (SIFT) features are extracted from a ROI in the feature-extraction module, these are matched with template SIFT features from the tattoo database. Experimentally, 24 tattoos with at least two tattoos from each of the eight classes of tattoos labeled in the ANSI/NIST-ITL 1-2000 standard are used. Each tattoo in the tattoo database has an average of 56 template SIFT features, so the tattoo database consists of 1,338 SIFT features. The

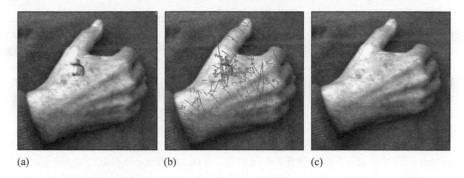

Figure 13.7   Tattoo de-identification; (a) an example of a still image obtained by a colour camera; (b) extracted SIFT features; (c) de-identified tattoo still frame. © 2014 IEEE. Reprinted, with permission, from [136]

de-identification process is performed in the skin-swapping module in such a way that the original tattoo's region is replaced by pixels from a surrounding, non-tattoo region. After replacement, a median filter is applied to the de-identified area. With this procedure, the authors try to hide the tattoo location and its visual appearance, and preserve the naturalness of the de-identified image (Figure 13.7).

The experiments have shown that tattoo localization based on SIFT features gave satisfactory results in well-controlled conditions, such as lighting, high tattoo resolution and no motion blur. For tattoos with a low-quality visual appearance, the SIFT features have to be combined with some region segmentation based on a combination of colour, gradient and/or texture methods. For surveillance applications, by using skin- and tattoo-area tracking based on a spatial and temporal correspondence between the frames, tattoo detection, localization and de-identification can be improved.

In [137], an approach is presented for tattoo de-identification which uses the ConvNet model of a deep convolutional neural network inspired by the VGGNet [138] for tattoo detection. De-identification of the tattoo area is performed by replacing the colours of the pixels belonging to the tattooed area with the colour of the surrounding skin area. In order to find surrounding skin pixels, a morphological dilate operation is applied on the previously found binary mask corresponding to the detected tattooed area.

13.6   Conclusion

In spite of the huge efforts of various academic research groups, institutions and companies, research in the field of de-identification of personal identifiers is still in its infancy. Due to recent advances in multisensor acquisition and recording devices and remote surveillance systems, there is a need for the research and development of multimodal de-identification methods that simultaneously hide, remove or substitute different types of personal identifiers from multimedia content. A solution to the problem of multimodal de-identification still remains a major challenge.

It is worth noting that important aspects of de-identification are metrics in measuring privacy protection, utility, intelligibility and/or naturalness of the de-identified data, as well as the evaluation protocol [49]. There is not yet a common framework for the evaluation and assessment of these components in de-identified contents. Researchers are primarily focusing on the evaluation of privacy protection, and the trade-off between privacy protection for privacy filters applied on face regions in images and video sequences (FERET database, PEViD-HD and PEViD-UHD datasets [34,38,139]). The evaluation of privacy protection and the trade-off between privacy protection and utility/intelligibility are usually performed by objective methods (PCA, Linear Discriminant Analysis (LDA) and Local Binary Pattern (LBP) automatic face recognition) and subjective evaluation [41,42] based on crowdsourcing [18], or by experts (video-analytics technology and privacy protection solution developers, or law enforcement personnel). Ongoing research activities regarding privacy protection and its evaluation in surveillance systems are presented in MediaEval workshops, established as an independent benchmarking initiative in 2010 [140].

The assessment of the de-identification of behavioural biometric identifiers is mainly devoted to privacy protection and to the intelligibility of de-identified speech [141].

In this chapter, we have tried to give an up-to-date review of de-identification methods for privacy protection based on de-identification of personal biometric identifiers; based on a proposed taxonomy of personal identifiers, we have presented de-identification of physiological, behavioural biometric identifiers and soft-biometric identifiers. Regarding the trends in the surveillance and biometric technologies, we have identified some new directions in the de-identification research: de-identification of iris and fingerprints captured at a distance, gait and gesture de-identification and multimodal de-identification which combines non-biometric, physiological, behavioural and soft-biometric identifiers.

The problem of selectively concealing or removing the context-sensitive information or objects from the environment which can be used to reveal the identity of a person is still open, too. This could be solved in the near future by using a knowledge-based approach for modelling a specific environment and situation to detect additional ROIs and to obscure them.

This chapter covers mainly the technical aspects of de-identification. But, due to the social, legal and political importance of privacy protection, we are aware that real solutions for de-identification, which are acceptable to both users and the law enforcement organizations in a networked society, will have to be based on the collective effort of not only technical experts, but also of experts from the fields of law, ethics, sociology and psychology.

Acknowledgements

This work has been supported by the Croatian Science Foundation under project 6733 De-identification for Privacy Protection in Surveillance Systems (DePPSS).

It is also the result of activities in COST Action IC1206 'De-identification for Privacy Protection in Multimedia Content'.

References

[1] Solove D. J. *Understanding Privacy*. Cambridge: Harvard University Press; 2008.

[2] Warren S. D., Brandeis, L. D. 'The Right to Privacy'. *Harvard Law Review*. 1890;**IV**(5):193–220. http://readingnewengland.org/app/books/righttoprivacy/ [Accessed 06 Feb 2014].

[3] Westin A. F. 'Social and Political Dimensions of Privacy'. *Journal of Social Issues*. 2003;**59**(2):431–453.

[4] Campisi P. 'Security and Privacy in Biometrics: Towards a Holistic Approach' in Campisi P. (ed.). *Privacy and Security in Biometrics*. London: Springer; 2013. pp. 1–23.

[5] Rane S. 'Standardization of Biometric Template Protection'. *IEEE MultiMedia*. 2014;**21**(4):94–99.

[6] ISO. *JTC1 SC27 IT Security Techniques; ISO/IEC24745: Biometric Information Protection*; 2011.

[7] Savvides M., Vijaya Kumar B. V. K., Khosla, P. K. 'Cancelable Biometric Filters for Face Recognition'. *Proc. Int. Conf. on Pattern Recognition (ICPR)*; **3**. 2004. pp. 922–925.

[8] Ratha N. K., Chikkerur S., Connell J. H., Bolle, R. M. 'Generating Cancelable Fingerprint Templates'. *IEEE Transactions on Pattern Analysis and Machine Intelligence*. 2007;**29**(4):561–572.

[9] Hao F., Anderson R., Daugman, J. 'Combining Cryptography with Biometrics Effectively'. *IEEE Transactions on Computers*. 2006;**55**(9):1081–1088.

[10] Xu W., He Q., Li Y., Li, T. 'Cancelable Voiceprint Templates Based on Knowledge Signatures'. *Proc. Int. Symposium on Electronic Commerce and Security*. 2008. pp. 412–415.

[11] Woodward J. D. 'Biometrics: Identifying Law & Policy Concerns' in Jain A. K., Bolle R., Pankanti, S. (eds.). *Biometrics, Personal Identification in Networked Society*. Boston:Springer; 1999. pp. 385–405.

[12] Solove D. J. *Nothing to Hide*. New Haven & London: Yale University Press; 2011.

[13] Angwin J. *Dragnet Nation*. New York: St. Martin's Press; 2015.

[14] Ribaric S., Ariyaeeinia A., Pavesic, N. 'De-identification for Privacy Protection in Multimedia Content: A Survey'. *Signal Processing: Image Communication*. 2016;**47**:131–151.

[15] Bhagwan V., Grandison T., Maltzahn, C. *Recommendation-Based De-Identification, A Practical Systems Approach Towards De-identification of Unstructured Text in Healthcare* [online]. 2012. Available from http://www.almaden.ibm.com/cs/people/tgrandison/SPE2012-ReDid.pdf [Accessed 15 Jul 2014].

[16]   IC1206 COST Action. *Memorandum of Understanding (MoU)* [online]. 2013. Available from http://w3.cost.eu/fileadmin/domain_files/ICT/Action_IC1206/mou/IC1206-e.pdf [Accessed 03 Dec 2015].

[17]   Padilla-Lopez J. R., Chaaraoui A. A., Florez-Revuelta, F. 'Visual Privacy Protection Methods: A Survey'. *Expert Systems with Applications*. 2015;**42**(9):4177–4195.

[18]   Korshunov P., Cai S., Ebrahimi, T. 'Crowdsourcing Approach for Evaluation of Privacy Filters in Video Surveillance'. *Proc. ACM Multimedia 2012 Workshop on Crowdsourcing for Multimedia (CrowdMM'12)*. 2012. pp. 35–40.

[19]   Jain A. K., Dass S. C., Nandakumar, K. 'Soft Biometric Traits for Personal Recognition Systems'. *Proc. Int. Conf. on Biometric Authentication*. 2004. pp. 731–738.

[20]   Reid D. A., Nixon M. S., Stevenage, S. V. 'Soft Biometrics; Human Identification Using Comparative Descriptors'. *IEEE Transactions on Pattern Analysis and Machine Intelligence*. 2014;**36**(6):1216–1228.

[21]   Li S. Z., Jain, A. K. (eds.). *Handbook of Face Recognition*. New York: Springer; 2005.

[22]   Boyle M., Edwards C., Greenberg, S. 'The Effects of Filtered Video on Awareness and Privacy'. *Proc. ACM Conf. on Computer Supported Cooperative Work*; Philadelphia, USA. 2000. pp. 1–10.

[23]   Neustaedter C., Greenberg S., Boyle, M. 'Blur Filtration Fails to Preserve Privacy for Home-Based Video Conferencing'. *ACM Transactions on Computer Human Interaction*. 2006;**13**(1):1–36.

[24]   Ribaric S., Pavesic, N. 'An Overview of Face De-identification in Still Images and Videos'. *Proc. 11th IEEE Int. Conf. and Workshops on Automatic Face and Gesture Recognition (FG)*. 2015. pp. 1–6.

[25]   Gross R., Sweeney L., de la Torre F., Baker, S. 'Model-Based Face De-identification'. *Proc. 2006 Conf. on Computer Vision and Pattern Recognition Workshop (CVPRW'06)*. 2006; pp. 161–169.

[26]   Phillips P. J., 'Privacy Operating Characteristic for Privacy Protection in Surveillance Applications' in Kanade T., Jain A. K., Ratha, N. K. (eds.). *Audio- and Video-Based Biometric Person Authentication, Lecture Notes in Computer Science (LNCS)*. Berlin & Heidelberg: Springer; 2005. pp. 869–878.

[27]   Newton E., Sweeney L., Malin, B. 'Preserving Privacy by De-identifying Facial Images'. *IEEE Transactions on Knowledge and Data Engineering*. 2005;**17**(2):232–243.

[28]   Gross R., Airoldi E., Malin B., Sweeney, L. 'Integrating Utility into Face De-identification' in Danezis G., Martin, D. (eds.). *PET – Privacy Enhancing Technologies 2005, Lecture Notes in Computer Science (LNCS)*. Berlin & Heidelberg: Springer; 2006. pp. 227–242.

[29]   Gross R., Sweeney L., Cohn J., de la Torre F., Baker, S. 'Face De-identification' in Senior A. (ed.). *Protecting Privacy in Video Surveillance*. Dordrecht: Springer; 2009. pp. 129–146.

[30] Sweeney L. 'k-Anonymity: A Model for Protecting Privacy'. *International Journal of Uncertainty, Fuzziness and Knowledge-Based Systems*. 2002;**10**(5):557–570.

[31] Meng L., Sun Z., Ariyaeeinia A., Bennett, K. L. 'Retaining Expressions on De-identified Faces'. *Proc. Special Session on Biometrics, Forensics, De-identification and Privacy Protection (BiForD)*. 2014. pp. 27–32.

[32] Meng L., Sun, Z. 'Face De-identification with Perfect Privacy Protection'. *Proc. Special Session on Biometrics, Forensics, De-identification and Privacy Protection (BiForD)*. 2014. pp. 9–14.

[33] Yuan L., Korshunov P., Ebrahimi, T. 'Privacy-Preserving Photo Sharing Based on a Secure JPEG'. *Proc. 3rd Int. Workshop on Security and Privacy in Big Data Security*. 2015. pp. 185–190.

[34] Korshunov P., Ebrahimi, T. 'Using Face Morphing to Protect Privacy'. *Proc. IEEE Int. Conf. on Advanced Video and Signal-Based Surveillance*. 2013. pp. 208–213.

[35] Chi H., Hu, Y. H. 'Face De-identification Using Facial Identity Preserving Features'. *Proc. 2015 IEEE Global Conf. on Signal and Information Processing (GlobalSIP)*. 2015. pp. 586–590.

[36] Letournel G., Bugeau A., Ta V.-T., Domenger, J.-P. 'Face De-identification with Expressions Preservation'. *Proc. 2015 IEEE Int. Conf. on Image Processing (ICIP)*. 2015. pp. 4366–4370.

[37] Zhu X., Ramannan, D. 'Face Detection, Pose Estimation, and Landmark Localization in the Wild'. *Proc. IEEE Conf. on Comput. Vision and Pattern Recogn*. 2014. pp. 2497–2504.

[38] Korshunov P., Ebrahimi, T. 'Towards Optimal Distortion-Based Visual Privacy Filter'. *Proc. IEEE Int. Conf. on Image Processing (ICIP)*. 2014. pp. 6051–6055.

[39] Erdely A., Barat T., Valet P., Winkler T., Rinner, B. 'Adaptive Cartooning for Privacy Protection in Camera Networks'. *Proc. 11th IEEE Int. Conf. on Advanced Video and Signal Based Surveillance (AVSS)*. 2014. pp. 26–29.

[40] Dufaux F., Ebrahimi, T. 'Scrambling for Privacy Protection in Video Surveillance Systems'. *IEEE Transactions on Circuits and Systems for Video Technology*. 2008;**18**(8):1168–1174.

[41] Dufaux F., Ebrahimi, T. A. 'Framework for the Validation of Privacy Protection Solutions in Video Surveillance'. *Proc. IEEE Int. Conf. on Multimedia and Expo (ICME)*. 2010. pp. 66–71.

[42] Sohn H., Lee D., De Neve W., Plataniotis K. N., Ro, Y. M. 'An Objective and Subjective Evaluation of Content-Based Privacy Protection of Face Images in Video Surveillance Systems Using JPEG XR' in Flammini F., Setola R., Franceschetti, G. (eds.). *Effective Surveillance for Homeland Security: Balancing Technology and Social Issues*. Boca Raton & London & New York: CRC Press/Taylor & Francis; 2013. pp. 111–140.

[43] Samarzija B., Ribaric, S. 'An Approach to the De-identification of Faces in Different Poses'. *Proc. Special Session on Biometrics, Forensics, De-identification and Privacy Protection (BiForD)*. 2014. pp. 21–26.

[44]   Agrawal P. 'De-identification for Privacy Protection in Surveillance Videos'. *Master of Science Thesis, Center for Visual Information Technology International Institute of Information Technology Hyderabad.* 2010. 49 pages.

[45]   Agrawal P., Narayanan, P. J. 'Person De-identification in Videos'. *IEEE Transactions on Circuits and Systems for Video Technology.* 2011;**21**(3):299–310.

[46]   Bonetto M., Korshunov P., Ramponi G., Ebrahimi, T. 'Privacy in Mini-drone Based Video Surveillance'. *Proc. 11th IEEE Int. Conf. and Workshops on Automatic Face and Gesture Recognition (FG)*; **4**. 2015. pp. 1–6.

[47]   Çiftçi S., Korshunov P., Akyüz A. O., Ebrahimi, T. 'MediaEval 2015 Drone Protect Task: Privacy Protection in Surveillance Systems Using False Coloring'. *Proc. MediaEval Workshop.* 2015. pp. 1–2.

[48]   Saini M., Atrey P. K., Mehrotra S., Kankanhalli, M. 'Adaptive Transformation for Robust Privacy Protection in Video Surveillance'. *Advances in Multimedia.* 2012;**2012**:1–14. Article ID 639649.

[49]   Winkler T., Rinner, B. 'Security and Privacy Protection in Visual Sensor Networks: A Survey'. *ACM Computing Surveys.* 2014;**47**(1):1–39.

[50]   Lio S., Jain A. K., Li, S. Z. 'A Fast and Accurate Unconstrained Face Detector'. *IEEE Transactions on Pattern Analysis and Machine Intelligence.* 2016;**38**(2):211–223.

[51]   Weber E. H. 'Tastsinn und Gemeingefühl [Sense of touch and common feeling]' in Wagner R. (ed.). *Handwörterbuch der Physiologie mit Rücksicht auf physiologische Pathologie. Band 3, Teil 2 [Concise dictionary of physiology with regard to physiological pathology. Vol. 3, Pt. 2].* Braunschweig: Vieweg; 1846. pp. 481–588.

[52]   Koestinger M., Wohlhart P., Roth P. M., Bischof, H. 'Annotated Facial Landmarks in the Wild: A Large-scale, Real-world Database for Facial Landmark Localization'. *Proc. 1st IEEE Int. Workshop Benchmarking Facial Image Anal. Technol.* 2011. pp. 2144–2151.

[53]   Jain V., Learned-Miller, E. *FDDB: A Benchmark for Face Detection in Unconstrained Settings.* Univ. Massachusetts, Amherst, MA, USA, Tech. Rep. UM-CS-2010-009, 2010.

[54]   Marčetić D., Hrkać T., Ribarić, S. 'Two-stage Cascade Model for Unconstrained Face Detection'. *Proc.1st Int. Workshop on Sensing, Processing and Learning for Intelligent Machines (SPLINE).* 2016. pp. 21–24.

[55]   Simonyan K., Zisserman, A. 'Very Deep Convolutional Networks for Large-Scale Image Recognition'. *Proc. Int. Conf. Learn. Represent (ICLR).* 2015. pp. 1–14.

[56]   Zhang X., Zou J., He K., Sun, J. 'Accelerating Very Deep Convolutional Networks for Classification and Detection'. *IEEE Transactions on Pattern Analysis and Machine Intelligence.* 2016;**38**(10):1943–1955.

[57]   Wen L., Lei Z., Lyu S., Li S. Z., Yang, M.-H. 'Exploring Hierarchical Dense Structures on Hypergraphs for Multi-Object Tracking'. *IEEE Transactions on Pattern Analysis and Machine Intelligence.* 2016;**38**(10): 1983–1996.

[58]   Qin Z., Shelton, C. R. 'Social Grouping for Multi-Target Tracking and Head Pose Estimation in Video'. *IEEE Transactions on Pattern Analysis and Machine Intelligence.* 2016;**38**(10):2082–2095.

[59]   MarketsandMarkets. *Next Generation Biometric Market-Forecasts & Analysis 2014–2020* [online]. 2014. Available from https://www.marketsand markets.com [Accessed 15 Nov 2015].

[60]   Swanson S. *Fingerprints Go the Distance* [online]. 2011. Available from https://www.technologyreview.com/s/422400/fingerprints-go-the-distance [Accessed 07 Dec 2015].

[61]   Maltoni D., Maio D., Jain A. K., Prabhakar, S. (eds.). *Handbook of Fingerprint Recognition.* New York: Springer; 2003.

[62]   Ratha N. K., Connell J. H., Bolle, R. M. 'Enhancing Security and Privacy in Biometrics-Based Authentication Systems'. *IBM Systems Journal.* 2001;**40**(3):614–634.

[63]   Jain A. K., Uludag, U. 'Hiding Biometric Data'. *IEEE Transactions on Pattern Analysis and Machine Intelligence.* 2003;**25**(11):1494–1498.

[64]   Sheng L., Kot, A. C. 'Privacy Protection of Fingerprint Database Using Lossless Data Hiding'. *Proc. IEEE Int. Conf. on Multimedia and Expo (ICME).* 2010. pp. 1293–1298.

[65]   Sheng L., Kot, A. C. 'Fingerprint Combination for Privacy Protection'. *IEEE Transactions on Information Forensics and Security.* 2013;**8**(2):350–360.

[66]   Ross A. *De-identifying Biometric Images for Enhancing Privacy and Security* [online]. 2014. Available from http://biometrics.nist.gov/cs_links/ibpc2014/presentations/08_tuesday_ross_VC-MIXING_ IBPC2014.pdf [Accessed 17 Nov 2016].

[67]   Lugini L., Marasco E., Cukic B., Dawson, J. 'Removing Gender Signature from Fingerprints'. *Proc. Special Session on Biometrics, Forensics, De-identifications and Privacy Protection (BiForD).* 2014. pp. 63–67.

[68]   Daugman J. 'How Iris Recognition Works'. *IEEE Transactions on Circuits and Systems for Video Technology.* 2004;**14**(1):21–31.

[69]   Wildes R. P. 'Iris Recognition: An Emerging Biometric Technology'. *Proceedings of the IEEE.* 1997;**85**(9):1348–1363.

[70]   George A. M., Durai, C. A. D. 'A Survey on Prominent Iris Recognition Systems'. *Proc. Int. Conf. on Information Communication and Embedded Systems (ICICES).* 2013. pp. 191–195.

[71]   Matey J. R., Naroditsky O., Hanna K., *et al.* 'Iris on the Move: Acquisition of Images for Iris Recognition in Less Constrained Environments'. *Proceedings of the IEEE.* 2006;**94**(11):1936–1947.

[72]   Morpho [online]. 2014. Available from http://www.morpho.com/en/media/20140311_iris-distance-power-behind-iris [Accessed 23 Jul 2015].

[73]   Fancourt C., Bogoni L., Hanna K., *et al.* 'Iris Recognition at a Distance'. *AVBPA 2005, Lecture Notes in Computer Science (LNCS).* Berlin & Heidelberg: Springer; 2005. pp. 1–13.

[74]   Dong W., Sun Z., Tan T., Qiu, X. 'Self-adaptive Iris Image Acquisition System'. *Proc. SPIE, Biometric Technology for Human Identification*; **6944**. 2008. pp. 6–14.

[75]   Wheeler F. W., Amitha Perera A. G., Abramovich G., Yu B., Tu, P. H. 'Stand-off Iris Recognition System'. *Proc. 2nd IEEE Int. Conf. on Biometrics: Theory, Applications and Systems (BTAS)*. 2008. pp. 1–7.

[76]   Bashir F., Casaverde P., Usher D., Friedman, M. 'Eagle-Eyes: A System for Iris Recognition at a Distance'. *Proc. IEEE Conf. on Technologies for Homeland Security*. 2008. pp. 426–431.

[77]   de Villar J. A., Ives R. W., Matey, J. R. 'Design and Implementation of a Long Range Iris Recognition System'. *Proc. Conf. Record of the 44th Asilomar Conf. on Signals, Systems and Computers (ASILOMAR)*. 2010. pp. 1770–1773.

[78]   Abiantun R., Savvides M., Khosla, P. K. 'Automatic Eye-level Height System for Face and Iris Recognition Systems'. *Proc. 4th IEEE Workshop on Automatic Identification Advanced Technologies*. 2005. pp. 155–159.

[79]   Lee D., Plataniotis, K. N. 'A Novel Eye Region Based Privacy Protection Scheme'. *Proc. IEEE Int. Conf. on Acoustics, Speech and Signal Processing (ICASSP)*. 2012. pp. 1845–1848.

[80]   Abaza A., Ross A., Hebert C., Harrison M. A. F., Nixon, M. S. 'A Survey on Ear Biometrics'. *ACM Computing Surveys*. 2013;**45**(2):1–35.

[81]   Pflug A., Busch, C. 'Ear Biometrics: A Survey of Detection, Feature Extraction and Recognition Methods'. *IET Biometrics*. 2012;**1**(2): 114–129.

[82]   Yuan L., Mu, Z.-C. 'Ear Detection Based on Skin-Color and Contour Information'. *Proc. 6th Int. Conf. on Machine Learning and Cybernetics*. 2007. pp. 2213–2217.

[83]   Kumar A., Hanmandlu M., Kuldeep M., Gupta, H. M. 'Automatic Ear Detection for Online Biometric Applications'. *Proc. 3rd National Conf. on Computer Vision, Pattern Recognition, Image Processing and Graphics*. 2011. pp. 146–149.

[84]   Abaza A., Hebert C., Harrison, M. A. F. 'Fast Learning Ear Detection for Real-Time Surveillance'. *Proc. 4th IEEE Int. Conf. on Biometrics: Theory Applications and Systems (BTAS)*. 2010. pp. 1–6.

[85]   Descartes Biometrics. *Software Developer Kit for Biometric Ear Recognition Technology* [online]. 2014. Available from http://www. descartesbiometrics.com [Accessed 07 Dec 2015].

[86]   Kinnunen T., Li, H. 'An Overview of Text-independent Speaker Recognition: From Features to Supervectors'. *Speech Communication*. 2010;**52**(1): 12–40.

[87]   Stylianou Y. 'Voice Transformation: A Survey'. *Proc. IEEE Int. Conf. on Acoustics, Speech and Signal Processing (ICASSP)*. 2009. pp. 3585–3588.

[88]   Muda L. B., Elamvazuthi, M. I. 'Voice Recognition Algorithms Using Mel Frequency Cepstral Coefficient (MFCC) and Dynamic Time Warping (DTW) Techniques'. *Journal of Computing*. 2010;**2**(3):138–143.

[89]   Sundermann D., Voice Conversion: State-of-the-Art and Future Work'. *Fortschritte der Akustik.* 2005;**31**(2):1–2.

[90]   Abe M., Nakamura S., Shikano K., Kuwabara, H. 'Voice Conversion through Vector Quantization'. *Proc. IEEE Int. Conf. on Acoustics, Speech and Signal Processing (ICASSP).* 1988. pp. 655–658.

[91]   Upperman G., Hutchinson M., Van Osdol B., Chen, J. *Methods for Voice Conversion* [online]. 2004. Available from http://ftpmirror.your.org/pub/misc/cd3wd/1006/Methods_for_Voice_Conversion_electr_physics_cnx_x10252_. pdf [Accessed 14 Dec 2014].

[92]   Sundermann D., Bonafonte A., Ney H., Hoge, H. 'A First Step Towards Text-Independent Voice Conversion'. *Proc. Int. Conf. on Spoken Language Processing (ICSLP).* 2004. pp. 1–4.

[93]   Sundermann D., Hoge H., Bonafonte A., Ney H., Hirschberg, J. 'Text-Independent Cross-Language Voice Conversion'. *Proc. IEEE Int. Conf. on Acoustics, Speech and Signal Processing (ICASSP)*; **1**. 2006. pp. 1–4.

[94]   Mouchtaris A., Van Spiegel J., Mueller, P. 'Non-Parallel Training for Voice Conversion by Maximum Likelihood Constrained Adoption'. *Proc. IEEE Int. Conf. on Acoustics, Speech and Signal Processing (ICASSP)*; **1**. 2004. pp. I-1–I-4.

[95]   Jin Q., Toth A. R., Schultz T., Black, A. W. 'Voice Converging: Speaker De-identification by Voice Transformation'. *Proc. IEEE Int. Conf. on Acoustics, Speech and Signal Processing (ICASSP).* 2009. pp. 3909–3912.

[96]   Jin Q., Toth A. R., Schultz T., Black, A. W. 'Speaker De-identification via Voice Transformation'. *Proc. IEEE Workshop on Automatic Speech Recognition & Understanding (ASRU).* 2009. pp. 529–533.

[97]   Pobar M., Ipsic, I. 'Online Speaker De-identification Using Voice Transformation'. *Proc. Special Session on Biometrics, Forensics, De-identification and Privacy Protection (BiForD).* 2014. pp. 33–36.

[98]   Justin T., Struc V., Dobrisek S., Vesnicer B., Ipsic I., Mihelic, F. 'Speaker De-identification using Diphone Recognition and Speech Synthesis'. *Proc. 11th IEEE Int. Conf. and Workshops on Automatic Face and Gesture Recognition (FG)*; **4**. 2015. pp. 1–7.

[99]   Pathak M. A., Raj, B. 'Privacy-preserving Speaker Verification as Password Matching, *Proc. IEEE Int. Conf. on Acoustics, Speech and Signal Processing (ICASSP).* 2012. pp. 1849–1852.

[100]   Zhang D. D. *Automated Biometrics – Technology and Systems.* New York: Kluwer Academic Publishers; 2000.

[101]   Nixon M. S., Carter J. N., Cunado D., Huang P. S., Stevenage, S. V. 'Automatic Gait Recognition' in Jain A. K., Bolle R., Pankanti, S. (eds.). *Biometrics, Personal Identification in Networked Society.* New York: Kluwer Academic Publishers; 1999. pp. 231–249.

[102]   Yoo J.-H., Hwang D., Nixon, M. S. 'Gender Classification in Human Gait Using Support Vector Machine'. *ACIVS 2005, Lecture Notes in Computer Science (LNCS)*; **3708**. Berlin & Heidelberg: Springer; 2005. pp. 138–145.

[103]   Lee L., Grimson, W. E. L. 'Gait Analysis for Recognition and Classification'. *Proc. IEEE Int. Conf. on Automatic Face and Gesture Recognition (FG)*. 2002. pp. 148–155.

[104]   Collins R. T., Gross R., Shi, J. 'Silhouette-Based Human Identification from Body Shape and Gait'. *Proc. 5th IEEE Int. Conf. on Automatic Face and Gesture Recognition (FG)*. 2002. pp. 351–356.

[105]   Tao D., Li X., Wu X., Maybank, S. J. 'General Tensor Discriminant Analysis and Gabor Features for Gait Recognition'. *IEEE Transactions on Pattern Analysis and Machine Intelligence*. 2007;**29**(10):1700–1715.

[106]   Veres G. V., Gordon L., Carter J. N., Nixon, M. S. 'What Image Information is Important in Silhouette-Based Gait Recognition?'. *Proc. IEEE Computer Society Conf. on Computer Vision and Pattern Recognition (CVPR)*. 2004. pp. II-776–II-782.

[107]   Wang L., Tan T., Ning H., Hu, W. Silhouette 'Analysis-Based Gait Recognition for Human Identification'. *IEEE Transactions on Pattern Analysis and Machine Intelligence*. 2003;**25**(12):1505–1518.

[108]   Sarkar S., Phillips P. J., Liu Z., Vega I. R., Grother P., Bowyer, K. W. 'The HumanID Gait Challenge Problem: Data Sets, Performance, and Analysis'. *IEEE Transactions on Pattern Analysis and Machine Intelligence*. 2005;**27**(2):162–177.

[109]   Boulgouris N. V., Hatzinakos D., Plataniotis, K. N. 'Gait Recognition: A Challenging Signal Processing Technology for Biometric Identification'. *IEEE Signal Processing Magazine*. 2005;**11**(6):78–90.

[110]   Baaziz N., Lolo N., Padilla O., Petngang, F. 'Security and Privacy Protection for Automated Video Surveillance'. *Proc. IEEE Int. Symposium on Signal Processing and Information Technology*. 2007. pp. 17–22.

[111]   Mitra S., Acharya, T. 'Gesture Recognition: A Survey'. *IEEE Transactions on Systems, Man and Cybernetics – Part C: Applications and Reviews*. 2007;**37**(3):311–324.

[112]   Abdallah M. B., Kallel M., Bouhlel, M. S. 'An Overview of Gesture Recognition'. *Proc. 6th Int. Conf. on Sciences of Electronics, Technologies of Information and Telecommunications (SETIT)*. 2012. pp. 20–24.

[113]   Lentsoane N. D., Kith K., Van Wyk B. J., Van Wyk, M. A. 'Identity Verification System Using Hand Gesture Information'. *Proc. 17th Int. Symposium of the Pattern Recognition Society of South Africa*. 2006. pp. 1–6 [online]. Available from http://www.prasa.org/proceedings/2006/prasa06-13.pdf [Accessed 12 Apr 2016].

[114]   Lentsoane N. D. 'Identity Verification System Using Hand Gesture Information'. *Magister Technologiae: Electronic Engineering, Department of Electrical Engineering. Faculty of Engineering, Tshwane University of Technology*. 2007. 202 pages.

[115]   Fong S., Zhuang Y., Fister, I. 'A Biometric Authentication Model Using Hand Gesture Images'. *BioMedical Engineering OnLine*. 2013:1–18 [online]. Available from https://biomedical-engineering-online.biomed central.com/articles/10.1186/1475-925X-12-111 [Accessed 24 Jun 2014].

[116] Yang S., Premaratne P., Vial, P. 'Hand Gesture Recognition: An Overview'. *Proc. IEEE Int. Conf. on Broadband Network & Multimedia Technology (BNMT)*. 2013. pp. 63–69.

[117] Tome P., Fierrez J., Vera-Rodriguez R., Nixon, M. S. 'Soft Biometrics and Their Application in Person Recognition at a Distance'. *IEEE Transactions on Information Forensics and Security*. 2014;**9**(1):464–475.

[118] Waymann J. L. 'Large-Scale Civilian Biometric Systems Issues and Feasibility'. *Proc. Card Tech/Secur. Tech ID*. 1997.

[119] Park U., Jain, A. K. 'Face Matching and Retrieval Using Soft Biometrics'. *IEEE Transactions on Information Forensics and Security*. 2010;**5**(3):406–415.

[120] Congl D.-N. T., Achard C., Khoudour, L. 'People Re-identification by Classification of Silhouettes Based on Sparse Representation'. *Proc. 2nd Int. Conf. on Image Processing Theory Tools and Applications (IPTA)*. 2010. pp. 60–65.

[121] Ivasic-Kos M., Iosifidis A., Tefas A., Pitas, I. 'Person De-identification in Activity Videos'. *Proc. Special Session on Biometrics, Forensics, De-identification and Privacy Protection (BiForD)*. 2014. pp. 63–68.

[122] Nodari A., Vanetti M., Gallo, I. 'Digital Privacy: Replacing Pedestrians from Google Street View Images'. *Proc. 21st Int. Conf. on Pattern Recognition (ICPR)*. 2012. pp. 2889–2893.

[123] Yang M., Yu, K. 'Adapting Gender and Age Recognition System for Mobile Platforms'. *Proc. 3rd Chinese Conf. on Intelligent Visual Surveillance (IVS)*. 2011. pp. 93–96.

[124] Lin H., Lu H., Zhang, L. 'A New Automatic Recognition System of Gender, Age and Ethnicity'. *The 6th World Congress on Intelligent Control and Automation (WCICA)*; **2**. 2006. pp. 9988–9991.

[125] Guo G., Mu G., Fu Y., Dyer C., Huang, T. 'A Study on Automatic Age Estimation Using a Large Database'. *Proc. 12th IEEE Int. Conf. on Computer Vision (ICCV)*. 2009. pp. 1986–1991.

[126] Chen D.-Y., Lin, K.-Y. 'Robust Gender Recognition for Real-Time Surveillance System'. *Proc. IEEE Int. Conf. on Multimedia and Expo (ICME)*. 2010. pp. 191–196.

[127] Muhammad G., Hussain M., Alenezy F., Mirza A. M., Bebis G., Aboalsamh, H. 'Race Recognition Using Local Descriptors'. *Proc. IEEE Int. Conf. on Acoustics, Speech and Signal Processing (ICASSP)*. 2012. pp. 1525–1528.

[128] Chen O. T.-C., Gu J. J., Lu P.-T., Ke, J.-Y. 'Emotion-Inspired Age and Gender Recognition Systems'. *Proc. 55th IEEE Int. Midwest Symposium on Circuits and Systems (MWSCAS)*. 2012. pp. 662–665.

[129] Tariq U., Hu Y., Huang, T. S. 'Gender and Ethnicity Identification from Silhouetted Face Profiles'. *Proc. 16th IEEE Int. Conf. on Image Processing (ICIP)*. 2009. pp. 2441–2444.

[130] Lee J.-E., Jain A. K., Jin, R. 'Scars, Marks and Tattoos (SMT): Soft Biometric for Suspect and Victim Identification'. *Proc. Biometrics Symposium (BSYM)*. 2008. pp. 1–8.

[131] Jain A. K., Park, U. 'Facial Marks: Soft Biometric for Face Recognition'. *Proc. 16th IEEE Int. Conf. on Image Processing (ICIP)*. 2009. pp. 37–40.

[132] Heflin B., Scheirer W., Boult, T. E. 'Detecting and Classifying Scars, Marks, and Tattoos Found in the Wild'. *Proc. 5th IEEE Int. Conf. on Biometrics: Theory, Applications and Systems (BTAS)*. 2012. pp. 31–38.

[133] Laumann A. E., Derick, A. J. 'Tattoos and Body Piercing in the United States: A National Dataset'. *Journal of the American Academy of Dermatology.* 2006;**55**(3):413–421.

[134] Manger D. 'Large-Scale Tattoo Image Retrieval'. *Proc. 9th Conf. on Computer and Robot Vision (CRV)*. 2012. pp. 454–459.

[135] Lee J.-E., Jin R., Jain, A. K. 'Image Retrieval in Forensics: Tattoo Image Database Application'. *IEEE MultiMedia*. 2011;**19**(1):40–49.

[136] Marcetic D., Ribaric S., Struc V., Pavesic, N. 'An Experimental Tattoo De-identification System for Privacy Protection in Still Images'. *Proc. Special Session on Biometrics, Forensics, De-identification and Privacy Protection (BiForD)*. 2014. pp. 57–62.

[137] Hrkać T., Brkić K., Ribarić S., Marčetić, D. 'Deep Learning Architectures for Tattoo Detection and De-identification'. *Proc. 2016 1st International Workshop on Sensing, Processing and Learning for Intelligent Machines (SPLINE)*. 2016. pp. 45–49.

[138] Simonyan K., Zisserman, A. 'Very Deep Convolutional Networks for Large-Scale Image Recognition'. *Proc. Int. Conf. Learning Representations (ICLR)*. 2015. pp. 1–14.

[139] Korshunov P., Melle A., Dugelay J.-L., Ebrahimi, T. 'Framework for Objective Evaluation of Privacy Filters'. *Proc. SPIE Optical Engineering + Applications*. 2013. pp. 88560T–88560T-12.

[140] *MediaEval Workshops* [online]. Available from http://www.multimediaeval.org/ [Accessed 24 Jun 2016].

[141] Justin T., Mihelič F., Dobrišek, S. 'Intelligibility Assessment of the De-Identified Speech Obtained Using Phoneme Recognition and Speech Synthesis Systems'. *Lecture Notes in Artificial Intelligence (LNAI)*; **8655**. Cham: Springer; 2014. pp. 529–536.

Chapter 14

Secure cognitive recognition: brain-based biometric cryptosystems using EEG

Emanuele Maiorana[1] and Patrizio Campisi[1]

Abstract

Cognitive biometric recognition systems, based on the exploitation of nervous tissues' responses as identifiers, have recently attracted an always-growing interest from the scientific community, thanks to the several advantages they could offer with respect to traditional biometric approaches based on physical or behavioral characteristics, such as fingerprint, face, signature, and so forth. Biosignals are in fact much more robust against presentation attacks, being hard, if not impossible, to covertly capture and then replicate them. Liveness detection is also inherently provided. Nevertheless, their usage could expose several sensitive information regarding people's health and capability, making the system prone to function creep issues. With the aim of guaranteeing proper privacy and security to the users of the such systems, different general cryptosystem architectures for cognitive biometric traits are therefore presented in this chapter. Their effectiveness is evaluated by applying the proposed approaches to brain signals sensed through electroencephalography (EEG). A multi-session EEG dataset comprising recordings taken in three distinct occasions from each of 50 subjects is employed to perform the reported experimental test.

14.1 Introduction

A notable vulnerability issue common to many traditional biometric identifiers, such as fingerprint, face, iris, and so forth, is the presentation attack: most biometric traits can be in fact covertly acquired and replicated at a later time, thus infringing the system security. Although liveness detection strategies can be designed to cope with such threat, their effectiveness needs to be improved for their adoption in practical biometric recognition systems with reliable performance. For high-security applications, it would be therefore preferable resorting to biometric modalities allowing to

[1]Department of Engineering, Roma Tre University, Italy

reduce or eliminate such threat by design. Cognitive biometrics [1], exploiting nervous tissues' responses collected either in correspondence of an external stimulus or during the execution of a task, have been recently investigated as potential identifiers not affected by the aforementioned concerns. In fact, body signals such as those collected through electroencephalography (EEG), electrocardiography (ECG), electrodermal response, or blood pulse volume necessarily require users' cooperation during the acquisition process, making covert acquisition more difficult than for traditional biometrics. Moreover, in addition to inherently guarantee liveness detection, they can also be easily exploited for both single-attempt and continuous recognition.

Most of the mentioned biosignals are commonly used in medical applications for the diagnosis of possible pathologies. Brain signals are for instance successfully employed to study diseases such as Alzheimer's, Parkinson's, and schizophrenia, among the others. Biosignals are potentially prone to function creep, that is, the possible use of the collected data for purposes beyond the intended ones, such as the extraction of sensitive personal information which may lead to discriminate people for hiring or to deny insurance to those with latent health problems, thus incurring critical privacy concerns [2]. It is therefore crucial to implement and adopt protected recognition systems, where the stored templates do not reveal any information regarding the original biometrics they are derived from. Such property is one of the main requirements to be taken into account when designing a template protection scheme, which typically needs also to satisfy the renewability and the performance requirement. This implies that the employed biometric templates should be revocable and able to guarantee recognition accuracies equal, or at least close, to those achieved in unprotected scenarios [3].

Such methods can be roughly described as the application of intentional and repeatable modifications to the original templates and are typically categorized into two major classes: cancelable biometrics and biometric cryptosystems [4]. The former is based on the adoption of non-invertible transformation functions [5]. However, cancelable biometric systems do not allow to define general metrics to be used to perform a proper robustness analysis, quantitatively evaluating the actual non-invertibility of the proposed methods. On the other hand, biometric cryptosystems [6] commonly rely on cryptographic protocols for which several metrics assessing the achievable security and privacy have been proposed in literature. These approaches can be further classified into key generation approaches, where cryptographic keys are directly created from biometric data, and key binding methods, which combine biometric templates with binary keys [7].

In this chapter, we present a key binding method for EEG template protection for biometric recognition. In more detail, different key binding architectures are here proposed to allow cognitive user recognition in a secure modality. The present work significantly extends our preliminary findings [8], where the issues related to the protection of cognitive biometric templates have been considered for the first time in literature. In more detail, in [8], the relevance of using multi-biometrics approaches based on information fusion for improving both the recognition and security performance of the considered cognitive biometric cryptosystems has been introduced. This chapter is organized as follows. In Section 14.2, a brief introduction about EEG biometric recognition, along with an analysis on the state of the art of this modality, is given. The schemes here employed to perform people recognition through cognitive biometrics, in both unprotected and protected ways, are then outlined in Section 14.3. An extensive set of experimental tests, performed on a large database comprising recordings taken from 50 healthy subjects during three distinct recording sessions spanning a period of one month, is then presented in Section 14.4, where the recognition rates and security levels achievable with the proposed system configurations are evaluated and discussed. Conclusions from the reported work are then drawn in Section 14.5.

14.2 EEG-based biometric recognition

Brain activity has interested researchers in the medical field since the beginning of the twentieth century, allowing to get several significant insights in the diagnosis and care of brain disorders such as epilepsy, spinal cord injuries, and stroke to cite a few. However, only recently it has been postulated that brain signals may have some potentials to be used for automatic people recognition [9]. In more detail, among the different modalities to sense the electrical activity of the brain, EEG has received a significant attention since it allows collecting brain information using portable and relatively inexpensive devices, which is a relevant advantage for making the adoption of such modality feasible in practical biometric recognition systems.

EEG signals are the result of the electrical fields generated in the brain by the synchronous firing of specific spatially aligned neurons of the cortex, i.e. the pyramidal neurons. The generated electrical activity can be measured by sensing the voltage differences between specific positions on the scalp surface. Wet electrodes currently represent the preferable choice to sense brain activity with the lowest possible noise level. Unfortunately, their use implies the adoption of electrolyte gel for reducing the impedance between the electrodes and the subjects' scalp, resulting in subject inconvenience and non-negligible time to set-up the recording process. Such issue has implied that most of the studies on EEG biometric trait have been carried out on datasets collected during a single session for each considered subject, with the number of involved participants usually in the order of a couple of dozens [10]. In general, different partitions of the same recording have been obtained, thus generating training and test samples belonging to the same acquisition session [11]. The reliability of such studies [12] may be therefore questionable, since it is hard to state whether the reported recognition rates are only dependant on the discriminative features of each user's brain characteristics, or if session-specific exogenous conditions, such as the capacitative coupling of electrodes and cables with lights or computer, induction loops created between the employed equipment and the body, power supply artefacts, and so on, may significantly differ between distinct acquisition sessions, thus affecting both inter- and intra-class variability of EEG recordings.

Only a handful of works in literature have instead compared EEG recordings taken in different days for recognition purposes. Specifically, data collected during three consecutive days, with four sessions each day, from nine subjects performing a motor imagery task have been exploited in [13]. Evaluating the recognition performance through the half-total error rate (HTER), defined as the average value between the false rejection rate (FRR) and the false acceptance rate (FAR) associated to a given operating point, the best result has been achieved at HTER = 19.3%. Signals from 20 people, recorded during two sessions at a median distance of 15 months, have been used in [14], where an identification scenario resulting in a correct recognition rate (CRR) of about 88% has been taken into account. A biometric system based on an imagination task performed by six subjects, whose EEG signals have been recorded in four different days, has been instead analysed in [15], where CRRs ranging from 78.6% to 99.8% have been reported. A database collected from nine subjects during two one-year-apart sessions has been considered in [16], achieving CRR = 87.1%. Signals from nine subjects have been recorded during two sessions spanning up to three weeks in [17] and exploited to achieve perfect identification rate for EEG data acquired in eyes-closed (EC) conditions, and an identification rate at 90.53% for the eyes-open (EO) scenario. Event-related potentials, obtained as responses to visual stimuli, have been exploited in [18] obtaining CRRs at 89.0% and 93.0%, respectively, for a database collected from 15 people during two sessions spanning one week, and a dataset comprising signals from eight persons acquired during two six-month-separated sessions. Signals recorded from four subjects in EC conditions during two one-week-apart sessions have been processed through continuous wavelet transform in [19] to guarantee a CRR at 92.58%. Visual-evoked potentials to both target and non-target stimuli have been evaluated in [20] to provide equal error rates (EERs), respectively, of about 18% and 13% over a database comprising signals acquired from 50 users during three sessions taken during a period of one month. Parsimonious representations in the frequency domain have been proposed in [21], where CRR = 87.9% and CRR = 75.4% have been respectively achieved in EC and EO conditions, to process EEG signals taken from 30 subjects during two recording sessions spanning one month. Eventually, the performance achievable when comparing data captured from 50 subjects during three different sessions spanning a 1-month period has been presented in [22]. A CRR at 90.8% has been reported when comparing signals captured in EC conditions, and a CRR = 85.6% has been achieved for the EO scenario, almost regardless of the sessions being compared out of the three available ones. The same dataset employed in [22] is exploited here to test the proposed biometric cryptosystem, therefore characterizing the reported results with a greater reliability with respect to those reported in [8], where EEG signals collected from 40 subjects, and only in EC conditions, have been considered.

14.3   Proposed cognitive biometric cryptosystems

The present section describes the proposed cognitive biometric cryptosystem. In order to assess its performance in Section 14.4.2, the reference unprotected framework is described here as well. In more detail, the preprocessing performed in both protected and unprotected scenarios is presented in Section 14.3.1. The considered unprotected

frameworks are then introduced in Section 14.3.2, while the proposed cognitive biometric cryptosystems are outlined in Section 14.3.3. The information fusion strategies
exploited to improve the performance of the described systems are then discussed in
Section 14.3.4.

14.3.1 Preprocessing

Let us assume that the cognitive biometric acquisition available for either enrolment or verification purposes can be expressed in terms of C time sequences $\mathbf{z}^{(c)}[t]$,
with $t = 1, \ldots, T$ representing the temporal index, and $c = 1, \ldots, C$ the number of
recorded biosignals. A generic cognitive biometrics can be in fact represented as a
time-dependent signal, which can be acquired through multiple sensors. In case of
EEG biometrics, C corresponds to the number of electrodes adopted in the employed
montage. In the performed implementation, the considered EEG signals $\mathbf{z}$ are obtained
from the raw acquisitions after being first processed through a spatial common average referencing filter [23], applied to the acquired EEG data in order to improve their
signal-to-noise ratio, by subtracting to each raw EEG signal the mean voltage sensed
over the entire scalp. The obtained signals are then band-pass filtered in order to isolate
the EEG sub-band whose discriminative properties have to be analysed. Specifically,
in the following, we always refer to EEG signals in the $\alpha - \beta = [8, 30]\,Hz$ sub-band,
having verified during the performed experimental tests that it carries the information guaranteeing the best achievable performance. Given the considered sub-band, if
EEG signals are originally acquired at a higher sampling frequency as in our case, the
filtered signals are downsampled at 64 Hz to reduce the computational complexity of
the subsequent processing.

Under the commonly assumed hypothesis of stationarity or quasi-stationarity of
the recorded signals [24], the available data can be decomposed into N consecutive
frames $\mathbf{f}_{(n)}^{(c)}[t_n]$, with $n = 1, \ldots, N$ and $t_n = 1, \ldots, T_n$, each representing a portion
of the original biometrics having total duration T_n. The stationarity assumption has
been often considered and evaluated for biosignals such as ECG and EEG [24,25],
although it may depend on the adopted acquisition protocol. The frame division
process performed on the considered EEG data is carried out selecting overlapping frames lasting 5s, with a 40% overlapping factor between each frame and the
previous one.

A set of R features can be then extracted from each processed frame, thus gen-
erating the representations $\mathbf{x}_{(n)}[r]$, with $r = 1, \ldots, R$. The template $\mathbf{x}_E$ conveying
the discriminative information of the considered biometrics can be then obtained
through the average of the information extracted from each frame: $\mathbf{x}_E = \sum_{n=1}^{N} \mathbf{x}_{(n)}$.
The template $\mathbf{x}_E$ can be stored as is in an unprotected system, while proper techniques
should be implemented to secure it in protected schemes. The entire process described
here is performed, during enrolment and verification modalities, in both the considered unprotected and protected frameworks, respectively, depicted in Figures 14.1
and 14.2.

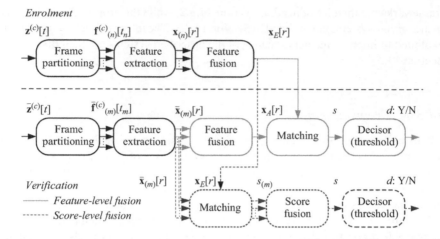

Figure 14.1   Frameworks for subject recognition without template protection

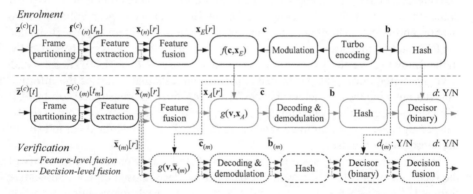

Figure 14.2   Frameworks for subject recognition with template protection

14.3.2   *Unprotected systems*

As already mentioned, the preprocessing steps described in Section 14.3.1 are performed for both the considered protected and unprotected systems. In more detail, the described preprocessing is applied to the available EEG data in both enrolment and verification stages. Yet, while during enrolment it is required to generate a single biometric template $\mathbf{x}_E$ from the N representations $\mathbf{x}_{(n)}$ obtained from each nth frame, $n = 1, \ldots, N$, different approaches can be taken to exploit the information contained in each frame during the verification phase. Specifically, in an unprotected system, it would be possible to combine such information using either feature or score-level fusion.

Having indicated with $\bar{\mathbf{z}}^{(c)}$ the probe sample provided for recognition purposes, and with $\bar{\mathbf{x}}_{(m)}$, $m = 1, \ldots, M$, the representations generated from each of the M frames extracted from the available data during verification, a feature-level fusion of the available information can be performed as during the enrolment phase, in order to generate a single template $\mathbf{x}_A = \sum_{m=1}^{M} \bar{\mathbf{x}}_{(m)}$ to be compared with $\mathbf{x}_E$. A single similarity score s would be thus determined during the verification process, and then compared against a system threshold to determine if the presented subject is actually who he/she declares to be.

Alternatively, the information extracted from the M available frames can be fused at the score level. In this case, each representation $\bar{\mathbf{x}}_{(m)}$ is compared against $\mathbf{x}_E$, thus generating M different similarity scores $s_{(m)}$, $m = 1, \ldots, M$. Different fusion rules can be then considered to combine these metrics into a single score s, for instance by selecting the mean, the maximum, or the minimum among the M outcomes. According to the performed experimental tests, the minimum rule can guarantee the best results among the mentioned methods, and therefore it will be used in Section 14.4.2. For similar reasons, the L1 distance is preferred over L2 when comparing enrolment and authentication templates. A decision on the identity can be eventually taken by comparing the obtained fused score s with a specific threshold.

The strategies for unprotected verification are depicted in Figure 14.1, which illustrates the frameworks employed to both perform feature-level fusion during authentication, as well as to implement a score-level integration of the information captured by different frames. It is worth remarking that, in both the considered cases, several operating points, providing different performance in terms of FRR and FAR, can be selected by setting the thresholds employed when taking decisions.

14.3.3 Protected system

Biometric cryptosystems can be employed to bind biometric information with cryptographic keys, in order to simultaneously protect both information. The cryptosystems proposed here, illustrated in Figure 14.2, are inspired by the approach in [7], able to guarantee high performance, in terms of both recognition rates and security, when dealing with traits characterized by a high intra-class variability, as it may happen with cognitive biometrics.

During enrolment, a cryptographic key is first randomly generated, and represented through a binary message $\mathbf{b}$ having length k. A turbo encoding process is then performed on $\mathbf{b}$ to generate an n-bit string, eventually modulated into a codeword $\mathbf{c}$ constituted by R symbols, belonging to a constellation with L points in the complex domain, with $R = \frac{n}{\log_2 L}$. Being R also the size of the obtained enrolment template $\mathbf{x}_E$, an helper data can be generated by binding $\mathbf{x}_E$ with $\mathbf{c}$ through an element-wise operator $f(\cdot, \cdot)$. In case the outputs of the binding function f provide no information on the function's arguments, $\mathbf{v} = f(\mathbf{c}, \mathbf{x}_E)$ can be stored in the system and made publicly available, together with an hashed version of the binary key $\mathbf{b}$.

With respect to a traditional fuzzy commitment method [26], the employed approach offers a wider choice for selecting the function used for binding the considered biometrics $\mathbf{x}_E$ with the cryptographic key $\mathbf{b}$. As in [7], a practical implementation of the employed protection scheme can be obtained by representing each codeword $\mathbf{c}$ through R symbols belonging to a set of L points lying on the unit circle in the complex domain, thus resorting to a phase-shift keying modulation. The binding function

$f(\mathbf{c}, \mathbf{x})$ is designed to first linearly map the range of admissible values of each element in $\mathbf{x}$ to the interval $[-\pi, \pi)$. In order to do this, once the inter-class mean μ and standard deviation σ of the used biometric representations are estimated, a clamped feature vector χ is obtained as

$$\chi[r] = \begin{cases} \pi \frac{\mathbf{x}[r]-\mu[r]}{\alpha\sigma[r]}, & -\alpha\sigma[r] \le \mathbf{x}[r] - \mu[r] \le \alpha\sigma[r] \\ -\pi & , \mathbf{x}[r] - \mu[r] < -\alpha\sigma[r] \\ \pi & , \mathbf{x}[r] - \mu[r] > \alpha\sigma[r] \end{cases} \qquad (14.1)$$

being α a system parameter, and $r = 1, \dots, R$. The coefficients in χ are then quantized to the closest element in the set with D terms $\mathscr{D} = \left\{\frac{\pi}{D} + \frac{2\pi}{D} \cdot w\right\}$, with $w = \left[-\frac{D}{2}, -\frac{D}{2} + 1, \dots, \frac{D}{2} - 1\right]$, thus deriving a quantized vector $\boldsymbol{\varphi}$. The binding of $\mathbf{x}$ and $\mathbf{c}$ is then performed by applying angular shifts depending on $\mathbf{x}$ to the original constellation points: $\mathbf{v} = f(\mathbf{c}, \mathbf{x}) = \mathbf{c} \cdot e^{i\varphi}$. In case D is greater than L, with $D = L \cdot 2^l$, where $l \in \mathbb{N}_0$ being $\mathbb{N}_0$ the set of non-negative integers, the performed processes make it hard to state which constellation point in $\mathbf{c}$ generates a given symbol in $\mathbf{v}$, thus preserving the security of the proposed system.

Once a biometric representation $\bar{\mathbf{x}}$ is made available during the verification phase, a string similar to the original codeword $\mathbf{c}$ can be retrieved as $\bar{\mathbf{c}} = g(\mathbf{v}, \bar{\mathbf{x}}) = \mathbf{v} \cdot e^{-i\bar{\varphi}}$, where $g(\cdot, \cdot)$ is an operator reciprocal to $f(\cdot, \cdot)$, such that $g(f(\mathbf{c}, \mathbf{x}), \mathbf{x}) = \mathbf{c}, \forall \mathbf{x}$, and $\bar{\varphi}$ is generated from $\bar{\mathbf{x}}$, just like φ is obtained from $\mathbf{x}$. A joint demodulation and decoding process is then applied to $\bar{\mathbf{c}}$ to obtain $\bar{\mathbf{b}}$. The hashed versions of $\mathbf{b}$ and $\bar{\mathbf{b}}$ are eventually compared to determine the binary outcome of the verification process.

It is however worth observing that it is possible to define $\bar{\mathbf{x}}$ in different ways, depending on the modality through which we want to exploit the information present in the M frames generated from the verification probe, represented through the vectors $\bar{\mathbf{x}}_{(m)}, m = 1, \dots, M$. Specifically, as in an unprotected scenario, a feature-level fusion approach can be followed to generate a single vector $\mathbf{x}_A$, then used to recover a potentially corrupted codeword $\bar{\mathbf{c}} = g(\mathbf{v}, \mathbf{x}_A)$. Comparing the hashed versions of such codeword, once decoded, with $\mathbf{b}$, results in the binary decision $d : Y/N(1/0)$ regarding the correspondence of the claimed identity with the one presented at the system. Differently from an unprotected scenario, it is instead impossible to perform a score-level fusion of the information extracted from the available frames in a protected scheme, since no score is here made available during the verification phase. It is however feasible resorting to a decision-level strategy, exploiting the M decisions which may be performed on each of the generated verification frames. Specifically, each of the M vectors $\bar{\mathbf{x}}_{(m)}, m = 1, \dots, M$, can be employed to generate a distinct codeword $\bar{\mathbf{c}}_{(m)}$ as $\bar{\mathbf{c}}_{(m)} = g(\mathbf{v}, \bar{\mathbf{x}}_{(m)})$. A decoding process is then applied to each string $\bar{\mathbf{c}}_{(m)}$ to derive M binary messages $\bar{\mathbf{b}}_{(m)}$, each compared through its hashed version with $\mathbf{b}$, thus generating M decisions $d_{(m)}, m = 1, \dots, M$. A subject is then accepted as who he/she declares to be in case at least a set percentage of the M obtained decisions $d_{(m)}$, $m = 1, \dots, M$, confirms such assumption.

The dependency of the proposed protected frameworks performance on the employed parameters L, D, and α is discussed in Section 14.4. It is worth remarking that, once the binding function and the code's error capability are set, the performance of the protected system relying on a feature-level fusion approach is expressed through a single operating point. Conversely, under the same conditions, a decision-level fusion strategy allows selecting different operating points, depending of the minimum percentage of correct matchings required to accept the presented subject as authentic.

14.3.3.1 Security evaluation

The robustness against brute-force attacks of the proposed cryptosystem can be expressed through the conditional entropy $H(\mathbf{b}|\mathbf{v})$, giving the uncertainty about the secret message $\mathbf{b}$ once the code-offset $\mathbf{v}$ has been made publicly available. Since the knowledge of $P = \frac{k}{\log_2 L}$ coefficients of φ would entirely reveal $\mathbf{b}$, once $\mathbf{v}$ is known and given the binding operators $f(\cdot, \cdot)$ and $g(\cdot, \cdot)$, it can be derived that $H(\mathbf{b}|\mathbf{v}) = \min_{\mathscr{P} \in \mathfrak{P}} \{H(\varphi^{\mathscr{P}}|\mathbf{v})\}$, being $\varphi^{\mathscr{P}}$ a string generated from φ by selecting only P coefficients out of the available R ones, with $\mathfrak{P}$ being the ensemble of all possible sets $\mathscr{P}$ of P coefficients, $\mathscr{P} \subset \{1, \ldots, R\}$. As in [7], the provided security $H(\mathbf{b}|\mathbf{v})$ is therefore empirically estimated by computing the minimum entropy $\min_{\mathscr{P} \in \mathfrak{P}} \{H(\varphi^{\mathscr{P}}|\mathbf{v})\}$ over the available EEG data, by resorting to an approximation based on second-order dependency trees. This approach is employed in Section 14.4.2 to express the robustness of the proposed biometric cryptosystems.

14.3.4 Exploiting multiple biometric representations: information fusion

In the previous sections, we have outlined how information individually extracted from each of the M frames available during verification can be exploited, in either an unprotected or a protected system, resorting to fusion strategies at the feature, score, and decision level. However, given the high intra-class variability typically affecting cognitive biometrics, the proposed cryptosystems may not be able to guarantee proper verification performance when a single parametric representation is adopted for describing the discriminative characteristics of the considered biometrics. Therefore, in this contribution, we evaluate the improvements which can be achieved when multiple biometric templates are exploited for representing the same biometric trait. In more detail, we consider two different sets of features to characterize EEG biometrics, based on auto-regressive (AR) reflection coefficients and mel-frequency cepstrum coefficients (MFCCs). Sections 14.3.4.1 and 14.3.4.2, respectively, introduce the employed representations, while Section 14.3.4.3 specifically deals with the strategies which could be exploited to fuse the discriminative information derived when separately using the two aforementioned models.

14.3.4.1 AR modelling

AR modelling is among the most commonly employed approaches for the analysis of EEG signals, and it has been often adopted for biometric recognition purposes too [9]. After the preprocessing stage, each channel $\mathbf{f}_{(n)}^{(c)}$ of a given nth frame, with $n = 1, \ldots, N$, can be modelled as a realization of an AR process of order Q.

The AR reflection coefficients characterizing the model can be obtained using the Levinson algorithm for solving the Yule–Walker equations. As in [22], the Burg method [27] is exploited here to derive the coefficients of an AR model of order Q

directly operating on the observed data $\mathbf{f}_{(n)}^{(c)}$, rather than going through the estimation of their autocorrelation functions. In more detail, having indicated with $K_{q(n)}^{(c)}$, $q = 1, \ldots, Q$, the qth reflection coefficient estimated for the AR model of $\mathbf{f}_{(n)}^{(c)}$, the nth frame of the considered signal can be represented by the feature vector:

$$^{(AR)}\mathbf{x}_{(n)} = [\underbrace{K_{1(n)}^{(1)}, K_{2(n)}^{(1)}, \ldots, K_{Q(n)}^{(1)}}_{\text{channel } 1}, \ldots, \underbrace{K_{1(n)}^{(C)}, K_{2(n)}^{(C)}, \ldots, K_{Q(n)}^{(C)}}_{\text{channel } C}] \tag{14.2}$$

composed of the $R = C \cdot Q$ computed AR reflection coefficients.

14.3.4.2 MFCC modelling

MFCCs have been commonly employed in speech recognition, while their use has been only recently spread to the biometric analysis of EEG signals [28]. In our implementation, a filterbank made of 27 mel-scaled triangular band-pass filters is first applied to the spectrum of the considered EEG channel $\mathbf{f}_{(n)}^{(c)}$ of a given nth frame. The natural logarithm of the resulting cepstral bins is then evaluated, in an attempt to perform an homomorphic filtering operation, able to separate the underlying neural activation signals from the effects caused by the propagation of these signals through the skull [29]. A discrete cosine transform is eventually performed on the obtained values, generating the desired MFCC representation through the selection of the first Q coefficients, with the exception of the DC component, from the obtained projections. Similarly to what is done for the AR representation, having indicated with $p_{q(n)}^{(c)}$, $q = 1, \ldots, Q$, the qth MFCC coefficient obtained from $\mathbf{f}_{(n)}^{(c)}$ through the aforementioned process, the nth frame of the considered EEG signal can be represented by the feature vector:

$$^{(MFCC)}\mathbf{x}_{(n)} = [\underbrace{p_{1(n)}^{(1)}, p_{2(n)}^{(1)}, \ldots, p_{Q(n)}^{(1)}}_{\text{channel } 1}, \ldots, \underbrace{p_{1(n)}^{(C)}, p_{2(n)}^{(C)}, \ldots, p_{Q(n)}^{(C)}}_{\text{channel } C}] \tag{14.3}$$

still composed of the $R = C \cdot Q$ computed MFCC coefficients.

14.3.4.3 Fusion strategies

Similarly to what has been proposed for combining the information extracted from each of the M frames available during verification, feature- and decision-level fusion approaches can be adopted for exploiting multiple biometric representations in order to improve the performance achievable in either unprotected or protected recognition systems. Specifically, a feature-level fusion can be straightforward implemented by combining, for each nth frame, both AR and MFCC representations into a single parametric vector, thus obtaining

$$\mathbf{x}_{(n)} = [^{(AR)}\mathbf{x}_{(n)}, ^{(MFCC)}\mathbf{x}_{(n)}]. \tag{14.4}$$

Such information fusion would be therefore performed, in both protected and unprotected systems, right after feature extraction, at the beginning of the described processing chain.

A decision-level fusion strategy can be also followed, combining the information obtained using AR and MFCC modelling at the very ending of the proposed processing chain. Specifically, once the decisions $^{(AR)}d$ and $^{(MFCC)}d$ have been taken for the presented verification probe, a final decision could be determined employing the AND rule, therefore accepting the present subject only in case both AR and MFCC analysis confirmed the authenticity of the queried identity. Two distinct keys **b** could be also exploited in this scenario, one for each adopted representation.

The described strategies for fusing information derived from multiple biometric representations can be applied to all the recognition systems described in Sections 14.3.2 and 14.3.3, being therefore jointly used with fusion methods employed to combine data extracted from multiple frames at either feature, score and decision level.

14.4 Experimental tests

The experimental tests carried out to verify the effectiveness of the proposed architectures are described here. Specifically, Section 14.4.1 outlines the employed experimental setup, detailing the EEG database employed in the performed tests, while Section 14.4.2 presents the results obtained when using either the described unprotected and protected systems for user recognition.

14.4.1 Experimental setup

The same database considered in [22], collected at the authors' institution, is exploited here for the performed tests. It comprises recordings taken from 50 healthy subjects, whose age ranges from 20 to 35 years, with an average of 25 years. During each EEG acquisition, subjects are comfortably seated on a chair in a dimly lit room. Signals are acquired using a GALILEO BE Light amplifier with $M = 19$ channels, placed on the scalp according to the 10–20 international systems [30] and an original sampling rate of 256 Hz. The electrical impedance of each electrode is kept under 10 kΩ using conductive gel at the beginning of each acquisition. The EEG signals are collected for each considered subject during three distinct acquisition sessions, indicated in the following as S1, S2, and S3. Specifically, the second session S2 is performed at an average temporal distance of about one week from the first session, while the third set of acquisitions is carried out at a distance from the first acquisitions having an average value of 34 days. During each session, the EEG signals of the considered subjects are acquired in two distinct baseline conditions: 4 min of recording are first taken in a relaxed state with EO, where the subject is asked to fix a light point on the screen, followed by other 4 min, still in resting state, during which the subjects remain with EC. It is worth pointing out that no restrictions are imposed on the activities the subjects performing the test may carry out before the acquisition, or on the diet they should follow, neither between an EEG acquisition and the following one, nor during the days of the recordings. This lack of restrictions is adopted in order to acquire data in conditions close to real life.

It has been shown in [22] that, within this temporal period and for both the considered acquisition protocols, comparing EEG data from S1 with signals from

S3 does not result in significantly different performance, with respect to comparing data from S1 with signals from S2, or data from S2 with signals from S3. Therefore, only S3 will be considered in the following to provide verification samples in the performed tests, while S1 and S2 will be employed for enrolment purposes.

14.4.2   Experimental results

We first evaluate the performance achievable with the proposed systems when selecting, for each user, 3 min of EEG recording, out of the 4 min available from the S1, for enrolment purposes. Authentication samples are obtained by selecting, for each subject, 1 min of EEG data captured during S3. A cross-validation procedure is performed by randomly selecting for three different times the EEG signals employed for enrolment from the considered training session, with ten random selections of the authentication probes performed at each run.

The recognition rates achievable with the unprotected frameworks described in Section 14.3.2 are reported in Figures 14.3 and 14.4, referring to EC and EO scenarios respectively. Specifically, considering EEG signals acquired in EC conditions, Figure 14.3(a) and (b) show the performance obtained when exploiting a feature-level approach for fusing information derived from multiple authentication frames, respectively using AR and MFCC features obtained for different Q values. The performance achievable exploiting a score-level fusion method for combining the information extracted from different frames are instead reported in Figure 14.3(c) and (d), referring to the use of AR and MFCC features respectively. Similar considerations can be done for the plots regarding the analysis of EO conditions, reported in Figure 14.4. It can be seen that lower recognition accuracy can be achieved collecting EEG signals with open eyes, with respect to a protocol requiring subjects having their eyes closed during EEG acquisition. From the obtained results, we can observe that, in unprotected scenarios, the adoption of a score-level fusion strategy with AR modelling allows achieving slightly better performance in EC conditions, while using a feature-level fusion method with MFCC representation provides better results in

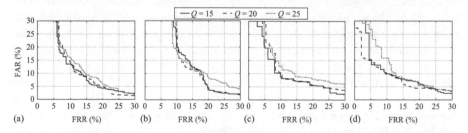

Figure 14.3   *Recognition performance in unprotected systems (EC resting state), using S1 for enrolment and S3 for testing purposes. (a) Feature-level fusion, AR modelling; (b) feature-level fusion, MFCC modelling; (c) score-level fusion, AR modelling; (d) score-level fusion, MFCC modelling*

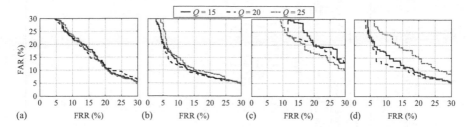

Figure 14.4 *Recognition performance in unprotected systems (EO resting state),*
using S1 for enrolment and S3 for testing purposes. (a) Feature-level
fusion, AR modelling; (b) feature-level fusion, MFCC modelling;
(c) score-level fusion, AR modelling; (d) score-level fusion, MFCC
modelling

EO scenarios. Specifically, the lowest EER is obtained in EC conditions when using
AR modelling with $Q = 15$ coefficients, resulting in an EER $= 10.1\%$. The best performance in EO scenarios is instead achieved using $Q = 20$ coefficients for MFCC
representation, with an EER $= 11.2\%$.

As for the proposed cognitive biometric cryptosystems, Tables 14.1 and 14.2
provides a summary of the performance, in terms of both recognition rates and
security, achievable when respectively considering EC and EO scenarios, still using
S1 for enrolment and S3 for testing purposes. The same error correction capability, roughly corresponding to a ratio $k/n = 1/15$, is considered for the turbo codes
employed in all the performed tests. As already commented, a single operating point,
expressed through its FRR and FAR, is obtained when considering a specific selection of the parameters L, D, and α for cryptosystems based on feature-level fusion.
The results reported for cryptosystems based on a decision-level fusion strategy are
instead referred to the lowest HTER, defined as HTER $= (FRR + FAR)/2$, achieved
when varying the threshold corresponding to the minimum percentage of correct
matchings required to accept a subject as genuine. The α parameter is chosen, for
each considered selection of L and D, as the one providing the lowest possible HTER
value. As already shown in [7], from the reported results, it can be seen that increasing the number of constellation points L improves the FAR and the provided security,
while worsening the FRR. The increase of the quantization level number D typically
improves the FRR, while the FAR and the security worsen. With respect to unprotected
scenarios, increasing the order Q of the employed models makes it harder to obtain
low HTERs values, even if higher values of Q guarantee higher security, since the
length k of the employed cryptographic keys can be longer. The best recognition accuracy is achieved in this case when performing decision-level fusion of information
derived from AR modelling with order $Q = 20$ in EC conditions, with an HTER at
about 5.0%. As for EO scenarios, the choice $Q = 15$ yields slightly better results than
$Q = 20$, with a best HTER obtained using decision-level fusion and AR modelling
at 9.2%. The recognition performance are therefore comparable with those of

Table 14.1 Recognition (in %) and security (in bits) performance for different values of Q, L, D and α, considering EEG signals acquired in EC conditions

				AR								MFCC							
				Recognition performance						Security		Recognition performance						Security	
				Feature-lev. fus.			Decision-lev. fus.			perform.		Feature-lev. fus.			Decision-lev. fus.			perform.	
Q	L	D	α	HTER	FRR	FAR	HTER	FRR	FAR	H(b|v)	k	HTER	FRR	FAR	HTER	FRR	FAR	H(b|v)	k
15	2	2	–	7.8	6.6	9.0	7.1	6.3	7.8	11.1		13.6	10.5	16.8	11.2	10.7	11.7	10.1	
		4	0.40	12.5	5.3	19.7	13.5	17.1	9.9	7.5	14	16.2	8.7	23.6	18.0	12.9	23.0	6.9	14
		8	0.40	21.4	6.2	36.6	25.4	30.0	20.8	4.6		26.2	9.2	43.3	32.9	21.7	44.1	4.1	
	4	4	0.35	9.8	11.0	8.7	11.9	13.0	10.8	21.3	32	14.8	16.2	13.5	18.0	14.8	21.1	20.3	32
		8	0.35	16.6	20.5	12.7	25.4	33.4	17.4	12.9		18.8	26.0	11.6	32.2	41.0	23.3	12.6	
		8	0.30	21.7	22.1	21.2	26.8	21.4	32.2	27.3	52	23.8	42.8	47.0	30.5	45.3	15.6	25.4	52
20	2	2	–	9.1	15.0	3.2	5.0	5.3	4.7	15.9		11.1	14.2	8.0	10.4	8.6	12.3	14.9	
		4	0.35	10.3	12.6	8.0	15.1	22.1	8.0	11.8	20	13.6	11.7	15.6	20.6	25.5	15.6	10.5	20
		8	0.35	16.5	21.2	11.9	31.6	24.1	39.1	7.1		18.0	17.8	18.3	35.5	32.1	38.9	6.6	
	4	4	0.25	14.5	24.7	4.3	15.4	17.4	13.3	31.3	46	17.0	24.4	9.6	20.6	23.3	17.8	27.8	46
		8	0.25	23.3	37.0	9.6	30.2	35.2	25.2	19.1		23.5	41.2	5.8	34.3	39.7	29.0	18.0	
		8	0.25	29.2	51.6	6.8	31.4	44.6	18.3	38.7	70	27.5	46.5	8.4	34.3	46.1	22.6	35.9	70
25	2	2	–	16.6	32.4	0.8	11.9	22.7	1.1	20.8		17.1	29.6	4.6	11.8	19.3	4.3	19.6	
		4	0.25	11.8	11.6	12.0	21.5	18.6	24.4	13.1	26	16.6	26.3	6.9	24.4	22.0	26.9	13.6	26
		8	0.25	19.7	31.2	8.3	32.6	39.5	25.7	5.9		20.0	32.5	7.5	38.5	16.9	60.1	8.0	
	4	4	0.25	22.6	44.0	1.1	19.4	30.2	8.6	37.8	58	23.8	43.3	4.4	24.2	34.2	14.2	36.8	58
		8	0.25	24.9	42.1	7.7	31.2	36.6	25.8	19.4		30.5	45.8	15.1	39.0	60.1	18.0	23.6	
		8	0.25	38.8	76.7	1.0	31.1	44.3	17.8	43.3	90	36.5	67.8	5.2	39.5	50.5	28.4	46.9	90

Table 14.2 Recognition (in %) and security (in bits) performance for different values of Q, L, D and α, considering EEG signals acquired in EO conditions

Q	L	D	α	AR								MFCC							
				Recognition performance						Security perform.		Recognition performance						Security perform.	
				Feature-lev. fus.			Decision-lev. fus.					Feature-lev. fus.			Decision-lev. fus.				
				HTER	FRR	FAR	HTER	FRR	FAR	H(b|v)	k	HTER	FRR	FAR	HTER	FRR	FAR	H(b|v)	k
15	2	2	–	10.1	13.5	6.6	9.2	9.8	8.6	11.3		13.8	17.2	10.4	12.3	7.2	17.4	10.8	
		4	0.30	16.8	8.7	24.8	15.1	7.1	23.2	7.4	14	19.2	14.2	24.2	17.4	22.1	12.6	7.7	14
		8	0.30	32.2	5.9	58.4	32.8	17.2	48.5	4.5		36.0	15.4	56.5	31.3	18.1	44.4	4.9	
	4	4	0.25	14.1	21.2	6.9	15.9	7.8	24.1	20.6	32	18.1	23.8	12.5	19.8	19.7	19.9	21.4	32
		8	0.25	25.2	35.8	14.7	28.7	32.2	25.1	12.5		25.0	33.3	16.8	31.2	32.4	30.0	13.1	
	8	8	0.25	29.6	50.2	9.0	32.4	37.2	27.6	28.0	52	29.4	47.0	11.9	36.1	40.4	31.8	27.7	52
20	2	2	–	16.6	31.1	2.2	9.6	18.0	1.3	15.6		16.4	27.2	5.6	16.5	27.2	5.9	15.8	
		4	0.30	14.9	25.7	41.0	20.7	18.7	22.7	11.0	20	18.6	17.7	19.6	22.0	21.8	22.3	10.6	20
		8	0.30	22.0	39.7	43.0	34.1	47.7	20.5	6.7		22.9	34.2	11.7	34.3	23.6	45.0	6.3	
	4	4	0.25	27.0	51.4	2.5	19.4	28.8	9.9	30.3	46	25.9	47.6	4.3	24.3	37.9	10.7	29.4	46
		8	0.25	30.6	47.9	13.3	36.2	39.4	33.0	18.9		29.6	46.5	12.6	36.1	17.1	54.6	17.8	
	8	8	0.25	37.6	73.2	2.0	35.0	44.4	25.5	40.3	70	38.7	75.4	2.0	38.4	54.0	22.9	39.3	70
25	2	2	–	25.6	51.1	1.0	24.3	48.3	0.2	20.8		21.7	41.3	2.0	26.8	52.6	1.0	21.0	
		4	0.25	18.6	31.2	6.1	23.8	31.8	15.9	13.9	26	20.3	32.1	8.6	23.7	13.9	33.4	15.0	26
		8	0.25	27.6	45.8	9.5	36.6	50.2	22.9	7.9		28.4	32.8	23.9	38.1	53.1	23.2	9.2	
	4	4	0.25	34.6	68.4	0.8	30.1	56.6	3.5	39.9	58	33.2	64.8	1.6	33.7	62.0	5.5	38.7	58
		8	0.25	36.2	69.1	3.2	38.3	56.8	19.7	24.8		35.6	67.2	4.0	40.6	49.5	31.8	25.7	
	8	8	0.25	44.3	88.3	0.2	40.9	74.1	7.7	50.3	90	44.0	87.6	0.4	42.8	75.5	10.1	53.6	90

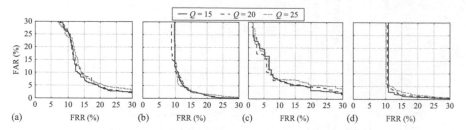

Figure 14.5   Recognition performance in unprotected systems (EC resting state), using S1 for enrolment and S3 for testing. (a) Feature-level frame fusion, feature-level fusion of AR and MFCC representations; (b) feature-level frame fusion, decision-level fusion of AR and MFCC representations; (c) score-level frame fusion, feature-level fusion of AR and MFCC representations; (d) score-level frame fusion, decision-level fusion of AR and MFCC representations

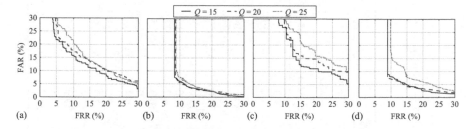

Figure 14.6   Recognition performance in unprotected systems (EO resting state), using S1 for enrolment and S3 for testing. (a) Feature-level frame fusion, feature-level fusion of AR and MFCC representations; (b) feature-level frame fusion, decision-level fusion of AR and MFCC representations; (c) score-level frame fusion, feature-level fusion of AR and MFCC representations; (d) score-level frame fusion, decision-level fusion of AR and MFCC representations

unprotected scenarios. The achievable security strongly depends on the number of coefficients employed for template representation, and on the selection of the parameters L and D. Values at about 15 bits and 11 bits are guaranteed in EC and EO conditions, respectively.

In order to further improve the achievable performance, the considered AR and MFCC models can be jointly used by resorting to either feature- or decision-level fusion strategies, as outlined in Section 14.3.4.3. When considering unprotected systems, the recognition rates obtained in EC and EO conditions are respectively shown in Figures 14.5 and 14.6. In more detail, for the EC scenario, Figure 14.5(a) and (b) reports the performance obtained when exploiting a feature-level approach for fusing information derived from multiple authentication frames, respectively combining AR and MFCC features obtained with the same Q values at either feature and decision level. Figure 14.5(c) and (d) refers to score-level fusion of the information derived from multiple authentication frames, combining AR and MFCC representations at feature and decision level respectively. With reference to the performance reported in Figures 14.3 and 14.4, a clear improvement can be observed for unprotected systems, able to achieve the best EER in EC conditions at 7.8% when using AR and MFCC models with $Q = 20$ coefficients at feature level, and exploiting score-level fusion of the information extracted from each verification frame. As for EO conditions, the best performance is obtained at EER $= 8.0\%$ when AR and MFCC models with $Q = 15$ coefficients are fused at the decision level, while combining at feature-level the information derived from each verification frame.

However, the proposed fusion approach of multiple-feature representations provides more interesting improvements when applied to the considered cognitive biometric cryptosystems, as shown by the results reported in Tables 14.3 and 14.4, respectively, referred to EC and EO scenarios. Specifically, for both the considered acquisition protocols, a notable improvement in recognition rates is achieved when exploiting a decision-level approach for fusing AR and MFCC information, guaranteeing a best EER $= 1.7\%$ for EC and EER $= 2.8\%$ for EO. On the other hand, a feature-based fusion method does not result in significant accuracy improvement, providing best EER $= 9.3\%$ in EC and EER $= 11.0\%$ in EO.

As for the achievable security performance, it is worth pointing out that we retain here a conservative approach, assuming that the entropy $H(\mathbf{b}|\mathbf{v})$ expressing the system robustness should be taken as the minimum value evaluated for either AR or MFCC representation, in both the considered fusion schemes at feature and decision level. In fact, we could speculate that recovering information related to only one of the adopted models already provides all the required insights on the original EEG signals, and therefore also on the alternative representation. For a feature-level fusion strategy, this would imply that recovering information on the coefficients of φ associated to one model may be already enough to argue on the whole combined representation. Similarly, in a decision-level fusion scheme, both the needed keys could be assumed to be retrieved once only one of the adopted representations φ is available. The reported k values are instead referred to the use of two keys for decision-level fusion schemes, and a single (longer) key for feature-level fusion approaches.

Inspired by the results reported in [22], we attempt to further improve the achievable recognition rates by exploiting both the available sessions S1 and S2 for enrolment purposes, taking $E/2$ frames from each of these recording sessions to train the AR and MFCC models of each considered subject, while S3 is used for testing purposes. Figures 14.7 and 14.8, respectively, show the recognition rates achievable when exploiting multi-session enrolment in EC and EO conditions, while exploiting both AR and MFCC features at either feature and decision level. A best EER $= 6.4\%$ can be thus obtained in the EC scenario with $Q = 20$, with a lowest EER $= 6.9\%$ achieved in EO conditions with $Q = 15$.

Tables 14.5 and 14.6 report the performance obtained in the proposed biometric cryptosystem when also exploiting multi-session enrolment. As already shown by the previous results, the best recognition rates are obtained when fusing at the decision

Table 14.3 Recognition (in %) and security (in bits) performance varying Q, L, D, and α, and fusing AR and MFCC representations at either feature or decision level, in EC conditions. S1 and S2 for enrolment and S3 for testing

				Recognition performance												Security perform.	
				AR & MFCC fused at feature level						AR & MFCC fused at decision level							
				Feature-lev. fus.			Decision-lev. fus.			Feature-lev. fus.			Decision-lev. fus.				
Q	L	D	α	HTER	FRR	FAR	HTER	FRR	FAR	HTER	FRR	FAR	HTER	FRR	FAR	H(b|v)	k
15	2	2	–	10.4	11.9	8.9	9.3	5.0	13.5	6.3	11.7	1.0	2.5	4.3	0.7	10.1	28
		4	0.35	13.6	10.2	17.1	14.7	13.0	16.3	7.8	11.0	4.6	5.8	5.5	6.1	6.9	
		8	0.35	19.1	13.5	26.3	30.2	22.9	37.6	15.1	12.0	18.3	17.1	22.7	11.6	4.1	
	4	4	0.30	12.3	19.0	5.6	15.0	15.9	14.1	7.7	12.6	2.8	4.9	6.8	3.1	20.3	64
		8	0.30	18.8	24.8	12.9	31.5	38.2	24.8	16.0	19.6	12.5	16.2	19.6	12.5	12.6	
	8	8	0.25	21.0	24.8	17.2	28.4	46.2	10.5	18.3	30.9	5.8	17.0	18.7	15.3	25.4	104
20	2	2	–	10.5	17.5	3.5	11.4	20.8	2.0	10.2	19.7	0.7	1.7	2.8	0.7	14.9	40
		4	0.30	12.2	11.1	13.2	16.5	20.0	12.9	8.4	13.2	3.7	7.3	7.8	6.8	10.5	
		8	0.30	22.3	11.3	33.3	35.4	31.1	39.7	15.1	26.7	3.4	22.4	17.5	27.3	6.6	
	4	4	0.25	14.2	24.2	4.1	17.7	19.4	16.0	17.5	34.6	0.4	6.2	4.4	8.0	27.8	92
		8	0.25	21.5	21.8	21.2	35.1	38.2	31.9	22.3	33.7	11.0	19.8	25.4	14.2	18.0	
	8	8	0.25	29.0	53.4	4.6	34.8	55.0	14.5	31.3	61.7	0.8	22.3	35.2	9.3	35.9	140
25	2	2	–	18.7	35.2	2.2	23.8	46.4	1.1	21.3	46.4	1.1	21.3	42.7	0.0	19.6	52
		4	0.25	13.6	16.9	10.4	22.7	31.2	14.2	11.9	20.9	2.9	11.9	13.5	10.2	13.6	
		8	0.25	21.8	41.0	2.6	36.5	29.0	44.0	21.2	40.7	1.6	24.1	33.3	14.8	8.0	
	4	4	0.25	26.4	51.3	1.4	25.5	44.3	6.8	27.3	54.4	0.1	13.1	24.2	2.0	36.8	116
		8	0.25	29.0	51.3	6.8	35.6	37.9	33.3	29.3	56.5	2.0	24.4	31.3	17.5	23.6	
	8	8	0.25	36.3	72.3	0.2	39.3	59.8	18.8	41.0	81.6	0.4	23.2	35.2	11.1	46.9	180

Table 14.4 Recognition (in %) and security (in bits) performance varying Q, L, D, and α, and fusing AR and MFCC representations at either feature or decision level, in EO conditions. S1 and S2 for enrolment and S3 for testing

Q	L	D	α	Recognition performance												Security perform.	
				AR & MFCC fused at feature level						AR & MFCC fused at decision level						$H(b|v)$	k
				Feature-lev. fus.			Decision-lev. fus.			Feature-lev. fus.			Decision-lev. fus.				
				HTER	FRR	FAR	HTER	FRR	FAR	HTER	FRR	FAR	HTER	FRR	FAR		
15	2	2	–	12.1	20.9	3.2	11.0	19.0	2.9	10.4	19.9	1.0	2.8	3.2	2.3	10.8	
		4	0.30	13.3	12.3	14.2	18.7	20.5	16.9	11.5	16.2	6.8	6.3	7.5	5.0	7.4	
		8	0.30	27.4	13.2	41.6	31.9	25.1	38.8	26.7	16.6	36.9	19.7	6.6	32.8	4.5	28
	4	4	0.25	16.5	25.7	7.4	16.8	15.1	18.6	15.7	30.5	1.0	8.3	9.0	7.5	20.6	
		8	0.25	21.2	31.4	11.0	23.7	24.2	23.2	21.2	24.8	17.7	17.4	18.6	16.3	12.5	64
	8	8	0.25	19.5	33.2	5.8	25.2	28.4	22.0	23.1	25.7	20.5	19.3	20.1	18.5	27.7	104
20	2	2	–	18.3	34.9	1.7	27.5	54.1	0.8	20.8	41.6	0.0	6.3	12.6	0.0	15.6	
		4	0.30	14.1	16.9	11.3	21.8	14.7	29.0	12.0	20.6	3.4	10.4	10.4	10.4	10.6	
		8	0.30	22.5	3.19	13.0	36.6	39.4	33.7	22.6	35.5	9.8	23.6	11.1	36.1	6.3	40
	4	4	0.25	28.7	56.2	1.3	25.2	46.1	4.3	31.6	63.0	0.2	11.9	20.9	2.9	29.4	
		8	0.25	31.3	56.1	6.5	36.5	40.3	32.7	32.3	61.3	3.4	26.4	10.8	42.1	17.8	92
	8	8	0.25	39.7	78.5	0.8	37.1	60.7	13.5	40.6	81.2	0.0	28.1	40.1	16.2	39.3	140
25	2	2	–	27.8	55.5	0.1	37.2	74.4	0.0	31.3	62.7	0.0	18.6	37.3	0.0	20.8	
		4	0.25	22.9	40.1	5.8	25.8	30.6	21.1	22.9	44.7	1.1	13.1	9.0	17.2	13.9	
		8	0.25	29.3	42.2	16.5	37.1	51.9	22.3	29.0	45.5	12.6	27.9	39.4	16.5	7.9	52
	4	4	0.25	38.6	77.0	0.1	38.8	75.7	1.9	38.4	76.9	0.0	25.2	49.1	1.4	38.7	
		8	0.25	40.0	78.2	1.7	40.6	56.5	24.7	38.5	76.3	0.7	30.0	34.9	25.1	24.8	116
	8	8	0.25	45.2	90.3	0.1	43.4	81.8	5.0	46.7	93.3	0.1	36.5	68.1	4.9	50.3	180

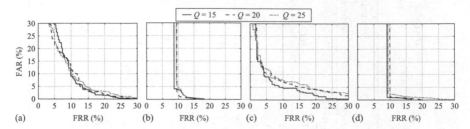

*Figure 14.7   Recognition performance in unprotected systems (EC resting state),
using S1 and S2 for enrolment and S3 for testing. (a) Feature-level
frame fusion, feature-level fusion of AR and MFCC representations;
(b) feature-level frame fusion, decision-level fusion of AR and MFCC
representations; (c) score-level frame fusion, feature-level fusion of
AR and MFCC representations; (d) score-level frame fusion,
decision-level fusion of AR and MFCC representations*

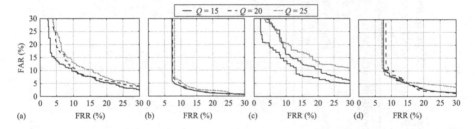

*Figure 14.8   Recognition performance in unprotected systems (EO resting state),
using S1 and S2 for enrolment and S3 for testing. (a) Feature-level
frame fusion, feature-level fusion of AR and MFCC representations;
(b) feature-level frame fusion, decision-level fusion of AR and MFCC
representations; (c) score-level frame fusion, feature-level fusion of
AR and MFCC representations; (d) score-level frame fusion,
decision-level fusion of AR and MFCC representations*

level AR and MFCC representations with $Q = 15$, and using $L = D = 2$, with
HTER $= 1.5\%$ for EC and HTER $= 3.2\%$ for EO. However, the security guaranteed
in these cases is quite low, being respectively 10.1 and 10.8 bits. In order to improve
the robustness of the proposed cryptosystem, a limited loss in verification accuracy
can be tolerated. Specifically, an HTER $= 5.5\%$ can be reached in EC conditions
fusing at the decision level both the R and the MFCC representations obtained with
$Q = 20$, while guaranteeing 27.8 bits of security. Using $Q = 15$ coefficients for each
representation allows achieving HTER $= 6.9\%$ with 20.6 bits of robustness in EO
scenarios. The same levels of security can be achieved through a feature-level fusion
of AR and MFCC templates, fusing frames at feature level as well, yet providing
lower recognition performance, at HTER $= 13.0\%$ and HTER $= 17.1\%$, respectively,
for the EC and EO scenarios.

Table 14.5 Recognition (in %) and security (in bits) performance varying Q, L, D, and α, and fusing AR and MFCC representations at either feature or decision level, in EC conditions. S1 and S2 for enrolment, S3 for testing

Q	L	D	α	Recognition performance												Security perform.	
				AR & MFCC fused at feature level						AR & MFCC fused at decision level							
				Feature-lev. fus.			Decision-lev. fus.			Feature-lev. fus.			Decision-lev. fus.				
				HTER	FRR	FAR	HTER	FRR	FAR	HTER	FRR	FAR	HTER	FRR	FAR	H(b|v)	k
15	2	2	–	6.8	5.8	7.8	6.3	4.7	8.0	5.4	8.6	2.2	1.5	2.2	0.8	10.1	28
		4	0.35	11.9	5.9	17.8	15.1	14.2	15.9	6.1	8.6	3.7	5.7	3.1	8.3	6.9	
		8	0.35	21.8	5.3	38.3	33.5	33.6	33.4	14.2	11.0	17.5	20.0	13.3	26.6	4.1	
	4	4	0.30	10.9	7.4	14.4	15.9	13.0	2.6	7.8	13.0	2.6	4.2	4.6	3.8	20.3	64
		8	0.30	18.1	23.5	12.7	30.4	40.3	20.5	15.6	27.2	40.0	17.4	8.1	26.6	12.6	
	8	8	0.30	19.1	33.4	4.7	28.9	47.1	10.7	19.9	31.2	8.6	18.9	25.1	12.7	25.4	104
20	2	2	–	7.2	10.5	4.0	8.1	13.3	2.9	7.3	13.9	0.7	1.7	1.9	1.4	14.9	40
		4	0.35	9.2	7.1	11.3	18.6	8.0	29.3	6.5	10.5	2.5	6.9	7.7	2.2	10.5	
		8	0.35	21.4	10.2	32.5	37.5	6.9	68.0	12.8	23.2	2.5	23.8	14.4	33.1	6.6	
	4	4	0.25	13.0	21.5	4.4	17.9	20.5	15.3	14.5	28.1	0.8	5.5	2.2	8.9	27.8	92
		8	0.25	21.2	34.5	8.0	36.9	45.3	28.5	21.6	32.7	10.5	23.6	12.9	34.3	18.0	
	8	8	0.25	27.9	51.1	4.6	36.0	39.1	33.0	31.2	62.2	0.2	22.3	16.5	28.1	35.9	140
25	2	2	–	13.0	24.8	1.3	23.0	45.2	0.8	17.3	34.5	1.0	4.6	9.3	0.0	19.6	52
		4	0.25	11.3	14.7	7.8	22.4	25.4	19.4	10.7	18.8	2.5	8.4	2.5	14.2	13.6	
		8	0.25	19.8	36.9	2.8	38.7	58.4	19.0	21.0	33.7	8.3	22.6	18.6	26.6	8.0	
	4	4	0.25	23.8	45.9	1.7	27.7	48.6	6.8	27.0	54.0	1.0	12.0	21.5	2.5	36.8	116
		8	0.25	28.7	49.8	7.7	38.0	66.0	10.1	30.1	57.8	2.3	23.5	10.2	36.7	23.6	
	8	8	0.25	36.3	71.5	1.0	37.9	58.9	16.9	41.2	82.2	1.0	29.0	46.2	11.9	46.9	180

Table 14.6 *Recognition (in %) and security (in bits) performance varying Q, L, D, and α, and fusing AR and MFCC representations at either feature or decision level, in EO conditions. S1 and S2 for enrolment, S3 for testing*

				Recognition performance												Security perform.	
				AR & MFCC fused at feature level						AR & MFCC fused at decision level							
				Feature-lev. fus.			Decision-lev. fus.			Feature-lev. fus.			Decision-lev. fus.				
Q	L	D	α	HTER	FRR	FAR	HTER	FRR	FAR	HTER	FRR	FAR	HTER	FRR	FAR	H(b|v)	k
15	2	2	–	11.6	18.7	4.6	12.0	19.1	4.9	9.4	17.8	1.0	3.2	5.3	1.1	10.8	
		4	0.30	18.3	9.2	27.3	20.9	23.5	18.3	11.0	15.1	6.9	7.6	10.1	5.2	7.4	28
		8	0.30	39.7	6.3	73.0	40.6	6.9	74.4	26.3	15.1	37.5	23.1	10.5	35.7	4.5	
	4	4	0.25	17.1	28.1	6.1	19.0	26.9	11.1	15.4	29.6	1.1	6.9	4.9	8.9	20.6	64
		8	0.25	25.6	24.2	27.0	32.5	33.6	31.5	22.7	40.3	5.2	18.0	18.6	17.5	12.5	
	8	8	0.25	17.5	31.2	3.8	23.2	26.4	20.0	21.2	22.8	18.6	17.4	18.2	16.4	27.7	104
20	2	2	–	16.1	30.9	1.3	21.9	42.8	1.0	18.1	35.8	0.4	6.1	11.6	0.5	15.6	
		4	0.25	14.5	18.7	10.2	22.2	26.3	18.1	12.4	20.2	4.7	11.3	6.2	16.5	10.6	40
		8	0.25	23.5	31.1	15.9	39.0	59.5	18.6	20.8	39.7	1.9	24.9	15.7	34.0	6.3	
	4	4	0.25	22.1	43.1	1.1	25.1	43.8	6.3	26.1	52.2	0.0	11.7	20.2	3.2	29.4	92
		8	0.25	28.3	48.2	8.4	38.2	63.0	13.3	28.6	53.8	3.4	27.2	8.1	46.2	17.8	
	8	8	0.25	36.4	71.7	1.1	38.9	74.5	3.2	37.3	74.7	0.0	26.7	36.0	17.4	39.3	140
25	2	2	–	26.2	52.2	0.2	34.8	69.7	0.0	27.4	54.9	0.0	16.1	32.2	0.0	20.8	
		4	0.25	22.4	39.8	5.0	23.1	26.6	19.6	21.6	42.1	1.1	14.1	9.8	18.4	13.9	52
		8	0.25	29.0	40.9	17.1	37.5	31.3	43.6	28.4	4.41	12.7	28.1	40.0	16.3	7.9	
	4	4	0.25	33.3	66.3	0.2	36.5	71.5	1.4	36.4	72.9	0.0	24.3	47.9	0.7	38.7	116
		8	0.25	39.0	77.2	0.7	39.6	55.2	23.9	37.9	75.1	0.7	27.9	32.7	23.0	24.8	
	8	8	0.25	44.2	88.7	0.0	42.4	80.8	4.1	45.7	91.5	0.0	34.3	64.4	4.1	50.3	180

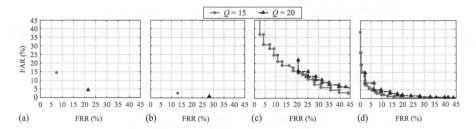

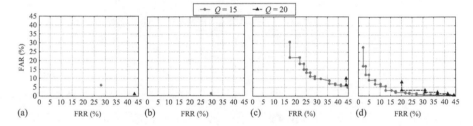

Figure 14.9 Recognition performance in protected systems (EC resting state), using S1 and S2 for enrolment and S3 for testing, $L = D = 4$.
(a) Feature-level frame fusion, feature-level fusion of AR and MFCC representations; (b) feature-level frame fusion, decision-level fusion of AR and MFCC representations; (c) decision-level frame fusion, feature-level fusion of AR and MFCC representations; (d) decision-level frame fusion, decision-level fusion of AR and MFCC representations

Figure 14.10 Recognition performance in protected systems (EO resting state), using S1 and S2 for enrolment and S3 for testing, $L = D = 4$.
(a) Feature-level frame fusion, feature-level fusion of AR and MFCC representations; (b) feature-level frame fusion, decision-level fusion of AR and MFCC representations; (c) decision-level frame fusion, feature-level fusion of AR and MFCC representations; (d) decision-level frame fusion, decision-level fusion of AR and MFCC representations

In order to better show the characteristics of the proposed feature- and decision-level fusion of information extracted from each frame, Figures 14.9 and 14.10 show in detail the recognition performance obtained for the protected systems in EC and EO scenarios, respectively. Fusion of AR and MFCC features at both feature- and decision-level are considered, with $L = D = 4$ and {S1,S2} jointly employed for enrolment purposes. Results obtained with $Q = 25$ are not reported since they are outside the acceptable performance range. As already remarked, the use of feature-level fusion of verification frames results in isolated operating points for each considered scenario, when using a single $k/n = 1/15$ ratio for setting the error correction capability of the employed turbo codes. On the other hand, decision-level fusion provides

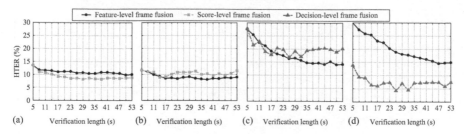

Figure 14.11   *Recognition performance versus verification time in unprotected and protected systems under EC conditions, using S1 and S2 for enrolment and S3 for testing, Q = 20 for AR and MFCC representations. (a) Unprotected system, feature-level fusion of AR and MFCC representations; (b) unprotected system, decision-level fusion of AR and MFCC representations; (c) protected system, feature-level fusion of AR and MFCC representations; (d) protected system, decision-level fusion of AR and MFCC representations*

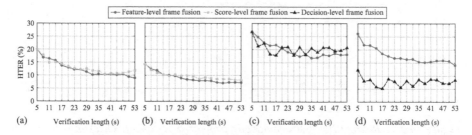

Figure 14.12   *Recognition performance versus verification time in unprotected and protected systems under EO conditions, using S1 and S2 for enrolment and S3 for testing, Q = 15 for AR and MFCC representations (a) unprotected system, feature-level fusion of AR and MFCC representations; (b) unprotected system, decision-level fusion of AR and MFCC representations; (c) protected system, feature-level fusion of AR and MFCC representations; (d) protected system, decision-level fusion of AR and MFCC representations*

a much greater flexibility in selecting the desired operating point, while keeping fixed the guaranteed security. As can be seen, in EC scenarios using $Q = 20$ provides verification performance close to $Q = 15$, while guaranteeing more security. In EO conditions, the gap in verification performance between $Q = 15$ and $Q = 20$ is much wider, therefore suggesting using $Q = 15$ as preferable choice.

Figures 14.11 and 14.12 eventually report the recognition performance achievable, in EC and EO conditions, by the considered unprotected and protected frameworks, for increasing durations of the temporal extension of the recognition phase. The employed model orders are with $Q = 20$ coefficients in EC conditions, and $Q = 15$ coefficients in EO scenarios, with $L = D = 4$ adopted for the protected cryptosystems. As it can be expected, increasing the number of frames employed for verification improves the performance of both unprotected and protected systems. Verification phases lasting less than 20s can be employed to provide HTERs close to 5.0% in both the considered acquisition protocols, for protected cryptosystems.

14.5 Conclusions

In this contribution we have presented several cognitive biometric cryptosystem architectures, performing EEG-based people recognition while providing security for the employed templates. The achievable performance has been evaluated for EEG signals captured in either EC and EO conditions, over a database comprising recordings taken from 50 subjects over three distinct sessions. Recognition rates very close to those of unprotected scenarios, with HTER $= 5.5\%$ and HTER $= 6.9\%$, respectively, in EC and EO conditions, are achieved with the considered secure systems, additionally providing security levels quantifiable in about 20–30 bits.

References

[1] Revett K, Deravi F, Sirlantzis K. Biosignals for user authentication – Towards cognitive biometrics? In: IEEE ICEST; 2010.

[2] Campisi P. Security and privacy in biometrics. Springer, London: Springer Publishing Company, Incorporated; 2013.

[3] Nandakumar K, Jain AK. Biometric template protection: Bridging the performance gap between theory and practice. IEEE Signal Processing Magazine. 2015;32(5):88–100.

[4] Rathgeb C, Uhl A. A survey on biometric cryptosystems and cancelable biometrics. EURASIP Journal on Information Security. 2011;3:1–25.

[5] Patel VM, Ratha NK, Chellappa R. Cancelable biometrics: A review. IEEE Signal Processing Magazine: Special Issue on Biometric Security and Privacy. 2015;32(5):54–65.

[6] Uludag U, Pankanti S, Jain AK. Biometric cryptosystems: Issues and challenges. Proceedings of IEEE. 2004;92(6):948–960.

[7] Maiorana E, Blasi D, Campisi P. Biometric template protection using turbo codes and modulation constellations. In: IEEE WIFS; 2012.

[8] Maiorana E, La Rocca D, Campisi P. Cognitive biometric cryptosystems A case study on EEG. In: IEEE International Conference on Systems, Signals and Image Processing (IWSSIP); 2015.

[9] Campisi P, La Rocca D. Brain waves for automatic biometric-based user recognition. IEEE Transactions on Information Forensics and Security. 2014; 9(5):782–800.

[10]   Almehmadi A, El-Khatib K. The state of the art in electroencephalogram and access control. In: International Conference on Communication and Information Technology (ICCIT); 2013.

[11]   Ruiz-Blondet MV, Jin Z, Laszlo S. CEREBRE: A novel method for very high accuracy event-related potential biometric identification. IEEE Transactions on Information Forensics and Security. 2016;11(7):1618–1629.

[12]   Chen Y, Atnafu AD, Schlattner I, *et al.* A high-security EEG-based login system with RSVP stimuli and dry electrodes. IEEE Transactions on Information Forensics and Security. 2016;11(12):2635–2647.

[13]   Marcel S, Millan JDR. Person authentication using brainwaves (EEG) and maximum a posteriori model adaptation. IEEE Transactions on Pattern Analysis and Machine Intelligence. 2006;29(4):743–748.

[14]   Näpflin M, Wildi M, Sarnthein J. Test-retest reliability of resting EEG spectra validates a statistical signature of persons. Clinical Neurophysiology. 2007;118(11):2519–2524.

[15]   Brigham K, Kumar BVKV. Subject identification from electroencephalogram (EEG) signals during imagined speech. In: Proceedings of the IEEE Fourth International Conference on Biometrics: Theory, Applications and Systems (BTAS'10); 2010.

[16]   Kostilek M, Stastny J. EEG biometric identification: Repeatability and influence of movement-related EEG. In: IEEE International Conference on Applied Electronics; 2012.

[17]   La Rocca D, Campisi P, Scarano G. On the repeatability of EEG features in a biometric recognition framework using a resting state protocol. In: BIOSIGNALS 2013; 2013. p. 419–428.

[18]   Armstrong BC, Ruiz-Blondet MV, Khalifian N, *et al.* Brainprint: Assessing the uniqueness, collectability, and permanence of a novel method for ERP biometrics. Neurocomputing. 2015;166:59–67.

[19]   Wang Y, Najafizadeh L. On the invariance of EEG-based signatures of individuality with application in biometric identification. In: IEEE International Conference of the Engineering in Medicine and Biology Society (EMBC); 2016.

[20]   Das R, Maiorana E, Campisi P. EEG biometrics using visual stimuli: A longitudinal study. IEEE Signal Processing Letters. 2016;23(3):341–345.

[21]   Maiorana E, La Rocca D, Campisi P. Eigenbrains and eigentensorbrains: Parsimonious bases for EEG biometrics. Neurocomputing. 2016;171: 638–648.

[22]   Maiorana E, La Rocca D, Campisi P. On the permanence of EEG signals for biometric recognition. IEEE Transactions on Information Forensics and Security. 2016;11(1):163–175.

[23]   McFarland D, McCane L, David S, *et al.* Spatial filter selection for EEG-based communication. Electroencephalography and Clinical Neurophysiology. 1997;103(3):386–394.

[24]   Blanco S, Garcia H, Quiroga RQ, *et al.* Stationarity of the EEG series. IEEE Engineering in Medicine and Biology Magazine. 1995;14(4):395–399.

[25] Xi Q, Sahakian AV, Ng J, *et al.* Stationarity of surface ECG atrial fibrillatory wave characteristics in the time and frequency domains in clinically stable patients. In: IEEE Computers in Cardiology; 2003.

[26] Juels A, Wattenberg M. A fuzzy commitment scheme. In: ACM Conf. on Computer and Communication Security; 1999.

[27] Kay SM. Modern Spectral Estimation. Theory and Applications. Upper Saddle River, NJ: Prentice-Hall; 1988.

[28] Nguyen P, Tran D, Huang X, *et al.* A proposed feature extraction method for EEG-based person identification. In: International Conference on Artificial Intelligence (ICAI); 2012.

[29] Davis P, Creusere CD, Kroger J. Subject identification based on EEG responses to video stimuli. In: IEEE International Conference on Image Processing (ICIP); 2015.

[30] Malmivuo J, Plonsey R. Bioelectromagnetism: Principles and Applications of Bioelectric and Biomagnetic Fields. New York: Oxford University Press; 1995.

Chapter 15
A multidisciplinary analysis of the implementation of biometric systems and their implications in society

Vassiliki Andronikou[1], Stefanos Xefteris[2], Theodora Varvarigou[1], and Panagiotis Bamidis[2]

With the advent of advanced technologies, including big data, Internet of Things (IoT) and cloud computing, biometric technologies enter in a whole new era. Great limitations regarding their applicability, robustness, effectiveness and efficiency can now be overcome bringing biometric systems much closer to large-scale deployment and application. Meanwhile, the major social issues related to biometrics stem from the irrevocable tight link between an individual's biometric traits and informational trails of this person, either in the form of records or within the biometric traits themselves, which raise serious debate and cause important restrictions in the realization of their full potential. This chapter will present Information and Communications Technology (ICT) developments, such as cloud computing, IoT and big data, which drive the way for the large-scale operation of robust biometric systems, as well as real world examples of biometric solutions. Moreover, it will analyse the privacy and ethical concerns and the societal implications of the introduction of biometrics systems for society as a whole and individually for citizens.

15.1 Introduction

Biometrics, the accompanying processes for their analysis and processing and their applications, have seen an increasing interest over the past decades. The need for deployment and use of large-scale applications requiring reliable authentication and authorization in critical and privacy sensitive fields, from banking and education to public safety and anti-terrorism, have further intensified this interest.

[1]National Technical University of Athens, School of Electrical and Computer Engineering, Greece
[2]Aristotle University of Thessaloniki, Greece, School of Medicine, Greece

The non-duplicability, non-transferability, uniqueness, distinctiveness, permanence, measurability and always-with-owner (and thus, never to lose, get stolen of – at least almost never – or disclosed) properties of biometrics constitute them as an irreplaceable means of identification and verification, but also profiling.

The real-world application of biometrics is highly determined by a series of factors, being:

1. population size,
2. biometric mode; i.e. identification, verification, profiling,
3. checks on the biometric system elements,
4. biometric system purpose in terms of security policy,
5. its organizational integration with the identity management system,
6. the infrastructural cost and the requirements including real-world operation need,
7. users level of familiarity with technology, among others.

The innate strengths of biometrics in terms of identification, verification and profiling and the fact that the biometric samples themselves can be an extremely rich informational source carrying much more information than their initial purpose of capturing, may lead to privacy-invasive misuse. In fact, as analysed in [1], there are three distinct and important types of risk in biometric systems: (i) an adversary may attack and compromise the biometric system, (ii) the latter may suffer itself from (serious) imperfections at hardware and software level leaving it vulnerable to intentional and unintentional security attacks, and (iii) the subject's privacy and overall rights may be encroached by the intentionally and flawlessly implemented capabilities of the system.

Recent advances in cloud computing, Internet of Things (IoT) including devices and processes, image processing techniques and big-data technologies are paving the way towards the realization of real-time, accurate, robust, acceptable and efficient biometric technologies. In fact, these advancements have a major bidirectional impact: biometric technologies have the potential to significantly benefit from cloud, IoT and big-data infrastructures as their enabling technologies which will allow the realization of their full potential while these infrastructures, given their penetration in several public and public sectors and their highly intensive identity management requirements, gradually turn to biometrics for ensuring reliable and robust authentication and authorization paving the way for biometrics to become the governmental and industrial standard. Within this context, industry is also seeking to benefit from such potential and focus their efforts in developing novel products incorporating and exploiting biometric solutions.

Several market analysts and consultants foresee that the biometrics market will impressively grow within the next years at a compound annual growth rate of more than 20 per cent from 2015 to 2020, with estimations of its expected size reaching $30 billion by 2021 [2] or even $41.5 billion by 2020 [3]. Conservative estimations regarding IoT connections of biometric sensors will reach 500 million by 2018 and 26 billion by 2020 [4]. Meanwhile, according to Gartner [5] in 2019 more than 57 per cent of smartphones available will support fingerprint and other biometrics contributing to a balance between security and user experience in daily activities while their embedding in a wide range of consumer devices, also being part of connected homes, will boost convenience [5]. In fact, what is expected is that by 2020 use of passwords will become obsolete and will be replaced by new biometric methods across 80 per cent of the market [6].

This gradual, seemingly irreversible shift to the biometrically enabled security and identity management overall in governmental and commercial applications involves profound threats to commonly accepted notions of privacy and security at individual and societal level. Going beyond the security and privacy risks of traditional means of authentication and authorization, the innate binding of biometrics with the individual's biological, psychological and social traits raises several concerns and strong debate is held regarding its serious ethical, legal, societal and cultural implications.

The rest of the chapter is organized as follows. Section 15.2 presents the overall technological landscape which promises to set the ground for the large-scale operation of biometric systems with high reliability and robustness. In Section 15.3, several biometric applications while Section 15.4 describes how the ICT advances can boost biometric systems, including currently running real world systems. Section 15.5 includes a thorough analysis of societal implications related to biometric systems, with particular focus on social media and secondary uses.

15.2 The ICT landscape for biometric systems

As already mentioned, within the past years, ICT advancements are extremely rapid and multidirectional driving economic growth, highly impacting and bringing disruptive change in daily life, social structure, business and governmental operations. Technological achievements are quite often highly interconnected, constituting a cornerstone for advancements in several other domains from medicine and astronomy to border security and financial transactions. Enabling technologies such as cloud, IoT and big data are driving the way for further boosting the advancements and their potential in reshaping the world within we live and work.

15.2.1 Internet of Things

The IoT refers to the use of sensors, actuators and data communications technology built into physical objects – from roadways to pacemakers – that enable those objects to be identified, tracked, coordinated, controlled and, eventually, used for several applications across a data network or the Internet. It is based on smart, self-configuring, widely distributed things which are interconnected in a dynamic network infrastructure. The underlying technologies of the IoT infrastructures vary from simple identification tags to complex sensors and actuators.

The core idea behind IoT is linking the physical and the digital worlds, leading to the pervasive presence of things with the ability to monitor, measure, infer and, in some cases, interact with, and potentially adapt the environment around people. The IoT not only orchestrates, interlinks and, eventually, uses things, from smart phones and cameras to smart clothing and traffic lights, for a common goal but also repurposes, quite often inexpensive, sensors and actuators allowing to serve different objectives and enable various applications in almost all domains of life.

Although still in its rather early stages of adoption, IoT has already a wide range of applications and deployments and is an important enabler for a transformative shift for everyday life, economy and the society. According to McKinsey Global Institute [7], the IoT has a total potential economic impact of $3.9 trillion to $11.1 trillion a year by 2025. In the biometric domain, in particular, ICT advancements in devices, wearables technology, signal processing and pattern matching techniques are only a small fragment of the IoT-driven impact in the potential of biometrics.

The major IoT contribution to the boosting of the scalability, performance and reliability of the biometric solutions, their potential wide adoption and the gradual unfolding of their full potential relies on **connectivity** and **mobility**. These two key aspects of IoT in fact allow for the full realization of one of the core IoT promises: *collecting live information from anywhere at any time*. Going a step back for the main prerequisites for connectivity and mobility, one of the strongest requirements posed is **interoperability**. In fact, according to McKinsey, '*of the total potential economic value the IoT enables, interoperability is required for 40 percent on average and for nearly 60 percent in some settings*' [7]. Lack of interoperability has raised lots of debate and worldwide efforts to harmonize existing and rising standards in the IoT stack, from devices and communication networks to IoT platforms overall but also goes beyond the pure technological domain and also covers relevant organizational and regulatory obstacles. Overcoming the fragmented development of IoT elements and solutions constitutes a major task which, however, promises that connectivity among people, processes, services and things will actually work no matter what technology and process is used. In order to capture to true impact of an interoperable IoT landscape estimates show that in 2025 remote monitoring could create as much as $1.1 trillion a year in value by improving the health of chronic-disease patients [7].

15.2.2   Cloud infrastructures for biometrics

Over the past years, there has been intensified interest around cloud technology. According to NIST [8] '*Cloud computing is a model for enabling ubiquitous, convenient, on-demand network access to a shared pool of configurable computing resources (e.g., networks, servers, storage, applications, and services) that can be rapidly provisioned and released with minimal management effort or service provider interaction*'. As the cloud provides a highly productive, flexible and simple to the user way of accessing powerful computational resources, it actually unlocks the potential of applications in all domains, including financial transactions, entertainment platforms, complicated business analytics, bioinformatics, etc.

Biometrics, in particular, can take advantage of cloud computing by leveraging its unbounded computational resources and important properties of flexibility, reliability and scalability in order to enhance their performance, expand their scope and boost their potential [9]. In fact, given that biometrics encapsulate several computational and data-intensive processes, from biometric sample processing and feature extraction to biometric matching, which have strong security and trust requirements, their deployment and use of cloud infrastructures seems like a one-way solution for their reliable large-scale operation. And going even a step further, by leveraging the everything as a Service paradigm being supported and further boosted by the cloud, several software elements and open source solutions of potential value for biometric solutions provide a valuable pool of resources to stakeholders developing such solutions, such as image processing tools, clustering mechanisms and matching techniques. Towards this direction, several research and commercial efforts are taking place worldwide which mitigate biometric solutions to the cloud and further develop them in order to exploit the full potential of its computational and informational resources. Hence, in [10], the development of an iris recognition system was based on Hadoop [11].

The proliferation of mobile devices, with more than 1.1 billion people currently using smartphones and tablets, and their intensive daily use in all contexts of human life has been taken into consideration by several development efforts in the cloud and the ICT field in general. Mobile computing has, in fact, introduced a whole new way of perception and interaction with the world and among the society. By linking people, services and infrastructures worldwide, it has brought individuals much closer to technology and computing as no other technological advancement has so far. Given that mobile devices currently are in most cases equipped with biometric sensors, such as cameras, microphones, fingerprint sensors, and applications which monitor the individual's physical and online activities, they are gradually becoming an important element also for biometrics systems. In fact, in combination with cloud technologies, mobile biometrics constitute a new era in the biometrics domain. In [12], Bommagani *et al.* showcase how they perform computationally intensive biometric recognition on a mobile device by offloading the actual recognition process to the cloud. MOCHA [13] is another example of an architecture which aims at coupling mobile computing with cloud infrastructure for performing face recognition with reduced latencies.

15.3   The new implementation potential

Biometrics technologies are constantly being augmented and upgraded with new features and capabilities both at software and hardware level. The breakthrough in service provision that cloud technologies and Service Oriented Architectures (SoA) brought is amalgamated with a plethora of new sensors, services, methods and algorithms, making biometrics systems adaptable, robust, modular, multimodal and more efficiently deployed as large-scale projects. Biometrics technologies are now being implemented as multimodal systems that are faster, accurate and more secure.

Through this boom in new technologies and fusion of modalities, biometric applications now encompass a range of application ranging from the 'traditional' localized or decentralized security applications to systems with widely different application fields such as Ambient Assisted Living systems [14–16].

As analysed in the previous paragraphs, the advent of cloud technologies ushered the shift to a novel paradigm: we now have the ability for on-demand and ubiquitous access to vast computing resources. Data storage, applications, servers and services are being rapidly provisioned on the fly with minimum effort or interaction with the provider. We can now use infrastructures, hardware and applications as services, while at the same time keep up a cost effective and secure framework that is easily deployed – or more accurately, designed to be – in large-scale schemes.

One of the important features that many biometrics implementations lacked in the previous years, was that of scalability. Cloud technologies and Service-oriented Architectures formulate a robust infrastructure for biometrics applications, ensuring ultra-scalability, offering a vast pool of distributed resources and storage as well as efficient processing. Smartphones and mobile devices in general act as access points/clients. This way, cloud and SoA bring forth the next generation of biometrics while upgrading and updating the application pool, broadening the spectrum of applications, boosting security and being cost-effective for the providers.

From the provider point of view, cloud based biometric technologies offer a variety of deployment possibilities, such as ambient intelligence applications, access control schemes, mobile platform applications, smart spaces, etc. At this point, locally deployed biometric systems are being more widely used and cloud based biometric solutions are still new.

Although this is the current state, there are existing solutions in the market, offered by various companies that lever the shift towards cloud based biometric solutions. Cloud implementations, Service Oriented Applications (SoAP) and RESTFul architectures are trending to become the current paradigm.

BioID [17] is a company that develops multi-modal biometric authentication solutions based on a cloud infrastructure of Microsoft Azure™ platform. The BioID web service is a face and voice biometric authentication service that offers SoAP and web API's for web, cloud and mobile applications. BioID integration is flexible and offers the choice of SoAP or RESTFul APIs.

HYPR [18] is a company that offers a range of multimodal biometric solutions that span software, firmware and hardware products. They make use of biometric tokenization to decentralize sensitive data storage and enhance security, integrate face recognition into the user's applications and embed voice recognition directly into the user's mobile devices. HYPR implementations extend biometric authentication even to the firmware level and thus enable the turn from smart IoT 'things' to biometric 'things'. HYPR also offer their own fingerprint sensor as a biometric token scanner. The integration of HYPR products enables password-less security through fingerprint, voice, face and eye recognition.

Security applications nowadays are heavily based on biometric systems. Advances in hardware and software enable sample enrolment at higher quality, more efficiently and more securely. New algorithms demonstrated ever lowering error rates for the various biometric modalities, such as face and iris recognition and fingerprinting. Fingerprint recognition algorithms have been vastly improved in the past decade. We now have the ability for high resolution fingerprint images for 'level 3' feature identification that can be improve performance even further. Application of novel techniques such as 3D face recognition already being implemented is overcoming some of the inherent problems in modern face recognition processes, namely variations in lighting, pose and facial expressions. Multimodal fusion, large scale and distributed deployment, semantic interoperability of biometric databases and real-time processing bring forth a completely new paradigm.

15.4 Evolution of biometric systems

In Figure 15.1, we showcase how advances in technology have augmented the evolution of biometric systems and extended their range of application in other fields too.

Early biometrics systems where based usually in one or fewer modalities based on a restricted pool of available algorithms.

Rapidly increasing citizens' mobility across countries, international security breaches and financial fraud posed – and still pose – a growing demand for large-scale efficient and effective identity management. Based on their innate characteristics, biometrics from their inception showed great potential for the deployment of large-scale solution. Nevertheless, performance and reliability of biometric systems failed to meet the requirements posed by their large-scale deployment; acceptable levels of accuracy, high demands on throughput, real-time service provision, privacy (both from the legal and ethical perspective), wide population coverage and scalability, among others.

In the effort to overcome the limitations posed by unimodal systems (poor population coverage high failure-to-enrol rates, etc.) the migration to multimodal biometric systems brought forth the ability of fusing information captured from different sources (multiple samples, various sensors, etc.). This brought forth the demand for improved algorithms and techniques for data pre- and post-processing that would overcome or take into account the technical caveats of the capturing sensors and broaden the system's flexibility and adaptability. Biometric applications had to move from deployment of a usually static set of sensors and algorithms with unconditioned usage among a spectrum of circumstances limiting performance to a more flexible paradigm of adaptive systems that increased reliability and performance.

Based on the key principles of Service Oriented Architectures, novel process flows were proposed, allowing for adaptability and scalability of biometric systems in large-scale scenarios [19], such as the metadata-aware architecture in Figure 15.2, that chooses and applies the most appropriate algorithm for a specific modality, taking into account the specific per-case difficulties and limitations for the collection of samples.

Such developments enabled also the expansion of biometric systems from unilateral application scenarios to a broader pool of use cases – making biometric systems business case driven.

15.4.1 Enabling advanced privacy

Recent advances in both software and hardware have also had a significant impact on the security and privacy aspects of biometric systems. The balance between sensitive private data availability and their protection from unwanted access is a delicate one

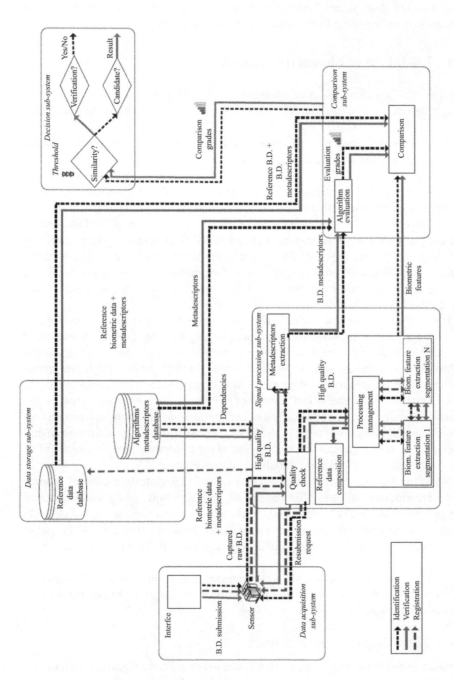

Figure 15.1 Evolution of biometric systems due to technology enablers

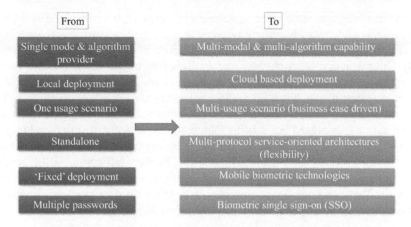

Figure 15.2   Metadata aware architecture of biometric systems

and numerous technologies are being developed, ensuring that such data is stored and accessed under a secure protection scheme, being also indecipherable in case of data loss.

The most used and ever-evolving examples of these technologies are cancellable biometrics and biometric cryptosystems [20]. The cancellable biometrics paradigm is a means of protecting biometric templates, alongside biometric cryptosystems. In order to protect user-specific sensitive data, the engineer employs a scheme of systematically repeatable distortion through transformation functions and making comparisons in the transformation domain, this making all data easily revocable [21]. The main objectives of a cancellable biometrics scheme are:

- *Diversity*: Different cancellable features are used for various applications so that the same biometric feature results in a large number of protected templates.
- *Reusability/Revocability*: Fast and easy revocation and reissue if the system is compromised.
- *Non-invertibility*: Transformations are non-invertible to heighten security.
- *Performance*: Transformations do not deteriorate the systems recognition performance.

Cancellable biometrics schemes ensure that inversion of transformed biometrics templates is not feasible by potential imposters, making use of ever evolving algorithms [22,23]. Rathgeb *et al.* [20] summarize the key characteristics of cancellable biometrics as follows:

> *In contrast to templates protected by standard encryption algorithms, transformed templates are never decrypted since the comparison of biometric templates is performed in transformed space which is the very essence of CB.*

Meanwhile, biometric cryptosystems are closely coupled with cloud based implementations of advanced biometric systems. In biometric cryptosystems, matching of templates is accomplished in the encrypted domain and stored data are never decrypted [24]. Biometric cryptosystems are mainly classified in two categories, key-binding or key-generation cryptosystems. In cryptosystems, we employ 'helper data', i.e. biometric-dependent public information to generate and retrieve encryption keys. The variance of biometric features makes them unsuitable for direct key extraction. Thus, helper data are employed in a scheme that retains significant information hidden but from which key reconstruction is easier. In a biometric cryptosystem, comparisons of biometric features are performed indirectly, by the verification of the generated keys' validity – the output of the authentication process is either a new key or a fail message. This exact verification process is in other words a biometric comparison but in an encrypted domain. Thus, template protection is ensured [20,25].

15.4.2   Ubiquitous biometrics

Ubiquitous biometrics is a novel notion in the field, referring to the identification of an individual anytime and anywhere by processing both biometric and non-biometric information. In a ubiquitous identification scheme 'traits' that are loosely connected to the person's identity will also be used to enhance identification. Social media profile and real life social interactions will be used in conjunction with the individual's location, on top of soft (behavioural) and hard biometric traits. Ubiquitous biometrics have the potential to become the basis of enhancing the trend for on-the-fly personalized services, but there are significant caveats in the design and deployment of such systems: the first obvious issue is that of privacy and ethical concerns. A rigid legal framework and ethics guidelines must be adapted to prevent abuse of the individual's biometric and behavioural data and technological safeguards must be in the core of ubiquitous biometric systems. Storage and use of enough data to clearly define and infer an individual's personality need to be heavily moderated and governed by as strict as possible legal frameworks. A second issue would be that of contextual application of a ubiquitous biometric system. A book purchase in a bookstore should be contextually defined as a situation that doesn't require a rigid identification framework, versus, for example, access in a secure environment or a high cost financial transaction [26].

15.4.3   Indian Unique ID project: a large-scale
implementation example

The Indian Unique ID (UID) [27] is an extremely large-scale biometric system that attempts to identify India's 1.2 billion people. The Indian UID [28] is the product of the Indian electronic government program and is intended for application in security and e-government applications such as tax payment, banking, welfare and citizen identification. In the near future, the UID Authority of India will issue a unique ID for each Indian citizen. Each citizen will be allocated a unique ID that will be resistant to forgeries and duplications, by linking collected biometric data into a central ID. Here, biometric modalities play a vital role: the system will collect features from the face, fingerprints of ten fingers and the irises of both eyes and check against duplications in previously recorded data. UID began in 2009 and the registration/matching of 200 million citizens was completed by the end of 2012. By the end of 2014, 600 million citizens were registered, almost half the population of India.

The largest issue in such a globally unprecedented large-scale project was the check for duplicate biometrics entries. Two steps were necessary for this check: essential management of the citizens' biometric information and the registration/matching based on biometric authentication. Considering the size of India's population, these two tasks were of a globally unprecedented scale.

The technical issues that needed to be addressed before the deployment of UID were:

1. **The accuracy of multimodal authentication**
   After solving the organizational issue of collecting biometric data from individuals in such a vast land and on such a population size, arose the issue of sample collection accuracy and data processing. High authentication accuracy presupposes high quality biometric information. Biometric information can be compromised depending on factors such as the individual's profession or ethnic/cultural background. For example in rural areas, most 'blue collar' workers had damaged fingers, so fingerprinting with traditional methods was difficult. Moreover, facial feature extraction was not always easy, since the traditional beard most men wear degrades biometric feature quality. So, matching algorithms had to be developed in order to face the specific degradation factors inherent in the collection process. More specifically image processing and face recognition algorithms adapted to the Indian people's facial and cultural characteristics (turbans, beards, etc.) where developed, fingerprint recognition algorithms took also into account age and deterioration through damage, etc.

2. **Optimization of multimodal processing algorithms**
   Processing huge amounts of information is a task that usually begins with design and development of some kind of search index that speeds up searching of – in our case – individuals. The use of biometrics induces a deterioration factor that may hinder searching, since biometric information collected at a time point a, can be quite different from biometric information collected later, due for example to aging. In the Indian UID, we have strict duplication checking without the ability to include demographic data to narrow down the check, so all that can be done is to check for duplicates by matching personal data to all of the registered data. To put things into scale, let's assume a sample population of 1.2 billion people and the simple operation of checking if a person that requests enrolment is already registered. The amount of comparisons needed rises to approximately 7.2×10^{17} operations. This number is calculated if we multiply the number of necessary searches for 1.2×10^9 subjects in a 1-to-N search scheme (approximately 1.44×10^{17}) times 5 – which is the number of different biometric feature streams (face, left eye, right eye, left hand, right hand) [27]. To enable operations on the initial registered population of 600 million, the biometric system should be

capable of processing data at the rate of 1.9 billion matches per second, without interruption.

So, the answer to this issue is not to use all the available faces or fingerprints or irises, but to use them optimally, using the minimum required number. The scheme employed made use of the false acceptance rate as a security index and false rejection rate as a usability index. The deployed scheme adjusts these to meet the required authentication accuracy, so that the number of operations/core/sec can be improved. Full analysis of the scheme surpasses the scope of this chapter, but it would be enough to note that with a multi-stage matching scheme deployed the system optimizes processing capability while maintaining a high authentication accuracy [29].

3. **System scalability**

    Obviously, even if individual operations are optimized, for such a large deployment scheme, system scalability poses great issue, since we need efficient response to matching requests. A distributed processing configuration is needed, in order to address this issue. Figure 15.3 showcases the authentication scheme deployed in this case:

    The process begins by an incoming matching request to the system.

    The system then creates the plan of the registered biometric information distribution and another plan for the execution of the comparison on the various servers, depending on the current load. Then it distributes the matching requests to the required servers, collects results from them and outputs the system's final response. We should note here that in such a scale, even the most minimal overhead delays will result in reduced efficiency or hamper the attainment of the scalability limit. This means that even if we add more matching servers, there is always a cut off threshold over which transaction throughput cannot be increased.

    The biggest technological issue to be resolved here was ensuring that the matching servers were provided with scale transparency and scale-out capability. Here, the whole does not surpass the sum of its parts as Aristotle would have it. This becomes more apparent as we add more servers and watch the matching improvement rate degrade. Thus, there is a threshold in server number over which the processing capability remains stagnant and we have zero improvement.

    The solution came by:

    (i)   Making the servers batch process the internal job transports and
    (ii) Managing transactions under a multiphase scheme

    Managing transactions using the matching servers induces a severe overhead if we consider the underlying scale, when everything (execution plans, placement plans, etc.) is individually aggregated. To ameliorate this, India UID servers manage transactions within a certain period with a single execution plan, thus optimizing CPU usage and internal communications.

4. **System design that ensures high availability of resources**

    The UID project comprises an extremely large number of servers. In such a paradigm obviously the fault rate can be high, especially when the servers in

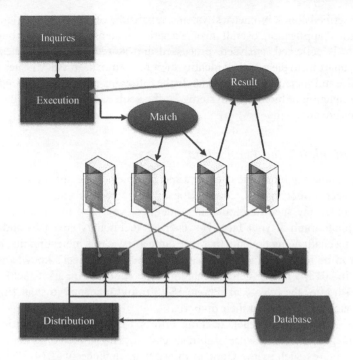

Figure 15.3   Distributed processing configuration

charge of matching are overused. Since this system is of national importance and the backbone for India's e-government plan, it therefore encounters the following major issues before deployment:

(i)   Requirement for zero loss of biometric information even in case of system failure
(ii) Undisrupted operation of the system in case of local failures
(iii) Grid and cloud based deployment of data centres as a failsafe against data loss and load distribution

15.5   The societal implications

The tremendous boosting of the capabilities and the rising of new systems and applications for making operations, transactions, travelling and everyday life both safer and more convenient stemming from the realization and operation of highly scalable, robust biometric solutions exploiting the technological wealth being continuously developed and advanced is only one part of the coin. The innate features of uniqueness, information richness, linkability, traceability and invasiveness that biometrics encapsulate create profound threats to commonly accepted notions of privacy and security.

As described above, biometrics systems, regardless of their goal, implementation and domain of application, overall involve a series of steps and phases through which the sample is collected, analysed, processed and stored. In each of these phases and steps, apart from the practical identity theft related risks, privacy issues affecting personal life and society as a whole exist. In the following paragraphs the implications of the introduction of biometrics systems for the society as a whole and individually for citizens are analysed.

15.5.1   Biometric data collection

In general, gathering biometric data for a specific purpose requires the participation of the subject, either through cooperation or coercion: for your face photo to be captured under any specific biometric guidelines, you have to let someone take a close-up photograph of your face. However, as technology advances and 'identity hacking' is a challenging domain, there are quite a few cases indicating that biometric samples can be retrieved and/or replicated without the subject's knowledge. As an example, in 2013, a hacker demonstrated how he spoofed Apple's TouchID sensors in only 24 h after the release of iPhone 5S [30] and managed to steal fingerprints without touching a person or their property.

Going a few steps further, hacking efforts for recreating biometric samples through images have been successful in several occasions, including demonstration on public figures such as the German minister of defence in 2014 [31] and German chancellor in 2015 [32]. In the first example, a few high-resolution images of the subject, collected via a standard digital camera from a 3 m distance, and using off-the-shelf software were enough for the reverse engineering of the subject's fingerprints. Till then research held on fingerprints spoofing [32] has shown that if the focus and lighting is proper, fingerprints can be recreated from images shot up to 3 m from the subject, indicating that peace signs actually expose fingerprints potentially allowing their spoofing. In the second example, the subject's high resolution photo from a press conference was used with the hacker stating that these images could be printed onto a contact lens.

In an effort to combat image and voice replay attacks, liveness detection techniques have been developed and incorporated in some biometrics solutions which aim at ensuring that the biometric traits are submitted to the biometric sensors by a live subject instead of a photo, recording or artificial biometric samples (such as gummy fingerprints). Still biometric scanners are not fully resistant to fake biometric forgeries with more sophisticated attacks being developed and applied which exploit material-specific noise, 3D face models mimicking movement, among others [33–35]. More details for approaches which aim at protecting biometric data collection can be found in other chapters of this book. In particular, the actual state of the art in inverse biometrics and privacy is discussed by Marta Gomez-Barrero and Javier Galbally, whereas techniques for protecting biometric templates are presented by Christian Rathgeb and Christoph Busch in 'Biometric Template Protection: State-of-the-Art, Issues, and Challenges'. Moreover, Slobodan Ribarić and Nikola Pavešić discuss de-identification approaches.

15.5.2   Biometrics in social media

In the new era of social media, individuals participating in the various social networks appearing in the Internet share traces of their biometrics in the form of images. This sharing either public or for a private group allows for other individuals and, consequently, ICT tools to have access to a wealth of biometrics-bearing pieces of information. In fact, researchers recently demonstrated that by leveraging a handful of pictures of an individual extracted from social media, facial reconstruction can be performed by building realistic, textured, 3D facial models which can trick liveness detectors into believing that the 3D model is a real human face [36].

Social media statistics reported clearly show that this risk will only be elevated as time goes. As an example, at present (Feb 2017), there are over 1.86 billion monthly active Facebook users, photo uploads are 300 million per day on average [37], the total number of uploaded photos reach 400 billion [38] while location-tagged photos are already 17 million [38]. In Instagram there are 75 million daily active users with nearly 85 million videos and photos people being uploaded on a daily basis [39].

Such large and – at least partly – openly accessible databases of identities with biometric traces pose three serious threats:

1.   **construction of biometric samples** *without the individual's consent or even knowledge* through the retrieval of published biometric traits and re-engineering of the biometric sample,
2.   **re-identification of individuals** in other online and offline sites and sources and, consequently, **linkability** of information and
3.   extraction of **soft biometrics** [40,41]

The first threat is also indicated in the example provided above in which faces can be reconstructed from online images and used for attacking face recognition systems even with liveness detection. And it also goes beyond the facial features, including biometric traits such as iris and fingerprints [30,31], which can be captured through an image or video given the necessary resolution. Having a public pool of such information, in fact, gives access to any impostor with the necessary ICT tools for feature extraction to process and extract the biometric traits which could be then used for accessing systems and applications which are using them as a basis for authentication and authorization.

This *stealing of identity* can have serious implications. Biometrics, unlike other means of identification such as passwords, *cannot be replaced* by the subject and thus constitute a permanent identification means which, if compromised, cannot be replaced. Hence, when an impostor replicates them, deprives the victim of using them unless the systems they have access to manage to include new methods of processing or additional means of identification which will be able to detect and reject artificial biometric samples. Such solutions include multimodal biometrics [41], which include more than one biometric traits as the basis of their identification, combination of biometric traits and traditional identification means (such as passwords or personal question answering) or challenge-response techniques [42]. In the latter case, voluntary (behavioural) or involuntary (reflex reactions) responses to external stimuli are requested. Hence, the user is asked to turn their head in specified random directions, smile or blink and perform motion analysis for verifying the response or human reflexive behaviour, such pupil contraction after a lighting event, is used as a basis [43]. Still, quite often such techniques suffer from the innate failure rates of the mechanisms applied (e.g., motion detection, pattern extraction, etc.) as well as limited user acceptance (the user might feel uncomfortable or unwilling to respond in a challenge such as grimacing at their mobile phone).

The *linkability potential* offered due to the widespread sharing of images is another important aspect. By extracting the biometric features from an image in one source (e.g., a social network, an online work profile, a blog or a personal web-page), an impostor can relate this person to other sources of their information with little or no overlap with the original one, as long as the biometric features match. For example, a person may have a Facebook profile with a photo of theirs and share no other identification-related information but post some daily activities or comments. This person can be matched to their Instagram account in which they may be shar-ing photos of their vacation and their Pinterest account at which they have uploaded many images of their new house, their LinkedIn account in which they present their educational and work related profile and a number of blogs in which they may be posting questions about new parents or symptoms of hypothyroidism. An impostor able to apply face recognition mechanisms, perform search by image and data mining, could in fact build a profile of this person based on the biometric traces the latter is leaving across the Internet. The information included in this profile is quite variable, including potentially information about their hobbies, daily activities, job, health, social connections and even location-tagging in real time.

It was only recently that an application called FindFace, initially launched as a dating application, had been using advanced facial recognition techniques to link photos of people to their social media profiles with a photo taken during daylight with an average smartphone being more than enough [44]. With its users reach-ing 500,000 in just the first 2 months of its operation, several concerns have been raised in fear of its use as a stalking tool as well as a means of finding informa-tion about people of whom their photos were taken sneakily in public places putting an end in anonymity. In fact, it was not long after its deployment that the applica-tion was abused by allowing users to unmask and harass local adult movie stars in Russia [45].

The *extraction of soft biometrics* is a rather tricky aspect which does not relate to directly identifying but rather to *profiling* the subject. Soft biometrics refer to measurable human characteristics that include information about the individual with no high inter-person variability, due to their lack in distinctiveness and permanence [40], such as gender, ethnicity and eye colour. Several techniques have been developed for the image-based extraction of gender [46–49], age [50–52], ethnicity [53–57] among others. Soft biometrics are in fact a proof that additional information about an individual apart from identification traits can be deduced from their biometric sample, in this case an image or video. When extracted without the person's knowledge and consent, this constitutes privacy leakage and given the type of information soft biometrics may refer to, such as race and ethnicity, ethical issues also rise in fear of biometric technologies assisting profiling practices which take race or ethnic origin into account.

Another important aspect which applies heavily on social media as a rich image-based source of biometric traits is ***picture-to-identity linking***, which aims at finding, and thus, linking together, the social network accounts which may belong to a specific person, who has shot a given photo. In [58] researchers have worked towards this direction by exploiting sensor pattern noise, that is, a unique 'fingerprint' left on a picture by the source camera sensor. The main answer they aimed at answering was: 'Given a picture P∗; how can I find an account belonging to the person who shot P∗?'. Although such a technological approach can have several beneficial applications, such as finding the digital traces of a stolen smartphone, combating cyber-bullying or identifying perpetrators of on-line child abuse, it still conceals some serious dangers. The first one goes back to the **linking of the online traces of a person across social media** and the Web as mentioned above without their knowledge and/or consent. The second one goes a step further and includes the linking of different individuals on the basis of the common device used for capturing their images. In other words, ***non-declared direct** (i.e. the two people have at least met) **or indirect** (i.e. the two people have been collocated at least once or share at least one common acquaintance) **social relations can be inferred*** on the assumption that some level of acquaintance must exist in order for two or more people to have their image or video captured through the same image-capturing device. This, in turn, builds a whole new set of information regarding the subject's social cycle which can be exploited in several ways, from targeted advertising to phishing and baiting attempts.

15.5.3 Other secondary uses of biometric data

As already mentioned above, the informational wealth that biometric samples encapsulate and the identification and linkability potential of biometric features set the ground for a series of threats to privacy and several ethical concerns. An important aspect lies in the secondary uses of biometric data. A privacy-preserving biometric system should operate on the *principle of purpose and proportionality*. This means that the purpose for which the biometric data are collected and processed should be clearly defined and a balance should exist between the purpose to be achieved and the risks related to the protection of individuals' fundamental rights and freedoms.

Given the innate tight link of biometrics to the subject's identity, activities and overall life as well as the multiple mapping potential of the biometric traits to much more than subject identification, when the original purpose for obtaining the biometric information is extended to include purposes other than the one originally stated without the informed and voluntary consent of the participants, the risk of serious privacy invasion is extremely high and is highly dependent, although not restricted by, the type of biometric information stored and processed, as illustrated in the earlier sections of this chapter.

When the extracted biometric features are stored rather than the captured biometric sample as a whole, as this is the common case for privacy and security reasons, the main threat related to secondary use of the biometric data lies in its *repurposing*, i.e. using the data for another purpose than the initially agreed upon and expected by the subject. In this case, data may be misused for several privacy-threating reasons. Assessment of the subject's physiological and psychological health condition or health-impacting lifestyle choices constitutes a major one. In [59], researchers show that they are able to detect health problems from gait analysis and in [60] health monitoring for chronic diseases which is based on gait analysis captured through mobile phones is presented. Other examples for medical analysis of biometric data are discussed by Marcos Faundez-Zanuy and Jiri Mekyska in their chapter 'Privacy of Online Handwriting Biometrics Related to Biomedical Analysis' of this book.

When an impostor has access to the point of biometric sample capturing, things can get even more complicated. Especially, given that biometrics are inherent to the individual and are tightly linked to their physiological and behavioural characteristics, researchers can often link medical predispositions, behavioural types, or other characteristics to particular biometric patterns. In [61], researchers announced how they manage to extract 'lifestyle intelligence' from a fingerprint. In fact, they indicate that simple analysis of the fingerprint can reveal information such as whether the subject has taken a particular type of drug, is a smoker or is using an aftershave. And going a step further, if the impostor manages to retrieve the biometric sample either by attacking and/or mediating in the collection process or by accessing stored samples, then a wealth of information may be revealed, regarding emotions [15,16,62,63] and health status. In [64], facial images are used as an alternative to blood tests for detecting health status, in [65], a new technique is introduced for assessing alcohol consumption based on pupillary reflex analysis, while in [66], skin images captured by a conventional digital camera of a mobile phone are used for performing skin status examination.

Although in all the above cases, different legitimate uses are considered as non-intrusive alternatives to traditional approaches, their relying on simple biometric samples easily collected without requiring the subject's consent and/or knowledge raises significant privacy and ethical concerns. Insurance companies could consider the extracted information as important aspects to consider when negotiating the terms of a health insurance contract, the subject's job seeking and career progress could be affected, and, even, social relationships could be disturbed.

15.5.4 Other concerns

Several other risks arise regarding physical integrity as well as individual perceptions of privacy. Physical or bodily integrity refers to the protection of physical selves against invasive procedures and emphasizes the importance of the human beings' self-determination over their own bodies [67]. Attacks at the point of biometrics' sample collection for either forcing the subject to enter their data or depriving the subject from the biometric trait (e.g., removing a finger or an ear) may prove to be significantly harmful for the subject's physical condition and psychology. Moreover, there are cases that objections are raised regarding hygiene with particular concern about the cleanness of the sensors. Given also the several risks accompanying the application of biometric systems along with their innate tight linking with the subject itself, religious objections also encountered [68], while as segments of the society associate some forms of biometrics, such as fingerprints, with law enforcement, the fear of social stigma further intensifies the perceived concerns over their application.

15.6 Conclusions

Recent and ongoing advancements in several ICT fields, such as IoT, big data, cloud infrastructures and computer vision, set a new promising ground for biometric technologies to flourish and realize their full potential in large-scale deployments and with complicated, highly impactful applications across the globe. The innate tight linking of biometrics with the subject itself raise several important privacy and ethical concerns that need to be dealt with at technological, societal and governmental level. The bursting of social media as a global infrastructure and network with an unprecedented wealth of biometric information further intensifies these concerns and prompts for the need of thorough analysis of the risks and possible measures to be taken for the benefit of individuals and society as a whole.

References

[1] Andronikou, V., Yannopoulos, A., Varvarigou, T.: 'Biometric profiling: Opportunities and risks', in 'Profiling the European Citizen' (Dordrecht: Springer, 2008), pp. 131–145.

[2] ABI Research: 'ABI research predicts fingerprint sensors, facial recognition, and biometric surveillance to propel the global biometrics industry to $30 billion by 2021', 2016. http://www.abiresearch.com/press/abi-research-predicts-fingerprint-sensors-facial-r/, accessed February 2017.

[3] News: 'Market researchers predict growth rates for biometrics' *Biometric Technology Today*, 2016, **2016**, (2), pp. 3–12.

[4] Rawlson, K.: '500 million biometric sensors projected for Internet of Things by 2018', 2014. http://www.biometricupdate.com/201401/500-million-biometric-sensors-projected-for-internet-of-things-by-2018, accessed February 2017.

[5] Lu, C.K.: 'Report highlight for market trends: Biometric technologies create a personalized experience and convenience in the connected home', 2016. https://www.gartner.com/doc/3184125/report-highlight-market-trends-biometric, accessed February 2017.

[6] Rivera, J., Van Der Meulen, R.: 'Gartner says the Internet of Things will drive device and user relationship requirements in 20 percent of new IAM implementations by 2016', 2014. http://www.gartner.com/newsroom/id/2944719, accessed February 2017.

[7] Manyika, J., Chui, M., Bisson, P., *et al.*: 'Unlocking the potential of the Internet of Things', 2015. http://www.mckinsey.com/business-functions/digital-mckinsey/our-insights/the-internet-of-things-the-value-of-digitizing-the-physical-world, accessed February 2017.

[8]   Mell, P., Grance, T.: 'Sp 800-145. The NIST definition of cloud computing' (National Institute of Standards & Technology, 2011).

[9]   Albahdal, A., Boult, T.: 'Problems and promises of using the cloud and biometrics', in 'Information Technology: New Generations (ITNG), 2014 11th International Conference on' (IEEE, 2014), pp. 293–300.

[10] Raghava, N.: 'Iris recognition on Hadoop: A biometrics system implementation on cloud computing', in 'Cloud Computing and Intelligence Systems (CCIS), 2011 IEEE International Conference on' (IEEE, 2011), pp. 482–485.

[11] Shvachko, K., Kuang, H., Radia, S.: 'The Hadoop distributed file system', in 'Mass Storage Systems and Technologies (MSST), 2010 IEEE 26th Symposium on' (IEEE, 2010), pp. 1–10.

[12] Bommagani, A., Valenti, M.: 'A framework for secure cloud-empowered mobile biometrics', in IEEE (Ed.): 'Military Communications Conference (MILCOM)' (2014), pp. 255–261.

[13] Soyata, T., Muraleedharan, R., Funai, C.: 'Cloud-vision: Real-time face recognition using a mobile-cloudlet-cloud acceleration architecture', in IEEE (Ed.): 'Computers and Communications (ISCC), 2012 IEEE Symposium on' (2012), pp. 59–66.

[14] Xefteris, S., Andronikou, V., Tserpes, K., Varvarigou, T.: 'Case-based approach using behavioural biometrics aimed at assisted living' *Journal of Ambient Intelligence and Humanized Computing*, 2010, **2**, (2), pp. 73–80.

[15] Xefteris, S., Doulamis, N., Andronikou, V., Varvarigou, T.: 'Behavioral biometrics in assisted living: A methodology for emotion recognition' *Engineering, Technology & Applied Science Research*, 2016, **6**, (4), pp. 1035–1044.

[16] Konstantinidis, E.I., Billis, A., Savvidis, T., Xefteris, S., Bamidis, P.D.: 'Emotion recognition in the wild: Results and limitations from active and healthy ageing cases in a living lab', in 'eHealth 360° International Summit on eHealth, Budapest, Hungary' (Springer, 2017), pp. 425–428.

[17] BioID: 'Face recognition & Voice authentication | BioID', 2016. https://www.bioid.com/.

[18] HYPR: 'HYPR Biometric Security', 2016. https://www.hypr.com/.

[19] Andronikou, V., Xefteris, S., Varvarigou, T.: 'A novel, algorithm metadata-aware architecture for biometric systems', in '2012 IEEE Workshop on Biometric Measurements and Systems for Security and Medical Applications (BIOMS) Proceedings' (IEEE, 2012), pp. 1–6.

[20] Rathgeb, C., Uhl, A., Jain, A., *et al.*: 'A survey on biometric cryptosystems and cancelable biometrics' *EURASIP Journal on Information Security*, 2011, **2011**, (1), p. 3.

[21] Ratha, N.K., Connell, J.H., Bolle, R.M.: 'Enhancing security and privacy in biometrics-based authentication systems' *IBM Systems Journal*, 2001, **40**, (3), pp. 614–634.

[22] Rathgeb, C., Busch, C.: 'Cancelable multi-biometrics: Mixing iris-codes based on adaptive bloom filters' *Computers & Security*, 2014, **42**, pp. 1–12.

[23] Sim, H.M., Asmuni, H., Hassan, R., Othman, R.M.: 'Multimodal biometrics: Weighted score level fusion based on non-ideal iris and face images' *Expert Systems with Applications*, 2014, **41**, (11), pp. 5390–5404.

[24] Bhattasali, T., Saeed, K., Chaki, N., Chaki, R.: 'A survey of security and privacy issues for biometrics based remote authentication in cloud', in 'IFIP International Conference on Computer Information Systems and Industrial Management' (2014), pp. 112–121.

[25] Nagar, A., Nandakumar, K., Jain, A.K.: 'A hybrid biometric cryptosystem for securing fingerprint minutiae templates' *Pattern Recognition Letters*, 2010, **31**, (8), pp. 733–741.

[26] Jain, A.K., Nandakumar, K., Ross, A.: '50 years of biometric research: Accomplishments, challenges, and opportunities' *Pattern Recognition Letters*, 2016, **79**, pp. 80–105.

[27] Leiming, S.U., Shizuo, S.: 'Data analysis platforms extremely-large-scale biometric authentication system – Its practical implementation' *NEC Technical Journal*, 2012, **7**, (2), pp. 56–60.

[28] Government of India, 'Unique Identification Authority of India (UIDA)', 2016. http://uidai.gov.in/, accessed December 2016.

[29] Mizoguchi, M., Hara, M.: 'Fingerprint/palmprint matching identification technology' *NEC Technical Journal*, 2010, **5**, (3), pp. 18–22.

[30] De, N.: 'Starbug's Touch ID Attack on Vimeo', 2013. https://vimeo.com/75324765, accessed February 2017.

[31] Hern, A.: 'Hacker fakes German minister's fingerprints using photos of her hands | Technology | The Guardian', 2014. http://www.theguardian.com/technology/2014/dec/30/hacker-fakes-german-ministers-fingerprints-using-photos-of-her-hands, accessed February 2017.

[32] McGoogan, C., Demetriou, D.: 'Peace sign selfies could let hackers copy your fingerprints', http://www.telegraph.co.uk/technology/2017/01/12/peace-sign-selfies-could-let-hackers-copy-fingerprints/, accessed February 2017.

[33] Bowden-Peters, E., Phan, R.C.-W., Whitley, J.N., Parish, D.J.: 'Fooling a liveness-detecting capacitive fingerprint scanner', in 'Cryptography and Security: From Theory to Applications' (Heidelberg: Springer, 2012), pp. 484–490.

[34] Chakraborty, S., Das, D.: 'An overview of face liveness detection' *arXiv preprint arXiv:1405.2227, International Journal on Information Theory (IJIT)*, 2014, **3**, (2), pp. 11–25.

[35] Solove, D.J.: 'A taxonomy of privacy' *University of Pennsylvania Law Review*, 2006, **154**, pp. 477–564.

[36] Xu, Y., Price, T., Frahm, J.-M., Monrose, F.: 'Virtual U: Defeating face liveness detection by building virtual models from your public photos', in '25th USENIX Security Symposium (USENIX Security 16)' (2016), pp. 497–512.

[37] Zephoria: 'Top 20 Facebook statistics – Updated January 2017', 2017. http://zephoria.com/top-15-valuable-facebook-statistics/, accessed February 2017.

[38] Smith, C.: '400 Facebook statistics and facts (February 2017)', 2017. http://expandedramblings.com/index.php/by-the-numbers-17-amazing-facebook-stats/14/, accessed February 2017.

[39] SocialPilot: '125 amazing social media statistics you should know in 2016', 2016. https://socialpilot.co/blog/125-amazing-social-media-statistics-know-2016/, accessed February 2017.

[40] Jain, A.K., Dass, S.C., Nandakumar, K.: 'Can soft biometric traits assist user recognition?', in 'Defense and Security' (2004), pp. 561–572.

[41] Ailisto, H., Lindholm, M., Mäkelä, S.-M., Vildjiounaite, E.: 'Unobtrusive user identification with light biometrics', in 'Proceedings of the Third Nordic Conference on Human-Computer Interaction' (2004), pp. 327–330.

[42] Khan, M.K., Zhang, J., Alghathbar, K.: 'Challenge-response-based biometric image scrambling for secure personal identification' *Future Generation Computer Systems*, 2011, **27**, (4), pp. 411–418.

[43] Sluganovic, I., Roeschlin, M., Rasmussen, K.B., Martinovic, I.: 'Using reflexive eye movements for fast challenge-response authentication', in 'Proceedings of the 2016 ACM SIGSAC Conference on Computer and Communications Security' (2016), pp. 1056–1067.

[44] FindFace: 'FindFace.PRO – Most accurate face recognition algorithm', 2016. http://findface.pro/en/, accessed February 2017.

[45] Kopstein, J.: 'Twitter bans Russian face recognition app used to harass porn stars – Vocativ', 2016. http://www.vocativ.com/384720/twitter-russian-face-recognition-porn-stars/, accessed February 2017.

[46] Moghaddam, B., Yang, M.-H.: 'Learning gender with support faces' *IEEE Transactions on Pattern Analysis and Machine Intelligence*, 2002, **24**, (5), pp. 707–711.

[47] Baluja, S., Rowley, H.A.: 'Boosting sex identification performance' *International Journal of Computer Vision*, 2007, **71**, (1), pp. 111–119.

[48] Guo, G., Dyer, C.R., Fu, Y., Huang, T.S.: 'Is gender recognition affected by age?', in 'Computer Vision Workshops (ICCV Workshops), 2009 IEEE 12th International Conference on' (2009), pp. 2032–2039.

[49] Makinen, E., Raisamo, R.: 'Evaluation of gender classification methods with automatically detected and aligned faces' *IEEE Transactions on Pattern Analysis and Machine Intelligence*, 2008, **30**, (3), pp. 541–547.

[50] Ramanathan, N., Chellappa, R.: 'Modeling age progression in young faces', in 'Computer Vision and Pattern Recognition, 2006 IEEE Computer Society Conference on' (2006), pp. 387–394.

[51] Lanitis, A., Taylor, C.J., Cootes, T.F.: 'Toward automatic simulation of aging effects on face images' *IEEE Transactions on Pattern Analysis and Machine Intelligence*, 2002, **24**, (4), pp. 442–455.

[52] Xia, B., Amor, B.B., Daoudi, M., Drira, H.: 'Can 3D shape of the face reveal your age?', in 'Computer Vision Theory and Applications (VISAPP), 2014 International Conference on' (2014), pp. 5–13.

[53] Xie, Y., Luu, K., Savvides, M.: 'A robust approach to facial ethnicity classification on large scale face databases', in 'Biometrics: Theory, Applications and Systems (BTAS), 2012 IEEE Fifth International Conference on' (2012), pp. 143–149.

[54] Tariq, U., Hu, Y., Huang, T.S.: 'Gender and ethnicity identification from silhouetted face profiles', in 'Image Processing (ICIP), 2009 16th IEEE International Conference on', IEEE (2009), pp. 2441–2444.

[55] Phillips, P.J., Jiang, F., Narvekar, A., Ayyad, J., O'Toole, A.J.: 'An other-race effect for face recognition algorithms' *ACM Transactions on Applied Perception (TAP)*, 2011, **8**, (2), p. 14.

[56] Lin, H., Lu, H., Zhang, L.: 'A new automatic recognition system of gender, age and ethnicity', in 'Intelligent Control and Automation, 2006. WCICA 2006. The Sixth World Congress on' (2006), pp. 9988–9991.

[57] Klare, B., Jain, A.K.: 'On a taxonomy of facial features', in 'Biometrics: Theory Applications and Systems (BTAS), 2010 Fourth IEEE International Conference on' (2010), pp. 1–8.

[58] Satta, R., Stirparo, P.: 'Picture-to-identity linking of social network accounts based on sensor pattern noise', in '5th International Conference on Imaging for Crime Detection and Prevention (ICDP 2013)' (IET, 2013), p. 15.

[59] Luo, J., Tang, J., Xiao, X.: 'Abnormal gait behavior detection for elderly based on enhanced Wigner-Ville analysis and cloud incremental SVM learning' *Journal of Sensors*, 2016, **2016**, pp. 1–18.

[60] Juen, J., Cheng, Q., Prieto-Centurion, V., Krishnan, J.A., Schatz, B.: 'Health monitors for chronic disease by gait analysis with mobile phones' *Telemedicine and e-Health*, 2014, **20**, (11), pp. 1035–1041.

[61] Sussman, P.: 'Fingertip diagnosis: Can fingerprints do more than just solve crimes? – CNN.com', 2007. http://edition.cnn.com/2007/TECH/science/07/05/fingerprint.diagnosis/, accessed March 2017.

[62] Datcu, D., Rothkrantz, L.: 'Semantic audiovisual data fusion for automatic emotion recognition' in *Emotion Recognition*, Hoboken, NJ: John Wiley & Sons, 2015, 2014, pp. 411–435.

[63] Majumder, A., Behera, L., Subramanian, V.: 'Emotion recognition from geometric facial features using self-organizing map' *Pattern Recognition*, 2014, **47**, (3), pp. 1282–1293.

[64] Shu, T., Zhang, B.: 'Non-invasive health status detection system using gabor filters based on facial block texture features' *Journal of Medical Systems*, 2015, **39**, (4), p. 41.

[65] Pinheiro, H.M., da Costa, R.M., Camilo, E.N.R., *et al.*: 'A new approach to detect use of alcohol through iris videos using computer vision', in 'International Conference on Image Analysis and Processing' (Springer, 2015), pp. 598–608.

[66] Chen, D., Chang, T., Cao, R.: 'The development of a skin inspection imaging system on an Android device', in 'Communications and Networking in China (CHINACOM), 2012 7th International ICST Conference on' (IEEE, 2012), pp. 653–658.

[67] Van der Ploeg, I.: 'Biometrics and the body as information', in 'Sorting: Privacy, Risk and Digital Discrimination' (2003).

[68] Anderson, R.: 'Security engineering – A guide to building dependable distributed systems' (Wiley, 2001).

Chapter 16

Biometrics, identity, recognition and the private sphere where we are, where we go[1]

*Emilio Mordini**

Abstract

The need for recognition schemes is inherent to human civilization itself. Each epoch has been characterized by different identification practices and has posed different challenges. Today we are confronted with "identification in the globalization age". Biometrics can be an important element of the answer to this challenge. With biometrics, for the first time in the history, human beings have really enhanced their capacity for personal recognition by amplifying their natural, physiological, recognition scheme, based on the appreciation of physical and behavioural appearances. Biometric technology can offer an identification scheme applicable at global level, indipendently of Nation States. Yet, when one speaks of global biometric identifiers, people immediately think of a nightmarish scenario, a unique world database, including billions of individuals, run by a global superpower. This is (bad) science fiction. We lack the technical and financial capacity, not to mention the international agreement, for creating such a database, which cannot exist today, and will hardly ever exist in the future. One could instead imagine a system based on many decentralized applications. An ongoing rhizome, made up of several distributed, interoperable, biometric databases, owned by local collaborative organizations and agencies. This system could increasingly support identity transactions on a global basis, at the beginning only in specific areas (e.g., refugees, migrants), siding traditional systems, and then, gradually, enlarging its scope and substituting old systems. This is expected to overturn many current ethical and privacy standards.

16.1 Introduction

In *Mimesis* [1], Erich Auerbach, the great German philologist, states that the whole Western civilization could be summarized in two great narratives, the tale of a man

*Responsible Technology, France
[1]This paper is partially based on a shorter, preparatory, work, "Mordini E., *Biometric identifiers for refugees*, Keesing Journal of Documents & Identity October 2016."

who spent 10 long years far from his homeland, travelling across the sea, and the tale of a man whom God once commanded *"Take your son, your only son, whom you love and go to the region of Moriah. Sacrifice him"* (*Gen.* 22,2).

Odysseus[2] and Abraham are the two fathers of the West, according to Auerbach. In his turn, Emanuel Levinas [2] argued that Odysseus' journey, a long journey home-coming, represents the essence of Western thought, say, *"un retour à l'île natale – une complaisance dans le Même, une méconnaissance de l'Autre"*[3] [2]. Levinas captures an important element, say, the relationship between journey, memory, and recognition. Recognition is essentially memory, and memory always implies nostalgia. To some extent, the Odyssey is anything but a long journey through the infinite nuances of memory and recognition. The poem starts with Odysseus, who incognito arrives among the Phaeacians as a refugee. Listening at a blind bard to sing a poem on the Trojan War, he cannot avoid bursting into tears, so revealing his identity. The hero tells as he wandered the sea for ten long years because of Gods' punishment, after blinding the Cyclops Polyphemus. Odysseus was captured by the cyclops; asked his name, he answered "Nobody". At the end, he succeeded in escaping, after blinding Polyphemus with a wooden stake. When, crying out and screaming, the cyclops asked for aid, other cyclops inquired "Who blinded you?" And Polyphemus obviously replied "Nobody", making them puzzled and unable of taking action. Recognition was at stake also in another, sinister and nocturnal, episode, when Odysseus evoked the deads, who appeared as ghosts, totally unable to recognize anyone. Only when spirits drank the fresh blood of sacrificed animals, they recalled their memories, becoming again capable to recognize living people. Finally, the concluding part of the poem is completely woven into many different forms of recognition, starting with Argos, Odysseus' faithful dog, which recognized his master, although the hero was disguised by beggar's features. Odysseus, always disguised as an old beggar, revealed his identity to his wife, the queen Penelope, by showing that he knew that their conjugal bed was carved in an olive tree and rooted in the ground. Similarly, he identified himself with his old father Laertes, by describing the orchard that Laertes had once given him. The episode of the recognition between Odysseus and his old nurse, Eurycleia, was still more intriguing. Although the hero tried to keep his identity secret, the house-keeper recognized him through an old scar on his leg. The next day then, the situation was reversed, it was the turn of Odysseus to identify himself voluntarily by showing the old scar to some servants.

Also, Abraham took his identity carved on his flesh, his body was "written" by the sign of the alliance with God, the circumcision. The circumcision was much more than a sign carved on the flesh, it symbolically hinted at Isaac' sacrifice. The momentous instant in which Abraham raised his knife over his son was already prefigured by the instant of the circumcision. There was an unavoidably short circuit of meanings between the two symbolic events. Words, flesh, and memory met themselves on Odysseus and Abraham's bodies.

---

[2] I will use the original Greek name, Odysseus, instead of the more common Latin translation, Ulysses.
[3] A return to his native island, a complacency with the same, a misrecognition of the other (my translation).

Memory, body, and recognition are sides of the same multifaceted prism, identity. In this chapter, I will explore some aspects of such a prism. My aim will be two-fold (1) to sketch how they interact with biometric technology and (2) to suggest that biometrics are legitimate to become the main identity technology of the global digital society. To be sure, this theme would require a book rather than a book chapter. It will not be possible to support my arguments as they would require, and to provide all the elements necessary to assess them as they would deserve. The current paper will be more a suggestive journey through problems, cases, ideas, than a organized, logic, construction. If, at the end of this paper, the reader will pose himself questions that he never asked before, my goal will be reached.

16.2 Identity, identification, recognition

Let's start our journey with a brief discussion of the core philosophical implications of the concept of "identity" [3]. Identity is the state of being the same, the "*sameness in all that constitutes the objective reality of a thing: oneness*" [4]; if an object x is identical with an object y, it follows as a matter of necessity, that x and y are one and the same object. To put it another way, if "A" is a name of x, and "B" a name of y, "A" and "B" name the same object: that object is named twice but should be counted once. In this sense, every object is identical to itself and to nothing else. This statement could seem a truism, which does not pose any major problem: yet, since Plato, who dealt with it in the *Cratilus*, philosophers know that this notion conceals many paradoxes. In what a sense one could state that an item is identical to itself? If one considers two items taken synchronically, say, in the same moment of time, this statement would mean that we were initially wrong, there was only one item instead of two. In other words, an identity statement would simply mean to declare a misperception or misjudgement, otherwise we should admit that two might be contemporaneously one, which is a nonsense. If one considers instead two items taken diachronically, say, in different moments of time, the idea of identity would imply that they share essential qualities, temporally invariant, although they might differ in accidental qualities. This poses the question whether any atemporal essence might ever exist. Rigorously speaking, only God is one, always identical to himself and to nothing else, independently of contingency and time. This is expressed by the concept of *Tawhid*, pure and absolute "oneness," which is the notion at the heart of Islam; similarly, in Judaism, God is the completely unmixed being, whose attributes entirely coincide with His essence, *Ehyeh asher Ehyeh* (Exodus 3:14), "I am *I am*." Except God, all other beings cannot "be" in an absolute sense, they instead "have" contingent qualities. In human experience, all beings are immersed in the ceaseless flow of time, their qualities are always contingent. In conclusion, there are two possibilities (1) either non-contingent qualities – the core identity, the essence – exist, but they escape our sensory perception; or (2) all beings possess only contingent attributes, and therefore core identity and essence are illusive. If (2) is true, then the common belief in an unitary identity, permanent in time, is deceiptive. Discussing these two, both legitimate, philosophical options is out of the scope of this chapter; it is worth noting, however, that none of them excludes that there could be qualities stable enough, in the relevant period of time, to be considered, for practical and operational purposes, permanent or essential.

This assumption is the starting point of second possible philosophical approach to "identity," which is based on assuming a weaker definition of identity, say, "*sameness of essential or generic qualities in different instances*" [4]. Philosophers call the strong definition of identity "numerical identity," to mean that it concerns the true uniqueness of an item; and "qualitative identity," the weaker one, to indicate that IT concerns only some attributes of an item, instead of the whole item. Two items "x" and "y" are *qualitatively* identical, when they are exactly similar with respect to some property they share. Monozygotic twins may be qualitatively identical in various respects but they are clearly two distinct individuals. It is worth drawing the attention on the fictional nature of qualitative identity. To keep on using the previous example, it is evident that two twins can be said "identical" only very roughly; at a more granular observation they differ in many essential aspects, including several physical features, like fingerprints and body symmetry. At the end, qualitative identity is a logic loophole invented to save the notion of identity, which would be otherwise untenable. The reason why we need to save the idea of "identity" is chiefly because we need to provide a foundation to the notion of "identification," which is not an abstract concept, but it is a term describing a specific human activity and practice. In other words, I argue that the idea of identification does not descend from the notion of identity, rather it is vice versa; from existing recognition practices (identification), we generated the theoretical and metaphysical notion of "identity".

"Identification" means treating an item as the same as another, consequently also the act of *recognizing* an item. The notion of identification always implies that something or someone, or some attributes, are treated as previously known (recognized). A totally unknown item, whose attributes are all totally unknown, cannot be identified because it could not be even thought. The idea of "absolute identification," picking out an item as completely distinct from every other, truly and rigorously unique, is purely metaphysical and abstract from human experience. When we identify, we always consider only "qualitative identity," the picking out of an object in terms of some property or quality that it possesses. Such qualities need not be indicative of the individuality or uniqueness of the individual in question. The distinction between identity and identification is, then, the distinction between who one is in essence (which is a metaphysical statement) and how (or in virtue of what) one may be recognized (which is a practical question).

Personal recognition is the process through which one identifies (recognizes) a human person; it has little to do with the (philosophical) notion of "personal identity". Persons are usually recognized through patterns of signs and attributes. None of them is totally distinctive and permanent. I am not referring only to long-term degradation of all attributes, including those which are treated by biometrics as though they were unique and stable (e.g., fingerprints, iris). I am referring to the notion of quality in itself. Qualities are distinct features only considered in abstract. In real world, a quality does not exist separated by all others and by the context. Qualities are mixed, they emerge only from the relationship between the whole of an object and a "sensor" (biological or mechanical). Sensor precision is always limited, both in biological sensory apparatuses and mechanical devises. It varies overtime, at different sites and according to different contextual conditions. This implies the practical impossibility that two or more human or machine observations might exactly match. Recognition is always by approximation and emerges from differences. The assumption that a unique identifier could ever exist is not only philosophically untenable, but it is also empirically impossible. In real life, we need a number of observations blended together to recognize an item, be a thing, or an animal or a person. This is the way in which human perceptive system works, and it is also the reason why biometric technology needs to generate and exploit "templates."

In conclusion of this chapter, there are two points that should be retained.

1. Identity is a notion full of metaphysical implications, it should be handled with great care and if possible avoided; this is the approach chosen by the *Standards Committee 37 of the International Organisation for Standardisation/International Electrotechnical Commission Joint Technical Committee 1* (ISO/IEC JTC1 SC37), which defines "biometrics" as *"automated recognition of individuals based on their biological and behavioural characteristics,"* avoiding any reference to identity [5];
2. The claim that biometric identification would be inherently demeaning because it turns "the human essence" into digits [6], is fallacious because personal identification does not ascertain personal identities – provided that they exist – rather it is a social practice with the pragmatic purpose of recognizing someone in a given context, for some given reasons.

16.3 Personal recognition through human history

In this chapter, I will outline the history of human identification, with the goal to show that radical changes in the socio-economic also affect the way in which personal identities and identification practices are conceptualized. My main focus will be on the bodily aspects of identification practices, because of their relevance to biometrics.

Along centuries, personal recognition has been based on various and different schemes. The need for specific methods for identification is presumably connected with the birth of the first urban societies during the so-called "Neolithic Revolution." Before this transition, human communities were probably made of few nomadic individuals, used to get their food by hunting, or fishing, and by gathering edible items, such as fruit and insects. Around 9,000 years ago, in the vast area once called "Fertile Crescent,"[4] the mankind started to develop a new economic model based on farming, cultivated crops and domesticated animals [7]. The transition from an economy based on hunting and gathering, to an economy based on farming, implied

---

[4]It is an area, including Iraq, Syria, Lebanon, Cyprus, Jordan, Israel, Egypt, as well as parts of Turkey and Iran.

many epochal consequences, including the emergence of sedentary dwelling. Human groups gave birth to sedentary communities organized in small villages and towns. Farming economy also meant the creation of food surpluses, which promoted trade of food and food related products (e.g., salt, which was probably one of first commodities because of its ability to preserve food). The growing societal complexity, alongside the development of trade, presumably generated the need for recognition schemes beyond interpersonal contacts within each community. The first categories of people who needed to be recognized and to recognize were probably those who travelled, say, merchants, sailors, caravan travellers, and soldiers. Recognition was necessary for establishing trustful relationships with unknown strangers. The Odyssey offers an indirect evidence of this novel situation. The Homeric poems provide a nice list of the main identifiers used by early human communities. Distinguished Italian Medievalist and Anglist, Piero Boitani, used to argue that the Odyssey is the greatest summa of recognition practices in western literature, including descriptions of physical appearance (e.g., body size and shape, skin and hair colour, face shape, physical deformities and particularities, wrinkles and scars, etc.), artificial body modifications (e.g., branding, tattooing, scarification, etc.), physical objects (e.g., passes, seals, rings, etc.), and mental tokens (e.g., memories, poems, music, recollection of family and tribal links, etc.). Some of these recognition tools warrant a further examination for at least two reasons, first, because they are still in use; second, because they provided the early model for today technology. It is thus possible to learn lessons from them, which could be fruitfully applied to the present.

Let's start with tattoos, which have a bizarre double-nature, of cosmetic and body writing. Tattoos started to be associate with identification thousands of years ago [8]. The word is a borrowing of the Samoan word *tatau*, meaning to "strike something," but also to "mark someone." In the Pacific cultures tattooing had a huge historic and cultural significance, splendidly evoked by Melville in *Moby Dick*'s character Queequeg, *"Many spare hours he spent, in carving the lid with all manner of grotesque figures and drawings; and it seemed that hereby he was striving, in his rude way, to copy parts of the twisted tattooing on his body. And this tattooing, had been the work of a departed prophet and seer of his island, who, by those hieroglyphic marks, had written out on his body a complete theory of the heavens and the earth, and a mystical treatise on the art of attaining truth; so that Queequeg in his own proper person was a riddle to unfold; a wondrous work in one volume; but whose mysteries not even himself could read, though his own live heart beat against them; and these mysteries were therefore destined in the end to moulder away with the living parchment whereon they were inscribed, and so be unsolved to the last."* (*Moby Dick, 110, 19*). It is difficult to escape the impression that in these few lines Melville was able, as only poets and artist know, to get closer to the heart of the mystery surrounding the "written body."

Tattoos have had, however, a quite different description in the West. Tattoos were banned by traditional Jewish culture. In Jewish law, any body modification is forbidden by *Deut. 4:9,15*, except those made for protecting or restoring health. Also, ancient Greek civilization did not love tattoos as well as any other mark on the skin [9]. Aristotle defined recognition through body signs *"the least artistic form"* (*Poet. 1453b*) and Greeks were proud that their civilization used cultural signs for people recognition rather than body marks. For them, the written body was almost a dead body. Greeks abhorred corpses, and artificial body signs were felt very close to corpse preparation for mortuary rituals. Greeks thought that living beings are by definition changing, only death is immutable. Thus, they were very suspicious towards any search for immutability. The observation of the impermanent nature of all human attributes and qualities was turned into a reflection on the fascinating, and ambiguous, relation between memory, body, and life. In Greek mythology, the deceased brought on his body the signs of the faults committed in life. When he arrived in the netherworld, his body was read by *Radamanthos*, the infernal judge, who wrote, still on the body, the punishment decreed [10]. The symbolism is transparent, our life is written on our bodies, through memories, wrinkles, scars, diseases, disabilities, and so; body are full of stories. The lesson to be learned is relevant also to us. Human bodies are never bare bodies, they have always, and unavoidably, a personal, biographical, dimension. Claiming that biometric technology strips human beings from their personal and biographical aspect, identifying people through their biological features, as they were animals [11], means to ignore the richness and complexity of human physicality.

Also, Romans despised tattooing, at least during the first centuries [8]. The Latin word for tattoo was "*stigma*," and Romans used to mark criminals and slaves, while citizens were identifiable through names. Names were extermely important in the Roman culture. The Roman Empire was the first society in the west providing for a universal identification system through a tripartite codified name scheme, connected to a comprehensive legal system of political citizenship [12]. The birth of the great empires, the Roman in the West and the Chinese in Asia, introduced new important drivers for personal identification, namely, taxation, conscription, and the administration of law [13]. In largely illiterate societies, it is unlikely that the birth of a class of public officers, capable for writing and tasked to tax, recruit, or bring to the court people, was happily accepted. The bad press that surrounds identification practices, and the deep rooted conviction that identification is always a way to curb people, is likely to date to this period.

The Modern Era, which started with the Reformation and the new geographic discoveries, was characterized by increased mobility of people and goods (the so-called first wave of globalization) associated with urbanization and, later on, industrialization. The need for different recognition schemes was due to urbanization and industrial work, two conditions that generated masses of people on the move, abandoning villages and the country, to move to cities and industries. With the industrial revolution, anonymity, almost impossible in medieval urban societies, and rural communities [14], became the standard condition for most individuals. The passage from the mediaeval identification scheme, based on community membership (e.g., family, guild, village, manor, parish, etc.), to a modern identification scheme, based on documents issued by the State, is crucial. Nation-States emerged from the political turmoil, which affected Europe for around a century. Nation-States had the primary need to affirm their authority, against transnational and global powers, such as the Roman Catholic Church and the cosmopolitan network of nobility. Nation-States' strategic move was to strip people from their communitarian identity. Communities were still largely controlled by religious authorities and nobility, it was then paramount that the new citizen was an unmarked person, enfranchised from any community. Being a member of a given category or community was no longer relevant before the State. France was the place where this strategy was more successfully and comprehensively experimented [15]. In a few years, the French Republican government enacted the first legislation in the West linking personal identities to birth registration; introduced the principle of the *jus soli,* according to which "*any individual born from foreign parents could claim French citizenship if he was at least 21 years old and declared his will to become a Frenchman and to settle in France*" [15]; created a national Committee of Public Safety (*Comité de salut public*) tasked to protect national borders from expatriated aristocrats (*les émigrés*); established a wide network of surveillance committees, tasked to supervise the rigorous application of republican principles, to deliver citizenship certificates, to deal with immigration matters at local level [16]. Religion, ethnicity, race, cast, social conditions, were declared irrelevant to identify individuals, at least in theory.[5] The Declaration of the Rights of Man and of the Citizen (*Déclaration des droits de l'homme et du citoyen*), passed by France's National Constituent Assembly in August 1789, provided the theoretical foundation for this new system. It is worth reading its first articles [17], because they describe the relevant attributes of the ideal citizen of the Nation-State,

Article I – Men are born and remain free and equal in rights. Social distinctions can be founded only on the common good.

Article II – The goal of any political association is the conservation of the natural and imprescriptible rights of man. These rights are liberty, property, safety and resistance against oppression.

Article III – The principle of any sovereignty resides essentially in the Nation. No body, no individual can exert authority which does not emanate expressly from it.

Such a citizen is recognized by using only three simple coordinates, allowing to localize him in space and time, (1) place of birth, which provides critical information about nationality; (2) date of birth, which complements the previous piece of information and allows establishing citizen's age, which is relevant to determine the rights that he actually possesses; (3) name, which is the token used in everyday life by the State to trace him. No further information is basically required, although further details can be added (e.g., residence, sex, profession, and so) but none of them is truly critical, as it is witnessed by the fact that only these three parameters are indispensable for birth registration almost everywhere, even today. Personal identification became necessary to demonstrate nationality, and consequently, to move across borders and to be fully entitled with political rights.

Finally, after World War II, a new powerful driver for personal identification emerged: the welfare state. The welfare state, which first appeared in north Europe, was based on the concept to provide all its citizens with a vast arrays of services,

---

[5]Actually, the history shows that most Nation-States kept on using identification schemes largely based also on racial, religious, ethnic, categories. The most horrible event of the twentieth century, the Shoah, was made possible chiefly by the existence of an effective bureaucratic apparatus for certifying racial identities.

free of charge, chiefly funded by the State via a redistributionist taxation. To be properly enforced, both redistributionist taxation and welfare provision needed robust and reliable systems for personal identification. As a consequence, welfare states progressively equipped themselves with an increasing number of identity and entitlement documents (e.g., social cards, social insurance numbers, etc.) and created huge bureaucratic apparatuses to manage public services and handle recognition systems. After civil and political rights, also social rights were progressively included among practices based on personal identification, hence making identification integral to most public services, including insurances, national health systems, tax offices, and so.

Here it is where we are, or, better, where we were till yesterday. There are two main concepts that one should retain,

1. History shows that personal recognition has always played an enabling role; it has allowed enforcing most social changes occurred in our civilization, including, 1) the naissance and development of long-distance trade and novel economic systems; 2) the birth of a legal system including civil and criminal codes; 3) the progress towards more social equity; 4) the enforcement of most liberty rights; 5) the promotion of social rights, e.g., the right to health, to employment, to housing, and so. To be sure, personal recognition has also partly contributed to the development of other, less positive, mechanisms, including more pervasive systems for social control and the proliferation of bureaucracies;

2. Interestingly enough, however, history also shows that very rarely personal recognition has been used to overpower and subjugate people. The misperception that personal recognition is integral to oppressive regimes, even to dictatorships, is chiefly based on a confusion between personal recognition and social categorization. Think of slaves in ancient Rome, they were tattooed and had no name, the tattoo was a sign of their category, it was not a true personal identifier. When a slave was freed, the first change in his life was to get a personal identifier, say, a name. Most crimes attributed to personal recognition, were perpetuated thanks to the development of parallel schemes for group categorization. This emerges with a compelling evidence from the period of French history immediately after the proclamation of the republic. In theory, the new definition of citizenship should have protected everybody against discriminating measures, and criminal law consented to prosecute people only on strict personal basis. Yet the system was structured for sorting out those who did not comply with the Republican government, notably the clergy still faithful to the Roman Church, and the landed gentry. These people were systematically exterminated, with the goal to eradicate their presence and cultural significance from the French society. The instruments used to sort out these "enemies of the revolution" were neither the new system for birth registration, nor the novel organization for delivering citizenship certification. They were identified by simply searching for people labelled as "priests" and "nobles" because of their way of dressing, or speaking, or their education [16].

16.4   Where are we going?

In this chapter, I will suggest that the global, digital, society needs an "identity revolution," similar to those occurred with the agricultural and the industrial revolutions. I will take some examples from two cases, refugees and digital economies, to suggest that biometrics could be a building block of such an identity revolution. Finally, I will argue that most reasons of ethical and privacy concerns related to this evolution are probably misplaced, although there are still reasons to carefully supervise these processes.

The world has reached a degree of interconnectedness never experienced before. Billion persons are moving each year across large geographic distances, and still more people are interconnected through the Internet; a new economy, based on synchronic communication, extensive financialization, and the exchange of data, is rapidly advancing. This situation dramatically transcends national control and state regulations, and has having momentous consequences for the identification scheme emerged from the French revolution. In such a context, biometric technology could be used either to support obsolete schemes for personal recognition, or to promote innovative approaches to global identity management.

16.4.1   Global mobility of people

Although people mobility within and across national borders is not at all a new phenomenon, what is surely novel is its current dimension. This is due to a variety of factors, including large availability of geographical information; increasing cultural homologation, which makes easier and easier to interact among people belonging to different cultures; global commerce and increasing number of networks of global retailers, which makes possible to find the same products and services almost everywhere; global communication systems at no cost, or almost no cost; possibility to travel at relative ease and at low costs; border porosity in some world areas; diffusion of electronic payments. Global people mobility includes three main categories of people,

- Economic migrants: they include "a wide array of people that move from one country to another to advance their economic and professional prospect" [18]. According to the United Nations *the number of international migrants worldwide has reached 244 million in 2015* (3.3% of the world's population) [19]. Most economic migrants are legal migrants, yet a part of them are not. Illegal aliens are foreigners who (1) have illegally entered a country or (2) have entered a country legally, but who no longer hold a valid residence permit. At least 50% of illegal aliens, both in the US and in the EU, are *undocumented immigrants*, say, foreigners who have crossed borders legally, but are resident under false pretences. For instance, they entered a country with a tourist visa although they intended to work; or their residence permit has expired and was not renewed; or they lost their refugee status. Undocumented immigrants are not necessarily all criminals, at worst, according to some jurisdictions, they are guilty of an administrative offense related to their failure to fulfil specific administrative requirements to stay in the country of destination.

- Refugees and displaced people: the term "refugee" refers to *"people who have had to abandon or flee their country of origin as a result of serious threat to their lives or freedom such as natural catastrophe, war or military occupation, fear of religions and racial or political persecution"* [20]. According to United Nations High Commissioner For Refugee (UNHCR) [21], there are in total 65.3 million people who had been forced to flee their homes worldwide, they include *internally displaced people*, who have been displaced within their country's borders (40.8 million in 2015); refugees, people to whom at has been granted the refugee status (21.3 million in 2015); *asylum seekers*, people who are seeking international protection but whose refugee status is yet to be determined (3.2 million in 2015); and *stateless people*, who (for various reasons) do not have the nationality of any country (10 million in 2015). Refugees and asylum seekers are legally defined by national legislations, in accordance with the international legal framework provided by the 1951 Geneva Convention Relating to the Status of Refugees and its 1967 Protocol [22]. At the core of this framework, there are two principles: (1) the right to asylum, conceived as a fundamental human right; (2) the principle of non-refoulement, which means that refugees cannot be forcibly returned to places where their lives or freedoms are threatened.
- Global travellers: they include tourists, business people, academics, students involved in international exchange programmes, people working in international organizations and NGOs, people working in multinational companies. According to the World Travel Monitor [23], in 2016 there have been 1 billion international trips, while in 2015 the World Tourist Organization has calculated around 1.2 billion international tourists [24].

16.4.1.1 Difficult challenges

Economics migrants, refugees, asylum seekers and alike are perceived by most politicians, the media, and part of the public as one of the most severe challenges posed by global mobility. This is only partially true. Migration is as intertwined with other problems (e.g., economic globalization, cultural homologation, international terrorism, internal security, and so) as to become an ideal "scapegoat." Yet, without 1 billion international travels per year, the global scenario could be quite different. When one seventh of the world population is on the move, this means that something epochal is happening, transcending the issue of immigration. Mobile people, no matter if they are tourists or refugees, must be provided with recognizable and reliable identifiers to guarantee them basic rights and empower them with basic capacities, e.g., using a credit card, signing a form, purchasing a flight ticket, etc. They must be protected from terrorism and criminal attacks, and, in the same time, terrorists, or other criminals should be preveted from inflitrating them. This huge amount of people is too large to be handled with traditional means, and they can be managed only by modern recognition technologies, such as biometrics. Large-scale biometric applications for border control, meet this need, being a way to automatize and speed up old, established, procedures. Yet, challenges posed by global mobility cannot be addressed within the traditional legal and political framework, they require a paradigm shift.

The population of migrants is very heterogenous and not so easily recognizable. Among them, asylum seekers should be distinguished from other categories of foreigners to grant them the refugee status. Although the 1951 Convention does not explicitly prescribe that asylum seekers' applications are assessed on an individual basis, in most cases there is no alternative. On the one hand, there are situations that can be evaluated only at individual level (e.g., someone who claims to be persecuted in his country of origin); on the other hand, applications could be rejected only at individual level to comply with the principle of nonrefoulement, which prevents collective deportation. Of course, there is always the possibility of collective admissions, which are not prevented by the 1951 Convention, but even in this case personal identification is inevitable. Think for instance of the collective admission of people fleeing from a war zone. They should prove at least that they truly come from the war zone. However, they could also be legitimately requested to demonstrate that they are not militants involved in the massacre of civilians. This is because the 1951 Convention does not apply to people who have committed a crime against peace, a war crime, a crime against humanity or a serious non-political crime. Moreover, the 1951 Convention bestows upon refugees some rights and obligations. Negative rights (e.g., freedom of speech, freedom of religion and freedom of movement) and obligations (e.g., respecting laws of the hosting country and complying with measures for the public order) could in principle also be applied to undocumented people, but positive rights (such as the right to work, to housing, to education and to access the courts) can be claimed only by people who are able to document their personal identity. Refugees also need to be able to document their identity to allow authorities track and monitor them, because the refugee status is not necessarily permanent. Finally, transnational criminals and terrorists could try to exploit the 1951 Convention to infiltrate a country. Making refugees personally identifiable could help prevent infiltration, expose criminals and monitor suspects.

However important the reasons to determine the personal identity of refugees, we are confronted with a number of difficulties. First of all, when refugees arrive, they may not have an identity document. They sometimes throw away their IDs to avoid disclosing their names, fearing for themselves and their relatives. People fleeing from war or military occupation may have the same problem. Refugees who have escaped natural disasters have sometimes lost their documents. Moreover, children are often undocumented because they were not registered at birth. According to UNICEF [25], 230 million children worldwide are unregistered, which means approximately one in three of all children under five around the world.

Finally, among "true" refugees there could be also "false" refugees, who are undocumented because they destroyed their IDs to hide their nationality, place of birth or age. However, even when refugees do hold a valid ID, this is often not enough. According to the UNHCR, 54% of all refugees worldwide came from three countries: Syria, Afghanistan, and Somalia. An ID is reliable as long as it is linked through an unbroken chain, granted by the civil register or a birth certificate. It is apparent that in Syria, Afghanistan and Somalia, as well as in many other countries, there is no longer an effective civil register system. Generally speaking, the civil register system is unreliable in most refugees' countries of origin, because building an effective system requires resources and bureaucratic structures often beyond the reach of low income countries. The recognition system originated from the French revolution is based on two pillars, (1) reliability of birth certification and (2) continuity and self-consistency of civil registration. Together, they imply that (1) babies are registered almost immediately after birth; (2) registration is performed in a reliable and transparent way, by trained personnel; (3) certificates are securely stored, and protected against any alteration over time; (4) civil registers are regularly updated and protected against alterations; (5) identity documents are checked for consistency with civil registers and birth certificates; (6) it is always possible to link back identity documents, civil registries, and birth certificates. Most of these conditions are not truly ensured by many low-income countries and there is little hope that they could be ever ensured in the future.

Lastly, there is also a minor issue that is worth mentioning. Many non-western languages use writing systems different from the Latin alphabet. This means names have to be transliterated. Transliteration is rarely unique; for instance, the name "Muhammad" can be transliterated from Arabic to the Latin alphabet in at least eight different ways. This may generate confusion and could enable the same person to submit multiple applications using differently transliterated original documents.

16.4.1.2 Providing refugees with biometric identifiers

The idea to use biometrics in refugee management dates to the late 1990s. The most obvious application concerns the use of biometrics at border checkpoints, where biometric data helps match entry and exit records and augments normal watch list used to provide IDs to undocumented refugees and refugees whose original IDs were not considered reliable. The UK and Dutch governments were among the first to use a central biometric database to manage the application, return and re-integration programmes for asylum seekers.

The effective management of refugees and asylum seekers is particularly difficult in Europe. The creation of a vast area of free movement of people (the 26 countries that abolished border control at mutual borders as per the year 2017) without a corresponding political and administrative union has resulted in an odd situation. This area of free movement, coupled with the lack of common criteria for assessing asylum applications, created so-called "asylum shopping." Asylum seekers, whose applications were rejected in a given EU country, applied in another EU country, and then in another, until they found a country that granted them refugee status. Once they were accepted in a country, they moved to other EU countries, including those that initially rejected their applications. This problem was addressed with the adoption of an EU-wide information technology system for the comparison of fingerprints of asylum seekers (European Dactyloscopy, EURODAC) in December 2000, which became operational in January 2003. EURODAC enables EU Member States to track foreigners who have already filed an asylum application in another Member State. The EURODAC regulation was recently amended to allow law enforcement access to the fingerprints database (according to the previous regulation, the system could be only searched for asylum purposes). EURODAC was not, however, a panacea. Collecting the fingerprints of all fingers of every applicant – as requested by the system – is hardly possible on the spot (think of boat people). Moreover, people could refuse to give their fingerprints and, according to 1951 Convention, they cannot be denied entry for just that reason. The practical difficulty in collecting biometric data on the spot and transmitting it to the Central Unit of EURODAC (which should be done within 72 h) inevitably delays the verification of an asylum seeker's data. This time gap could be – and has been – exploited to escape controls [26]. During this so-called European migrant crisis [27] EU leaders proposed to create a web of biometric hotspots for the swift identification, registration and fingerprinting of migrants, suggesting national border authorities use proportionate coercion in case of refusal to give fingerprints.

In the meantime, also international organizations have considered using biometrics to document refugees' personal identity. In 2014, the World Bank Group presented the Identification for Development Initiative, aiming "*to help countries reach the recently adopted Sustainable Development Goal target of 'providing legal identity for all, including birth registration, by 2030'.*" [28] The UNHCR launched a plan to provide displaced persons with biometric ID cards. After initial experiments in Senegal and Malaysia, in 2015 the UNHCR announced they had developed the Biometric Identity Management System (BIMS). This system for scanning and registering fingerprints and facial and iris images is already available in Apple and Android app stores [29]. These three biometrics are stored with biographic data on cards and in a central database in Geneva.

The UNHCR's BIMS could become the blueprint for a global ID system. A global ID scheme should not necessarily replace national ID systems, which could still work locally, but it could become a supranational scheme that facilitates cross border mobility, while raising security standards. Moreover, a similar scheme applied to countries whose birth registration system is not reliable could provide a solution to this problem. The Indian Aadhaar Biometric Identification Program [30], although with different goals and managed at State level, is extremely interesting and deserves to be carefully studied.

16.4.2 Digital economy

Together with global mobility, digital economy is the second powerful driver, leading to new recognition schemes. Libraries have been written in the last decades on digital economy, and it would not be worth repeating concepts better explored by other authors [31]. There is, however, an issue worth mentioning.

It concerns the very heart of the digital revolution, say, data. Immateriality of production is one of the main features of the information society. Knowledge, information, communication, relationships, emotional responses are the intangible, immaterial, financial assets of the information society. Personal information is no longer private but has become detachable and marketable through the use of Information Communication Technology (ICT). These processes correspond to the birth of a new commodity, data. What is data? Data is measured information. New ICT has allowed developing technological devices and metrics to turn qualitative information into quantitative information, say, data. The digital nature of data makes it easily storable, transferable, and marketable. During the agricultural revolution, and the industrial revolution, the economic transition was led by the "commodification" of a natural item, turned into a marketable product. Animals and plants existed well before the Neolithic revolution, but only when humans developed the technology to domesticate animals and cultivate plants, they became commodities. The commodification of animals and plants took millennia to spread from the Fertile Crescent to the rest of the world, generating myriads of dramatic transformation, affecting almost each aspect of human civilization. Something similar happened with the industrial revolution. Human labour had always existed, and humans had always sold their job. Yet, before the industrial revolution, no one used to sell his working time, which is the new commodity "invented" by the industrial revolution. The fundamental difference between a medieval and an industrial worker, is that medieval workers sold their skills, their products, their labour, while the industrial worker sells his working time. The enabling technology that supported the industrial revolution was the mechanical clock, which turned time into a measurable item, and into a commodity. This contributed to generate new concepts of human work, life, and identity, providing the foundation to that cultural transition, which got to completion with the French revolution. It took centuries to realize the transition from the agricultural to the industrial civilization. Today, we are witnessing transformations that are rapidly affecting people all over the world, spreading in few days from Paris to Tokyo, from Los Angeles to Delhi. The whole world economy is changing its structure, undergoing to increasing financialization and virtualization. National currencies are becoming more and more abstract, they are mathematical formula devoid of materiality, backed by a web of electronic transactions. They are no longer the expression of the richness of a Nation. The new virtual economy still lacks its gold standard. Identities could become such standard because they are the most valuable commodity of the digital world. This implies the need new definitions of identity, identity data, personal data, private sphere, privacy rights and data protection.

16.5 Privacy, person and human dignity

Privacy is the third epochal driver demading an identity revolution. The debate surrounding biometrics has often focused on personal data protection, I am not convinced that this is the real challenge. To be sure, it is extremely important to protect biometrics data, which are personal and – according to the General Data Protection Regulation of the EU – *by default* sensitive. Yet, the crucial point is the redefinition of the cultural notion of privacy. Privacy is often conceptualised in normative (legal and ethical) and political terms. In this chapter, I will focus on psychological aspects of privacy. My goal is to ground the concept of privacy on the notion of human (bodily, psychologically, socially) integrity instead of on personal data.

The earliest form of polarity between public and private spheres can be probably traced to early infancy. In early developmental stages infants hardly distinguish between themselves and the environment [32]. States of wholeness, timelessness and oneness alternate with states in which the awareness of space, time and separateness, slowly emerge. Through mother's body, the infant starts exploring the world and perceiving a distinction between the inward and the outward. The inward is what is evident only to the subject, and can become evident to others only if it is communicated. This is likely to be the earliest experience of what will become later the notion of privacy, which therefore implies that (1) some experiences can be kept separated, even hidden; and (2) the inner, private, world is bridged with the outer, public, world. Inward and outward are in a mutual, ongoing, dynamic communication, and the main difference between private and public spheres does not dwell in any specific content but in the different rules that govern the two realms. A total discontinuity between private and public experiences would coincide with autism, which is indeed a true "pathology of the private sphere" [33].

Growing up, the polarity between private and public is often described in spatial terns. Space (physical and virtual) is variously segmented, also according to the degree of control or influence exercised over it by each individual. Ethologic studies show that most animals tend to have outside boundaries of their movement during the course of their everyday activities. Also human beings tend to segment the territory around them [34]. In the inner circle there is an area that is perceived to be private, which is commonly called "personal space." Most people feel discomfort when their personal space is violated, and personal spaces can be trespassed only in particular circumstances and only by selected others. This mechanism is rooted in neurophysiology, as it has been demonstrated by individuals who lack reactions to personal space violations and also show lesions of a small cerebral region involved in emotional learning and memory modulation, the amygdale [35].

A third, important step, in the internal construction of the private sphere, is the naissance of feelings of intimacy, shame and modesty. These feelings are deeply influenced by the cultural context, yet they are universal. They could regard completely different, even opposite, behaviours, body parts, and social situations, but in their elementary structure they are always similar. Everywhere offending intimacy, shame and modesty causes intense emotional reactions. There is a very strict connection between these sentiments, the perception of a private sphere, and the idea of dignity. All experiences in which intimacy, shame or modesty are offended and not respected, imply some components of degradation, Primo Levi, in *The Drowned and the Saved*, argues that degradation destroys the sense of worth and self-esteem generating humiliation. *"To be humiliated means to have your personal boundaries violated and your personal space invaded."* My argument runs as follows, 1) psychologically speaking, the main difference between private and public spheres is in their internal rules; insisting too much on data contents could be misleading; 2) the tension between private and public spheres is essential to personal recognition and self-recognition of individuals; 3) any breach in personal integrity (physical, psychological, social) is a privacy breach; 4) psychologically speaking, privacy breaches always entail humilation; humilation and privacy breaches are the two sides of the same coin; 5) data protection – including personal and biometric data protection – affects property rights, while respect for privacy concerns the fundamental right to dignity.

The *EU Charter of Fundamental Rights* [36] has captured quite well this idea. Privacy is primarily addressed in Articles 7 and 8, which explicitly concern privacy, family life, and data protection. Yet, from a careful reading of the Charter, it emerges that the protection of privacy is addressed also in the first, most important, chapter devoted to Human Dignity. The Article 3 on the *Right to the integrity of the person*, reads:

1. Everyone has the right to respect for his or her physical and mental integrity
2. In the fields of medicine and biology, the following must be respected in particular: the free and informed consent of the person concerned, according to the procedures laid down by law [...]

The context in which Article 3 is collocated points out that *"the dignity principle should be regarded as a tool to identify the cases in which the body should be absolutely inviolable"* [35] and that consequently *"the principle of inviolability of the body and physical and psychological integrity set out in Article 3 of the Charter of Fundamental Rights rules out any activity that may jeopardise integrity in whole or in part – even with the data subject's consent."* [37] Body integrity is violated any time that an undue and unsolicited intrusion "penetrates" the individual's private sphere, independently from whether such an intrusion is tactile, visual, acoustic, psychological, informational, etc. or whether it produces physical injuries or dissemination of personal information.

16.5.1   Is biometric inherently demeaning?

Since the early 2000s, scholars, philosophers, ethical committees have warned against the risk that biometrics could irreparably offend human dignity. For instance, in 2007, the French National Ethics Committee published an opinion on biometrics, which reads inter alia, *"Do the various biometric data that we have just considered constitute authentic human identification? Or do they contribute on the contrary to instrumentalizing the body and in a way dehumanizing it by reducing a person to an assortment of biometric measurements?"* [38].

This question was first formulated by the prominent philosopher Giorgio Agamben, who argued that gathering biometric data from refugees and other vulnerable groups is a form of tattooing, akin to the tattooing of Jewish prisoners in Auschwitz [39]. Agamben argued that what makes human life (*bios* in ancient Greek) different from animal life (*zoe*, in ancient Greek) is its historical, biographical dimension. There are times, he argued, when rulers create indistinct zones between *bios* and *zoe*. In these areas, humans are stripped of everything except the fact that they have a bare life. *"No human condition is more miserable than this, nor could it conceivably be so. Nothing belongs to us anymore; they have taken away our clothes, our shoes, even our hair; if we speak, they will not listen to us, and if they listen, they will not understand. They will even take away our name."* [39]. In Auschwitz, prisoners' names were substituted by numbers as brands on livestock. To Agamben, this is the deeper sense behind the adoption of biometric recognition schemes at global scale.

Refugees, Agamben suggested, are only at the first step in a process that is going to affect everybody. The global citizen would be progressively treated as, and consequently turned into, a branded beast – maybe satisfied and well-fed – but destined to slaughter.

Agamben's argument is weak. The distinction between biographical and bare life is a fallacy, bare life does not exist in humans (and this author suspects that it does not exist even in animals). Human bodies are words made flesh, they are embodied biographies, as I tried to suggest along all this chapter. Second, it is true that biometrics digitalize (quantify) human attributes, once considered only from a qualitative point of view. Yet, if this meant "de-humanizing" a person, stripping him from his biographical dimension, most current medical diagnostic would do the same. Finally, the comparison with Nazi tattoos and extermination camps is abusive. As Mesnard and Kahan demonstrated [40], Agamben speciously selected some elements of life in camps, neglecting others, to demonstrate his thesis. It is true that prisoners were only identified by numbers, and this was an extreme humiliation imposed to them, but it is enough to read Primo Levi to understand that the perverted mechanism of the extermination camp was much deeper and subtler. Agamben overlooked the specific horror of the extermination camp, which could not – and should never – be compared to trivial situations, such as getting the US VISA, the event which stimulated his reflections. In a sense Agamben was, however, right, biometric identification can be used to humiliate people. Systems can be intrusive, physically or psychologically or socially, with the result to violate people feelings of intimacy in the widest sense. Yet, this is often due to operational procedures, rather than to technology per se. Very rarely devices are designed in such a way to be intrusive by default. More often, they are operated by operators who use, or set, them in intrusive and humiliating ways (e.g., by regulating the system in a way which obliges the subject to take an uncomfortable, or ridicolous, or degrading, physical position). Unfortunately, most privacy impact assessments ignore operational procedures, only focusing on technology.

Another reason for concern surrounding biometrics, regards the creation of centralized databases. It is true that centralized databases may increase security risks. If they are compromised, the entire identification system is threatened. Moreover, large centralized biometric databases are an easy target for hackers and other malicious entities, also because designers – in order to prevent system failure – often build in high redundancy in parallel systems and mirrors, thereby adding further vulnerabilities. Centralized databases also raise the risk of function creep, which is the term used to describe the expansion of a process or system, where data collected for one specific purpose is subsequently used for another unintended or unauthorized purpose. When function creep results from a deliberate criminal intention, it represents a serious ethical, sometimes also legal, offence. Even in democratic societies, there is always the risk that a public authority uses a central biometric database for its hidden agenda, such a monitoring specific religious or ethnic groups. This risk would be magnified in case of a global ID scheme, a nightmarish scenario for most privacy advocates. The words devoted to this issue by a UNESCO Report [41] well summarize this point,

If the international system did embrace extensive use of biometrics or another globally unique identifier, the move could signal the effective end of anonymity. It would become feasible to compile a complete profile of a person's activities (…) This death to anonymity would meanwhile be coupled with asymmetry in information: the individual's every move could be monitored, yet he may not have any knowledge of this surveillance. Beyond privacy, such a state of affairs does not bode well for the exercise of other fundamental freedoms such as the right to associate or to seek, receive, and impart information – especially as the intimidation of surveillance can serve as a very restrictive force.

Technical concerns regarding the creation of centralized biometric databases could be partly overcome using blockchain technology. The blockchain technology is chiefly known as the technology that underpins Bitcoin and other cryptocurrencies. A blockchain is a distributed database that keeps an ongoing list of transaction records protected against manipulation and revision. It consists of data blocks, which each contain batches of transactions, a timestamp, and information linking it to a previous block. Blocks can be used to prove ownership of a document at a certain time, by including a one-way hash of that document in the transaction. Any electronic transaction can be certified through a blockchain. Individuals could record any document that proves their identity on the blockchain. The structure of the blockchain guarantees that the chain between the original document and the transaction based on that document remains unbroken. Information in the blockchain does not need to be guaranteed by a third party (i.e., local community, state bureaucracy, international organizations) nor does it need to be stored in a centralized database. There are currently several applications in development, which couple biometrics and blockchain technology [42]. Most of them use the blockchain to secure biometric information and prove its integrity. Once secured, encrypted and unusable for any other purpose, biometric data is definitely more manageable. In 2015, BITNATION, a decentralized, open-source movement, launched the *"BITNATION Refugee Emergency Response,"* an application still in its embryonic stages that would provide refugees and asylum seekers with a "Blockchain Emergency ID" and "Bitcoin Debit Card" [43]. An individual's biometrics would be included in the blockchain and then used to issue both an ID and a debit card. The goal is to provide refugees with a political and financial solution without relying on hosting country resources or on international organizations. *"(We) out-compete governments by providing the same services cheaper and better through the blockchain"* according to Susanne Templehof, founder of BITNATION [44].

The blockchain network is trustless and decentralized. National or international institutions, global corporations and any other central authority, would not be any longer necessary to manage biometric large scale applications and the very idea of "large scale application" would be overturned. Biometric information would be secured within the blockchain, making theft and forgery of biometric details highly unlikely. Finally, the blockchain could provide higher data granularity, which is a

precious feature in databases that store multiple biometrics and non-biometric information. Higher data granularity means that one could operate on smaller, and more
focused, pieces of information, thereby limiting the risk of function creep and data
leakage. One could imagine a system based on many decentralized, interoperable, applications. An ongoing rhizome, made up of several distributed, biometric
databases, owned by local collaborative organizations and agencies. This dispersed,
non-hierarchical, system could increasingly support identity transactions on a global
basis, at the beginning only in specific sectors (e.g., refugees, migrants, international travellers), siding traditional systems, and then, gradually, enlarging its scope,
substituting old systems.

16.6   Conclusions

When one speaks of global biometric identifiers, people immediately think of a nightmarish scenario, a unique world database, including billions of individuals, run by a
global superpower. There is a conspiracy theory, which claims that the United Nations
have a plan to mark through biometrics the whole world population with the devilish
number 666, the "mark of the beast". Even the emergence of such a conspiracy theory demonstrates that a novel, global, recognition scheme is urgent. In a world with
7.3 billion people, one exists only if someone else recognizes you; "*Esse est agnitum*,"
"being is being recognized," paraphrasing a famous nineteenth century philosopher,
George Berkeley. People obsessed by conspiracy theory are people who cannot put up
with the idea of being totally ignored, of being only anonymous individuals in a huge,
anonymous, global crowd. They need to fantasize that the Big Brother is putting them
under surveillance, ignoring that in Orwell's novel only 1% population, the members
of the party, were constantly monitored. The majority of the population was "simply"
manipulated by pervasive propaganda and fake news.

   This global crowd is now on the move, across nations and continents and the
whole architecture of the human fabric is in turmoil. Still more people are interconnected through electronic highways, where they do business, meet other people, find
new jobs, learn foreigner languages, search for books, friends, and restaurants, and
review them publicly. People use their mobiles for paying, PayPal is more used than
American Express, money travels at the light speed across the world, whole countries
are abolishing cash money. This new, global, nomadic, population needs to establish trustful relationships, to travel, to make business, to purchase online, to resettle
themselves after a disaster, or fleeing from a war zone. They all want to minimize
risks, to maximize opportunities. Their perception of what is their private sphere, of
what is public and what is private, is rapidly changing, also because private information has become a marketable commodity. They demand to be respected and never
humiliated. Advanced personal reocgnition schemes are fundamental to achieve these
goals. There are thus three main challenges, 1) global nomadism; 2) digitalization
and economic virtualization; 3) human dignity and integrity. Biometric could provide
the technical framework to address all of them. A huge question mark still remains,
who will ever provide the political framework?

References

[1] Auerbach E. Mimesis: The Representation of Reality in Western Literature. Princeton, NJ: Princeton UP; 1963.

[2] Levinas E. Humanisme de l'autre homme. Paris: Fata Morgana; 1972.

[3] Harold N, Curtis B. Identity. In Zalta EN, editor. The Stanford Encyclopedia of Philosophy. Spring Edition; Stanford; 2017.

[4] Merriam Webster. Dictionary Merriam Webster. 2015. [Online]. [cited 2017 June 5]. Available from: https://www.merriam-webster.com/dictionary/identity.

[5] Wayman J, Mciver R, Waggett P, Clarke S, Mizoguchi M, Busch C. Vocabulary harmonisation for biometrics: the development of ISO/IEC 2382 Part 37. IET Biometrics. 2014; 3(1): p. 1–8.

[6] European Group on Ethics in Science and New Technologies to the European Commission. Opinion N.28 – Ethics of Security and Surveillance Technologies. Brussels; 2014.

[7] Bellwood P. First Farmers: The Origins of Agricultural Societies. London: Blackwell; 2004.

[8] Gilbert S. Tattoo History. A Source Book. New York: Juno Books; 2001.

[9] Andrieu B. Les cultes du corps. Paris: L'Harmattan; 1994.

[10] Chesson M. Social memory, identity, and death: anthropological perspectives on mortuary rituals. Archeological Papers of the American Anthropological Association. 2001; 10(1): p. 1–10.

[11] Agamben G. Homo Sacer. Il potere sovrano e la vita nuda. Torino: Einaudi; 1995.

[12] Mordini E. Life in a Jar. In Mordini E, Green M, editors. Identity, Security, and Democracy. Brussels: IOS Press; 2008. p. vii–xv.

[13] Scheidel W, editor. State Power in Ancient China and Rome. Oxford Studies in Early Empires. New York: Oxford University Press; 2015.

[14] Caplan J, Torpey J, editors. Documenting Individual Identity: The Development of State Practices in the Modern World. Princeton: Princeton UP; 2001.

[15] Berdah JF. Citizenship and National Identity in France from the French Revolution to the Present. In Ellis S, Eßer R, editors. Frontiers, Regions and Identities in Europe; Edizioni Plus. Pisa University Press. Pisa: 2009. p. 141–153.

[16] Lapied M. Le rôle des comités de surveillance dans la circulation de l'information, à partir de l'étude des comités du Sud-Est. Annales historiques de la Révolution française. 2002; 330: p. 29–29.

[17] Conseil Constitutionnel. Declaration of Human and Civic Rights of 26 August 1789. [Online]. 2002 [cited 2017 May 13]. Available from: http://www.conseil-constitutionnel.fr/conseil-constitutionnel/root/bank_mm/anglais/cst2.pdf.

[18] Semmelroggen J. The difference between asylum seekers, refugees and economic migrants. The Independent. 2015 August 18.

[19] United Nations. International Migration Report 2015. Department of Economic and Social Affairs, Population Division; 2016. Report No.: ST/ESA/SER.A/375.

[20]   Humburg J. Refugee. In Bolaffi G, editor. Dictionary of Race, Ethnicity and Culture. London: Sage; 2003.

[21]   UNHCR. Global Trends. Forced Displacement in 2015. Geneva: United Nations High Commissioner for Refugees; 2016.

[22]   UNHCR. The 1951 Refugee Convention. [Online]. 2017 [cited 2017 April 4]. Available from: http://www.unhcr.org/1951-refugee-convention.html.

[23]   IPK International. World Travel Monito. [Online]. 2017 [cited 2017 June 5]. Available from: http://www.ipkinternational.com/en/world-travel-monitor.

[24]   CNN. International Tourists Hit Record 1.2 billion in 2015, says UNWTO. [Online]. 2016 [cited 2017 June 3]. Available from: http://edition.cnn.com/travel/article/international-tourists-2015/index. html.

[25]   UNICEF. Every Child's Birth Right: Inequities and Trends in Birth Registration. [Online]. 2013 [cited 2017 February 21]. Available from: http://www.unicef.org/mena/MENA-Birth_Registration_report_low_res-01.pdf.

[26]   Frontex. Risk Analysis for 2017 – Europa EU. [Online]. 2017 [cited 2017 May 5]. Available from: http://frontex.europa.eu/assets/Publications/Risk_Analysis/Annual_Risk_Analysis_2017.pdf.

[27]   Teitelbaum M. Europe's migration dilemmas. Foreign Affairs. 2017 May 11.

[28]   World Bank. Identification for Development. [Online]. 2015 [cited 2017 May 9]. Available from: http://www.worldbank.org/.

[29]   UNHCR. Biometric Identity Management System. Enhancing Registration and Data Management. [Online]. 2015 [cited 2017 June 5]. Available from: http://www.unhcr.org/550c304c9.pdf.

[30]   Parussini G. India's Massive Aadhaar Biometric Identification Progra. [Online]. 2017 [cited 2017 June 20]. Available from: https://blogs.wsj.com/briefly/2017/01/13/indias-massive-aadhaar-biometric-identification-program-the-numbers/.

[31]   Goldfarth A, Greenstein S, Tucker C, editors. Economic Analysis of the Digita Economy. Chicago: Chicago UP; 2015.

[32]   Winnicott D. Human Nature. London: Free Association Books; 1988.

[33]   Blanco IM. The Unconscious as Infinite Sets. London: Duckworth; 1975.

[34]   Hall E. The Hidden Dimension. London: Anchor Books; 1966.

[35]   Kennedy PD, Gläscher J, Tyszka J, Adolphs R. Personal space regulation by the human amygdala. Nature Neuroscience. 2009; 12: p. 1226–1227.

[36]   European Parliament – The Council and The Commission. European Charter of Fundamental Rights, 2000/C 364/01. Brussels; 2000.

[37]   European Group on Ethics in Science and New Technologies to the European Commission. Op. N° 20, Ethical Aspects of ICT Implants in The Human Body. Adopted on 16/03/2005; Brussels: 2005.

[38]   CNB. Biometrics, identifying data and human rights, available at http://www.ccne-ethique.fr/docs/en/avis098.pdf: OPINION N° 98; Paris: 2007.

[39]   Agamben G. No to bio-political tattooing. Communication and Critical/Cultural Studies. 2008; 5(2): p. 201–202.

[40] Mesnard P, Kahan C. Giorgio Agamben à l'épreuve d'Auschwitz. Paris: Kimé; 2001.

[41] UNESCO. Information for All Programme. 2007.

[42] Amit. 12 Companies Leveraging Blockchain for Identification and Authentication. [Online]. 2016 [cited 2016 December 17]. Available from: https://letstalkpayments.com/12-companies-leveraging-blockchain-for-identificationand-authentication.

[43] BITNATION. Bitnation Refugee Emergency Response (BRER). [Online]. 2016 [cited 2016 December 17]. Available from: https://refugees.bitnation.co/.

[44] Allison J. Decentralised Government Project Bitnation Offers Refugees Blockchain IDs and Bitcoin Debit Cards. [Online]. 2016 [cited 2017 January 22]. Available from: http://www.ibtimes.co.uk/decentralised-government project-bitnation-offers-refugees-blockchain-ids-bitcoindebit-cards-1526547.

Index